Ecological Research Monographs

Series Editor: Yoh Iwasa

For further volumes:
http://www.springer.com/series/8852

Norio Yamamura · Noboru Fujita
Ai Maekawa
Editors

The Mongolian Ecosystem Network

Environmental Issues Under Climate and Social Changes

Springer

Editors
Norio Yamamura
Research Institute for Humanity and Nature
Motoyama, Kamigamo, Kita-ku
Kyoto, Japan

Ai Maekawa
Utrecht
The Netherlands

Noboru Fujita
Research Institute for Humanity and Nature
Motoyama, Kamigamo, Kita-ku
Kyoto, Japan

ISSN 2191-0707 ISSN 2191-0715 (electronic)
ISBN 978-4-431-54051-9 ISBN 978-4-431-54052-6 (eBook)
DOI 10.1007/978-4-431-54052-6
Springer Tokyo Heidelberg New York Dordrecht London

Library of Congress Control Number: 2012944983

Cover illustration: Front cover: A pastoralist in the Mongolian grassland. Back cover: *Left*: A herd of sheep and goats; *Center*: A GPS being attached to a sheep; *Right*: Measurement of shrub roots.

Printed on acid-free paper

Springer is part of Springer Science+Business Media (www.springer.com)

Preface

Most ecosystems on the planet, including grasslands and forests, have been seriously degraded by climate changes and human activities, particularly the overuse of natural resources, and are now in critical condition, with a loss of biodiversity and reduction in ecosystem functions. This situation is widely accepted as one of the most serious global environmental problems. Human activities that use resources are usually embedded in complex social-ecological systems and involve numerous interacting subsystems.

The Research Institute for Humanity and Nature (RIHN), Kyoto, Japan, promotes "integrated cooperative research toward the solution of global environmental problems" and has helped to create the field of global environmental studies, concentrating on interactions between people and nature. One RHIN project, "Collapse and Restoration of Ecosystem Networks with Human Activity," examines environmental problems associated with ecosystem reductions, using the key concept of ecosystem networks. In February 2010, the project hosted an international symposium on grassland reduction in Mongolia, which is one of the main topics of the project. The main environmental issue addressed in Mongolian grasslands was the reduction and loss of usable pastures for nomadic pastoralism.

The aim of this book is to describe what is currently occurring in the Mongolian grasslands, to analyze how various factors creating environmental problems interact with each other, and to suggest solutions for sustainable management of the grasslands. These analyses and discussions are based on the results of the symposium and other studies within the host project. The title, *The Mongolian Ecosystem Network*, indicates that environmental issues are analyzed from both natural and social perspectives.

This book is divided into three parts. Part I is an introduction, explaining the key concept of an ecosystem network within the project (Chap. 1) and providing background information on the general features of Mongolian nomadic pastoralism (Chap. 2). The distribution of vegetation in Mongolian grasslands is also reviewed, as a fundamental factor in the environment and for pastoralism (Chap. 3).

Part II describes the effects of natural environmental factors and nomadic activities on grassland conditions. Climate change caused by global warming, especially changes in precipitation, has affected the grassland vegetation (Chaps. 3, 4). Water dynamics that maintain the grassland system are analyzed in a steppe region with shrubs (Chap. 5) and in a forest-steppe region with trees (Chap. 6). Furthermore, the effects of grazing livestock on Mongolian ecosystems are described in relationship to forests (Chap. 7), soils (Chaps. 8, 9), and grassland vegetation (Chaps. 10, 11, 12, 13).

Part III describes the effects of economic and social factors on land use and the livelihood of herders. After the introduction of a market economy to Mongolia in the early 1990s, nomadic people tended to increase their livestock numbers, especially goats, and to move closer to large cities for economic advantages. This shift has led to the degradation of pastures by overgrazing. Rules for pastureland possession, designed to minimize overgrazing, are examined (Chap. 14), and a model representing movement between rural and urban sites is presented (Chap. 15). The social factors driving the increase in goat herds that produce economically profitable products are analyzed (Chaps. 16, 17). Recently, pastureland has also been used intensively for dairy farming, agriculture, and mining. These human activities are described, and their effects on grassland degradation are discussed (Chaps. 18, 19, 20). Finally, the impacts of global warming and globalization on the Mongolian society and ecosystem are examined (Chap. 21).

The chapters in Parts II and III address current environmental problems in Mongolian grasslands and suggest solutions for the sustainable use of grasslands, from both ecological and social perspectives, based on studies within the project. This book analyzes the environmental problems faced in Mongolian grasslands, but the contents contribute to the identification of environmental problems and consideration of sustainable pasture use in grassland areas throughout the world.

We acknowledge the Institutes of Geoecology, Biology, and Botany, Mongolian Academy of Science, and the Institute of Meteorology and Hydrology, Mongolian Ministry of Nature and Environment, for cooperative research in the project. We also thank RHIN, Japan, for financial support for the project (D-04) and the symposium.

Kyoto, Japan Norio Yamamura
Kyoto, Japan Noboru Fujita
Utrecht, The Netherlands Ai Maekawa

Contents

Contributors

Narantsetsegiin Amartuvshin (Chapters 3, 11, 13) Institute of Botany, Mongolian Academy of Sciences, Ulaanbaatar, Mongolia

Erdenegerel Ariunbold (Chapters 11, 13) Institute of Geoecology, Mongolian Academy of Sciences, Ulaanbaatar, Mongolia

Dorjgotov Ariungerel (Chapter 10) Mercy Corps, Gobi Forage Project, Ulaanbaatar, Mongolia

Maki Asano (Chapter 8) Graduate School of Life and Environmental Sciences, University of Tsukuba, Tsukuba, Japan

Zamba Batjargal (Chapter 21) World Meteorological Organization UN, New York, NY, USA

Batjargal Enkhjargal (Chapter 21) World Meteorological Organization UN, New York, NY, USA

Noboru Fujita (Chapters 3, 6, 11, 13) Research Institute for Humanity and Nature, Kyoto, Japan

Muneto Hirobe (Chapter 9) Laboratory of Forest Ecology, Graduate School of Environmental Science, Okayama University, Okayama, Japan

Reiichiro Ishii (Chapters 5, 6, 15) Research Institute for Global Change, Japan Agency for Marine-Earth Science and Technology, Yokosuka, Japan

Undarmaa Jamsran (Chapter 8) Center for Ecosystem Study, Mongolian State University of Agriculture, Ulaanbaatar, Mongolia

Luvsandorjiin Jargalsaikhan (Chapter 4) Institute of Botany, Mongolian Academy of Sciences, Ulaanbaatar, Mongolia

Akira Kamimura (Chapter 14) Tokyo University of Foreign Studies, Tokyo, Japan

Yutaka Kobayashi (Chapter 15) Center for Ecological Researches, Kyoto University, Kyoto, Japan

Yuki Konagaya (Chapters 2, 19) National Museum of Ethology, Suita, Japan

Junji Kondo (Chapter 9) Laboratory of Forest Ecology, Graduate School of Environmental Science, Okayama University, Okayama, Japan

Ai Maekawa (Chapters 2, 17) Utrecht, The Netherlands

Shinichi Mori (Chapter 18) International Management Group, Inc., Yokohama, Japan

Urianhai Galzuud Nachinshonhor (Chapter 12) Okayama University Graduate School of Environmental Science, Okayama, Japan

Yasuhisa Saito (Chapter 15) Department of Mathematics, Ajou University, Suwan, Republic of Korea

Izuru Saizen (Chapter 16) Graduate School of Global Environmental Studies, Kyoto University, Kyoto, Japan

Takashi Satoh (Chapter 5) Graduate School of Life and Environmental Sciences, University of Tsukuba, Tsukuba, Japan

Michiaki Sugita (Chapter 5) Faculty of Life and Environmental Sciences, University of Tsukuba, Tsukuba, Japan

Yukio Suzuki (Chapter 20) Ministry of Agriculture, Forestry and Fisheries, Tokyo, Japan

Kenji Tamura (Chapter 8) Graduate School of Life and Environmental Sciences, University of Tsukuba, Tsukuba, Japan

Jamsran Tsogtbaatar (Chapter 7) Institute of Geoecology, Mongolian Academy of Sciences, Ulaanbaatar, Mongolia

Maki Tsujimura (Chapter 5) Faculty of Life and Environmental Sciences, University of Tsukuba, Tsukuba, Japan

Indree Tuvshintogtokh (Chapter 10) Institute of Botany, Mongolian Academy of Sciences, Ulaanbaatar, Mongolia

Norio Yamamura (Chapters 1, 15) Research Institute for Humanity and Nature, Kyoto, Japan

Tsutomu Yamanaka (Chapter 5) Faculty of Life and Environmental Sciences, University of Tsukuba, Tsukuba, Japan

Part I
Ecosystem Network and Mongolian Nomadism

Chapter 1
Ecosystem Networks as Social–Ecological Systems

Norio Yamamura

Keywords Ecosystem network • Land cover • Land use • Mongolian grassland • Nomadic pastoralism • Social–ecological systems

1.1 Introduction

Most ecosystems on the planet, including grasslands and forests, have been seriously degraded by human activities, especially through the overuse of natural resources, and have reached critical situations, leading to a loss of biodiversity and reduction in ecosystem functions. This crisis is widely accepted as one of the most serious global environmental problems (Millennium Ecosystem Assessment 2005). Human activities that use natural resources are often embedded in complex social–ecological systems in which various subsystems interact with each other.

Recently, several groups have proposed research frameworks to address the complex social–ecological systems that have resulted in ecosystem degradation (GLP 2005; Folke et al. 2004; Ostrom 2009). Here, we propose another research framework for understanding ecosystem degradation, stressing the fact that social–ecological systems comprise a network of subsystems. In a project titled "Collapse and Restoration of Ecosystem Networks with Human Activity," conducted at the Research Institute for Humanity and Nature (RIHN), we designated the entire system as an ecosystem network, which we defined as a social–ecological system in which a network of ecological subsystems, characterized by land cover or land use, interacts with a human network, affecting the quality of and transitions among the subsystems.

N. Yamamura (✉)
Research Institute for Humanity and Nature, Kyoto, Japan
e-mail: yamamura@chikyu.ac.jp

N. Yamamura et al. (eds.), *The Mongolian Ecosystem Network: Environmental Issues Under Climate and Social Changes*, Ecological Research Monographs,
DOI 10.1007/978-4-431-54052-6_1,

The usefulness of this concept lies in its ability to forecast extensive human impacts through changes in land cover, for which data are generally available from area statistics, field surveys, and satellite information. The primary environmental issue in Mongolian grasslands is the reduction and loss of usable pastures for nomadic pastoralism that results from livestock overgrazing and agricultural cultivation (Hirano et al. 2006; Okayasu et al. 2007; Mearns 2004). Recently, extensive mine development to extract rich underground resources has also contributed to reduction in pasture areas (see Chap. 20, this volume). By explicitly depicting the structure of the ecosystem network, we can understand the network structure of environmental issues in Mongolian grasslands, including both ecological and social perspectives. We can also use this concept to examine how the network structure produces environmental problems and to determine which links should be modified to resolve or mitigate the problems.

1.2 Ecosystem Network for Mongolian Grasslands

For our project titled "Collapse and Restoration of Ecosystem Networks with Human Activity" (http://www.chikyu.ac.jp/rihn_e/project/D-04.html), conducted at the Research Institute for Humanity and Nature (http://www.chikyu.ac. jp/index_e. html), the field research takes place in the tropical rainforests of Sarawak, Malaysia, and in the grasslands of Mongolia. During the past few decades, social and environmental conditions in both areas have been profoundly affected by resource extraction, which has recently intensified in response to the demands of the global economy. Local people have used natural products from the forests in Sarawak and have lived by grazing livestock on the grasses in Mongolia. Although the ecological characteristics of their natural environments, such as the regeneration time of vegetation and the position of humans in the food web, are quite different, the livelihoods of many inhabitants in these regions are strongly dependent on services from natural ecosystems.

We hypothesized that the ecosystem network for Mongolian grasslands can be described as a network of social and ecological subsystems, as shown in Fig. 1.1. Subsystems and interactions are indicated by nodes and links, respectively, following the terminology of network theory. In the bottom row, we draw ecological subsystems as characterized by land cover or land use, including suitable pasture, degraded pasture, agricultural fields, and mining areas. Pastures interact with forests in the Forest-Steppe zones and often include shrubs in the Steppe and Desert-Steppe zones. These ecological subsystems are transformed into one another through human activities such as logging, planting, and overgrazing by livestock. In the middle row of Fig. 1.1, we place social subsystems, including nomads with their livestock, governments, and markets and enterprises, which affect the quality of and transitions among the ecological subsystems. The social subsystems also interact with each other. In the top row, we include external factors such as climate change, nongovernmental organizations (NGOs), international agencies, and foreign enterprises that affect the internal subsystems and links in the ecosystem network.

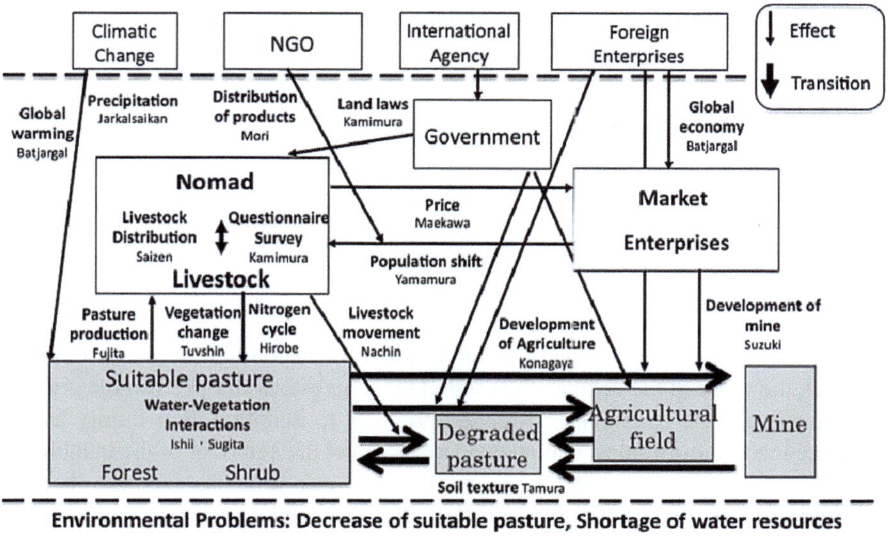

Fig. 1.1 Ecosystem network in Mongolia, where environmental problems are resulting in decreased availability of suitable pastures and shortages of water resources. The network of ecological subsystems, characterized by land cover or land use, interacts with the human social network, thereby affecting the quality of and transitions among subsystems. *NGO*, nongovernmental organization

Parts II and III of this book describe individual studies that clarify the strengths of the links among subsystems, the effects of subsystems on each other, and the interactions within subsystems in the ecosystem network. The links and nodes in the network (Fig. 1.1) reflect the contributions of numerous studies described in the chapters of this book. These ecological, economic, and social studies use various data sources, including field surveys, statistical data from the Mongolian government, and satellite data, to delineate land-cover distribution and its changes. Based on the network structure, we can understand the relationships among the individual studies in this book and can examine the emergence and prevalence of environmental problems related to grassland degradation.

The left side of the ecosystem network shown in Fig. 1.1 is related mostly to relationships between natural environments and grassland ecosystems. The chapters that correspond to links in this part of the network address mainly meteorological and ecological issues and are included in Part II, "Natural Environment and the Ecosystem Network." The right side of the network is related primarily to economic and social factors that affect land use and the livelihood of herders, and the related chapters are in Part III, "Socioeconomic Activity and the Ecosystem Network." Individual chapters discuss not only the corresponding link but also nearby links and subsystems as well as various effects throughout the network structure.

We also present a definite ecosystem network for the forest system in Sarawak. A comparison between ecosystem network structures further illuminates the ways in which interactions among social and ecological subsystems result in environmental problems that lead to ecosystem reductions.

1.3 Simplification of the Ecosystem Network and Implications for Conservation

We can simplify the structure of the entire ecosystem network shown in Fig. 1.1 by confining our attention to a specific environmental problem. Here, we focus on two important problems in Mongolian grasslands that result in the loss and deterioration of pastures used for nomadic pastoralism, specifically, overgrazing by livestock and mine development by foreign enterprises (see Fig. 1.2). The expansion and abandonment of agricultural fields have network structures and impacts similar to those of mine development.

In Mongolia, the vegetation (grasses) has no direct economic value for humans. Instead, the value of the vegetation is stored in the livestock that feed on the grasses, as humans use the livestock and related products to maintain their family and to produce income from sales. Global economics affect the behavior of the inhabitants of Mongolia: when international prices for livestock products increase, herders work to increase the number of livestock in the herd, which leads to overgrazing on vegetation and the subsequent degradation of grasslands (Fig. 1.2a). Climate change, especially with respect to precipitation, also has an impact on grasslands.

A reduction in the supply of livestock feed resulting from grassland degradation adversely affects the inhabitants. When grassland begins to degrade, the inhabitants may refrain from overuse by decreasing the number of livestock. Thus, through a type of negative feedback loop, inhabitant behavior and grassland condition can reach a steady state, and the system will tend to be inherently stable. The environmental problem of grassland degradation caused by overgrazing occurs when various factors disturb this negative feedback. Severely degraded grasslands may ultimately become deserts, contributing to the global environmental problem of yellow sands. Therefore, effective technological solutions or political policies are needed to address inhabitant behavior to restore the negative feedback loop.

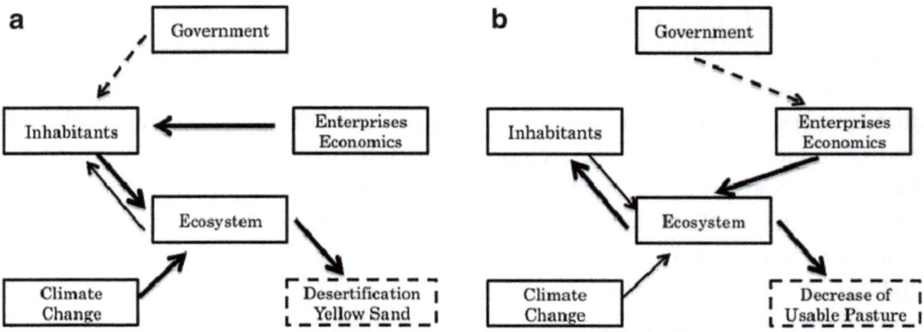

Fig. 1.2 Simplified ecosystem network in Mongolia. (**a**) A network with a negative feedback loop, represented by overgrazing. (**b**) A network with a positive feedback loop, represented by mine development

In contrast, mine developments are made by foreign enterprises and directly destroy wide areas of grassland with the permission of the government (Fig. 1.2b). The developed area can no longer be used by nomadic people and their livestock, and areas surrounding the mine must often be avoided because of water shortages and pollution. This process is almost independent of climate change. Enterprises can expand mining areas into unoccupied land that was traditionally used for pastoralism. Therefore, mine development and the withdrawal of pastoralism may constitute a positive feedback loop. Once mine development begins in an area, it easily expands to encompass wider areas. It is critical that effective solutions or political policies for the conservation of grasslands include limited mine development by enterprises and the government.

These two feedback patterns of overgrazing and mines in the ecosystem network structure (Fig. 1.2) appear to be typical in environmental problems of ecosystem degradation throughout the world. Some ecosystem networks may include both network patterns with different proportions. Therefore, a particular environmental problem could be placed at some point in the spectrum of network structures that has the overgrazing type and the mining time at both ends. Potential solutions and proposed political policies should be considered in the context of the ecosystem network structure. In conclusion, the ecosystem network concept may provide useful general guidelines for recognizing the structure of problems and devising sustainable ecosystem conservation in areas where the inhabitants depend strongly on natural ecosystems and are strongly influenced by economic globalization.

References

Folke C., Carpenter S., Walker B., Scheffer M., Elmqvist T., Gunderson L., Holling C.S. (2004) Regime shift, resilience, and biodiversity in ecosystem management. Annu Rev Ecol Evol Syst 35:557–581

GLP (2005) Science Plan and Implementation Strategy. IGBP Report No. 53/IHDP. Report No. 19. IGBP Secretariat, Stockholm. Available on the Internet: http://www.globallandproject.org.

Hirano A., Toriyama K., Komiyama, H. (2006) Spatiotemporal characterization of Mongolian grassland based on vegetation trend analysis. Asian Association on Remote Sensing-Proceedings A2–A8

Mearns (2004) Sustaining livelihood on Mongolia's pastoral commons: insight from a participatory poverty assessment. Dev Change 35:107–138

Millennium Ecosystem Assessment (2005) Synthesis. Island Press, Washington, DC. Available on the Internet at http://www.maweb.org

Okayasu T., Muto M., Jamsram U., Takeuchi K. (2007) Spatially heterogeneous impacts on rangeland after social change in Mongolia. Land Degrad Dev 18:555–566

Ostrom E (2009) A general framework for analyzing sustainability of social-ecological systems. Science 325:419–422

Chapter 2
Characteristics and Transformation of the Pastoral System in Mongolia

Yuki Konagaya and Ai Maekawa

Keywords Market economy • Mongolia • Nomadic pastoralism • Pastoral system • Socialism

2.1 Introduction

This chapter provides conceptual and historical contextualization of the keys to understand the Mongolian pastoral system.[1] Specifically, it details the structure of the pastoral system in the Republic of Mongolia (which is characterized by its high mobility), explains the socialist modernization process, and offers further analysis of the country's contemporary transformation into a market economy. Nomadism is sometimes treated as a legacy of romanticism; however, it is a contemporary phenomenon. Therefore, this chapter is designed to trace the process of modernization.

Researchers who have pointed out the importance of socioeconomic aspects, for example, Simukov (1935), Bazargur et al. (1989), and Erdenebaatar (1996), tried to elucidate the mechanism of seasonal movement but did not focus on the natural deterioration caused by the nomads' management strategy under a market economy. This chapter gives overall perspectives to understand the impacts of nomadic activities on the natural environment. It also aims to clarify the discussions of the chapters that follow.

[1] This article is based on the paper in E-journal *GEO* published by The Association of Japanese Geographers, vol. 2 (1)34–42, 2007, in Japanese.

Y. Konagaya (✉)
National Museum of Ethology, Suita, Japan
e-mail: yuki@idc.minpaku.ac.jp

A. Maekawa
Utrecht, The Netherlands
e-mail: ai@inmongolia.com

N. Yamamura et al. (eds.), *The Mongolian Ecosystem Network: Environmental Issues Under Climate and Social Changes*, Ecological Research Monographs,
DOI 10.1007/978-4-431-54052-6_2, © Springer Japan 2013

2.2 Historical Descriptions of Nomadic Pastoralists on the Mongolian Plateau

On the Mongolian Plateau, groups known by various names, such as the Xiongnu (Huns), Wuhuan, Xianbei, Rouran, Gaoche, and Tujue (Turks), have developed their strength by utilizing livestock since ancient times. Those societies adjacent to them often recorded their characteristically high degree of mobility. For example, a description by Sima Qian in the *Shiji* (*Records of the Grand Historian*), an account of the Xiongnu, reveals how "their animals eat grass and drink water, and they migrate regularly." Similar descriptions appear repeatedly in many later works. For example, the *Hou Han Shu* (*History of the Later Han*) describes the Wuhuan thus: "They pasture their animals widely across the grasslands, do not have permanent dwellings, and form encampments using yurts (*qionglu*)." A description contained in the *Beishi* (*History of Northern Dynasties*) observes of the Tujue (Turks) that "they live in yurts made of felt carpets, and move in order to seek for water and grass." These examples demonstrate how the mobility of the northern pastoralists was noticed and recorded from early times in China.

2.3 Adaptation to the Environment

The characteristics of, and adaptations to, the natural and social environments explain the Mongolian pastoral system.

2.3.1 Adaptation to the Natural Environment

In the mid-latitudes of the Eurasian continent, grasslands extend broadly in a belt shape from east to west. In the case of the Mongolian Plateau, evaporation and transpiration rates are reduced, although the aridity is comparatively mild because average temperatures are low. Additionally, rainfall is concentrated in the summer months. Amid the abundance of the summer season, female animals conceive, give birth, and lactate. The summer food of the nomads is derived from setting aside some of the milk, which is shared with the animals' young offspring. In addition, the animals fatten up in the summer and are thus able to withstand the winter. During the winter, the animals generally lose weight and deteriorate, and there are some that will not survive the winter. Nomads slaughter livestock that are unlikely to survive the winter to secure a supply of food. This system is a form of livelihood that utilizes the behavior of their animals according to the seasons, and as such can be regarded as an example of environmental adaptation (Konagaya 2005: 54–74).

The most serious uncertainties associated with arid regions are the disparities in rainfall in both time and space. Nomads may also endure extreme weather conditions across the entire nation (Shinoda and Morinaga 2005). If they raised their herds of animals in a fixed location under such uncertain precipitation, it would place a fatal burden on the plant life and vegetation cycle. Therefore, it becomes necessary to regulate the burden of livestock grazing on plant life according to the differences in rainfall over time and space. A pastoral system that supports this regulation through migration is referred to as nomadic pastoralism.

In the case of Mongolia, the level of mobility is noteworthy in the following respects. First, encampments are changed seasonally; second, there are multiple candidates for seasonal encampments concurrently; third, in times of natural disaster, such as extremely harsh winters, migration is allowed that goes beyond the territory which is normally utilized; and fourth, along with migration to the encampments, the composition of the camp group may change.

Furthermore, recent research has revealed that the high degree of mobility is not only a passive reaction to the natural environment by nomads but also helps to actively maintain the natural environment (Fujita 2003). Chapters 12 and 13 in this volume provide further information on the effects of this interaction between grazing and plants.

2.3.2 *Social Environment Relative to Natural Conditions*

One significant feature of the society on the Mongolian Plateau was the absence of permanent settlements based on an oasis. There are spring waters at the foot of the Altai Mountains in the middle of the Gobi Desert, but historically oases did not develop into permanent settlements. In other words, people did not concentrate on areas with springs, even though they were living in an extremely arid climate. Consequently, a unique social environment was formed that lacked permanent exchange partners and locations. Because they have a pastoral system characterized by a high mobility adaption to the natural environment, the people chose not to develop fixed settlements. Historically, Mongolian nomadic peoples did obtain luxury goods and some commodities from merchant caravans passing through the oasis cities to the west. Hence, some systematic exchanges took place in periodic markets set up in border areas with the agricultural regions to the south, as is known, for example, in the trade of tea and horses in the Ming Dynasty. However, neither of these habits constituted permanent markets.

The absence of constant trading partners resulted in the proportion of castrated males being higher in herds. The next section discusses this point.

2.4 Characteristics of Pastoral Management

The management of animal husbandry has two distinctive characteristics: one is that the herds consist of many castrated livestock, and the second is the usage of a mixed herd. Although these are characteristics from the perspective of animal husbandry management, they also reflect the features of both the social and natural environments.

Table 2.1 Livestock numbers and proportion of adult females

	Total (thousands)	Adult females (thousands)	Proportion of females (%)
Camels	254.2	76.3	30.0
Horses	2029.1	569.7	28.1
Cattle	1963.6	764.3	28.9
Sheep	12884.5	5751.8	44.6
Goats	13267.2	5721.8	43.1
Net totals	30398.6	12883.9	42.4

Source: National Statistical Office (2006) Mongolian Statistical Yearbook 2005

2.4.1 Pastoral System Featured by Castrated Male Livestock

The sheep herds raised in the Mediterranean region are composed mostly of females, as a majority of the males are slaughtered at a young age. Moreover, according to Dahl and Hjort's (1976: 88 ff.) exhaustive compilation of records on various livestock-rearing peoples from around the world, herds of sheep and goats consist mostly of females, and this finding is consistent with other kinds of livestock.

In contrast, adult females in comprise less than half of herds in Mongolia (Table 2.1). Male livestock are infertile because they are castrated before they reach maturity. They survive in the absence of any concerted attempt to cull their numbers. The fact that many male herd members are castrated and kept alive is the most distinctive ecological feature of the Mongolian pastoral system. This aspect of herd management is also intended to avoid the splitting up of the herds that would otherwise occur when females mate with the males after they reach maturity.

Generally, if trading is active, or if the transportation to markets is unproblematic, nomads can sell male livestock at a young age that have not undergone castration. Thus, the high ratio of castrated males in the herd indicates the low level of commercialization and the high degree of self-sufficiency in this economy. However, it does not necessarily mean that herders have to struggle to barely maintain a subsistence economy. Quite the contrary; they keep a large number of animals that they cannot utilize by themselves. We can categorize this pastoral management as the "castrated male keeping type."

According to Tani, during the 1970s, the proportion of males in the sheep herds of the transhumance pastoral system in Italy was 1–3% (Tani 1984). Matsubara reported the goatherds of the Turkish nomads (i.e., the Yoruk) accounted for 2–6% in the late 1970s (Matsubara 1983). These numbers are considerably lower than those of their Mongolian counterparts. Such herds do not contain castrated males, but only fertile males, that is, rams. Newly born sheep are eaten or sold in those pastoral societies. It is possible to posit that limited pasturelands do not allow them to keep infertile livestock, so these nomads live close to the consumers who do not own livestock. This idea suggests it is possible to discern between different pastoral systems.

Sneath discussed how, before the advent of socialism, there were a range of pastoral systems in Central Asia until the time of socialism, from the yield-focused type to the subsistence type (Humphrey and Sneath 1999: 218). It is possible to add the compositional characteristics of herds to his argument; that is to say, the "yield" can be converted to the "castrated males" and the variation of "yield -focused" type becomes synonymous with breeding herds with a large number of castrated males. It follows that subsistence economy means raising mostly females.

This castrated male keeping type verifies the state's historical power over the Mongolian plateau, as castrated males could be converted into military strength when needed. In former times, the fastest conveyance in the world was by horse, the most powerful draft engines were camels and oxen, and any castrated males were used as beasts of burden. Although sheep are not employed as beasts of burden, they were a mobile food resource. Therefore, we can consider sheep as contributors to military strength.

2.4.2 *Diversified Growing Pattern*

Pastoralism from Central Asia through Mongolia generally involves five species of livestock: sheep, goats, horses, cattle (including yaks and mixed cattle and yaks), and camels (Umesao 1976: 122). In contrast, pastoralism in other regions of the world breeds only one or two species.

Two types of farming can be posited: one in which a single crop variety is grown extensively (bumper crop) and another where many varieties are grown to some extent (diverse crop). In the same way, we can posit two types of pastoralism. The pastoralism of the Mongolian Plateau clearly belongs to the latter diverse pattern. In addition, the usage of livestock and their products is broad. The use of males and females is a characteristic feature. The development of multifarious uses has created a highly self-sufficient system.

Specifically, large livestock species (including horses, cattle, and camels) are necessary to guarantee a high degree of mobility. During the summer, people ride horses and use cattle to pull carts. During the winter months, camels are used for these purposes. In whichever variety of animals, milk is taken and made into precious dairy products. Smaller animals (such as sheep and goats) are important as a food supply. Moreover, felt is made from sheep's wool; however, herders do not spin the wool into thread to weave clothing. Sheepskins are used as coverlets or are tanned and made into leather clothing.

2.5 Transformations of the Socialist Period

From the time when the Chinese Revolution of 1911 liberated Mongolia from the rule of the Qing Dynasty, up until the Peoples' Revolution in 1921, a government headed by the Living Buddha Boghda Khaan, a leading Tibetan Buddhist lama, was inaugurated in Mongolia. Later, in 1924 when the Living Buddha had passed away, Mongolia became the world's second socialist country, which was christened the Peoples' Republic of Mongolia, after the Soviet Union. The democratic movement that began in 1989 led to the abandonment of the Peoples' Revolutionary Party dictatorship in 1990. As a result, a new constitution was established and the country's name was changed to Mongolia in 1992. Mongolia embarked on a course of modernization through socialism under the leadership of the Soviet Union for approximately 70 years.

Among the various aspects of modernization, the three following significant changes can be cited in the industrial field: first, the nomadic pastoralism was altered; second, an agricultural sector was created; and third, cities were built and a heavy industrial sector was created (Konagaya 2004). Women were encouraged to give birth to many children to provide workers for the new industrial development (Konagaya 1999). As a result, the population rapidly increased, and most of this increase was absorbed into the cities (Konagaya 2003: 522).

As for the modernizing changes in the pastoral system, I would like to highlight the following three points: first, socialist collectivization was carried out; second, livestock were bred and sold as raw materials for industries; and third, there was a heightened dependency on facilities in fixed locations. The following sections detail these three points.

2.5.1 *Socialist Collectivization*

In the days before socialism, society consisted of small numbers of wealthy nomads and large numbers of poor nomads (Vreeland 1957: 32–35; Goto 1968:249–264). To a certain extent, the Tibetan Buddhist clergy played a significant role in the redistribution of wealth. According to a survey in 1918, the number of livestock owned by Buddhist institutions was 19% of the national total (Maiskii 1921: 289). Each monastery held property known as *jas*, and those herds were called *jas* livestock. They were entrusted to poor nomads who lived in the vicinity for pasturing. Various conditions were placed on this outsourcing of animal herding (Toshimitsu 1986), but in principle, usage rights were allowed without transfer of livestock ownership, and a portion of the animals was paid as rent in kind. Consequently, the herders entrusted with these animals could support their livelihood, even though they owned no animals.

With the onset of the socialist period, the animals of poor herders were gradually gathered and ownership was transferred to a livestock collective union. Ownership of the *jas* livestock was transferred en masse to the livestock collectives, which became responsible for the redistribution of wealth.

By 1957, the socialist collectivization of livestock had spread to 30.1% of households and included 42.5% of the total number of animals (Foreign Ministry Asia Bureau China Section 1962: 153, 127), with 100% collectivization achieved by 1959.

2.5.2 *Industrialization of Livestock Rearing*

Through the modernization program under socialism, the pastoral system changed from being a lifestyle to support individual people to an industry to support the nation. It evolved to supply the raw materials of the livestock industry to the capital city to firmly establish the herding sector as an important national industry.

From the late 1930s onward, under the slogan "wool is gold," the cadres themselves engaged in wool-shearing projects and eventually wool shearing spread throughout the country (Konagaya 2004: 112). Besides its use in the manufacturing of felt in factories built in the center of each region, wool provided the material needed to make textiles and carpets at factories in the capital. None of these were products that had traditionally been produced on the Mongolian Plateau. The utilization of wool was thereby added to the pastoral system.

For milk products, the degree of freshness is important, and dairies were built on the outskirts of the cities. Surrounding the capital, 11 dairies were built under the organization of state farms. Women working at these dairies as milking specialists migrated to this area from all over the country.

In relationship to meat, a long-distance migration known as *tuuvar* is carried out in the fall, providing large amounts of meat to processing plants in the capital. Additionally, specialized farms have been built for the purpose of fattening up and holding livestock that have been driven to the capital's outskirts before slaughtering them to fill in the seasonal gaps in the meat supply. As a result, the proportion of castrated males in herds has dropped off drastically subsequent to the *tuuvar* migration. The castrated male animals that were once the source of military power, as already mentioned, have turned into raw material for the meat industry. Consequently, the Mongolian diet has also significantly changed (Konagaya 2005).

2.5.3 *Rise of Solidified Facilities*

In general, sedentarization has increased during the socialist period. The Sum (township) is the government areal unit that limits the extent of migration, and, generally, the territory of migration has become smaller than that of the banners under the Qing Dynasty. In addition, facilities for protection against the cold have been established at spring and winter encampments, which have contributed to the solidification of migration bases. During times of natural disasters, people once sought refuge from catastrophe through long-distance migrations, but these happenings have become less prevalent, because herders used the wheat and barley bran produced as animal feed as a result of the progress of agriculture, and this practice has reduced the risk to herding.

Furthermore, people have also come to live in clusters near established bases to take advantage of social services such as schools and hospitals. The lifestyle has moved toward the sedentary mode overall. However, mobility in herding has at times been encouraged as a means to have higher productivity at the same time

(Toshimitsu 1983). For example, the migration known as *otor* used to be a subsidiary for the horse herds pasturing far from the encamping site, or for conducting a march of escape from natural and social hazards. However, *otor* has encouraged an intensification of regular utilization of the grasslands in the fall, on a scale even larger than before. Accordingly, the socialist period created a new migration pattern in herding. This change typically meant having seasonal encampments at various times through the four seasons, which progressively moved by the fall. Over time, this became the standard model.

Also, excavations of wells were promoted by the state through the country. Consequently, it has become possible to use pastureland that had groundwater resources, and the potentially usable area has expanded.

In the ways just described, the pastoral system has been transformed in response to modernization (industrialization) policies by using the original characteristics of the Mongolian form of pastoralism. As a result, the number of livestock has increased threefold (Konagaya 2003: 523).

2.6 Changes Since Democratization

In 1989, the democratization movement became active, and Mongolia was pushed into a transition to a market economy. Public subsidies were no longer given to any industry because of acceptance of the economic policy known as shock treatment. All industries suffered great impacts. The most far-reaching changes to the herding system were the appearance of regional and household economic disparities.

2.6.1 Appearance of Regional Disparities

The cooperatives collected livestock products and engaged with the national markets during the socialist period. Although herders were required to fulfill a demanding production quota, they received a monthly cash salary. However, following the advent of marketization, the collective channel to transport livestock products stopped functioning. Herders in remote areas were placed at a significant disadvantage when it came to selling their livestock products because of their distance from the market. Chapter 17 by Maekawa discusses this issue in detail.

Regional disparities have therefore arisen: these are discernible in the regional difference in the ratio of adult female livestock according to the national statistics for 2005 (maps 1 through 5 in Fig. 2.1). If herders have better access to the markets, such as those located in the capital city and trading cities along the border, castrated male livestock are easily sold. Therefore, it is assumed that the ratio of female livestock reflects regional marketability. However, camels are an exception, given that there is a high ratio of females in the Gobi region where the marketability is indeed limited and the region is ecologically arid, because camels are a source of milk

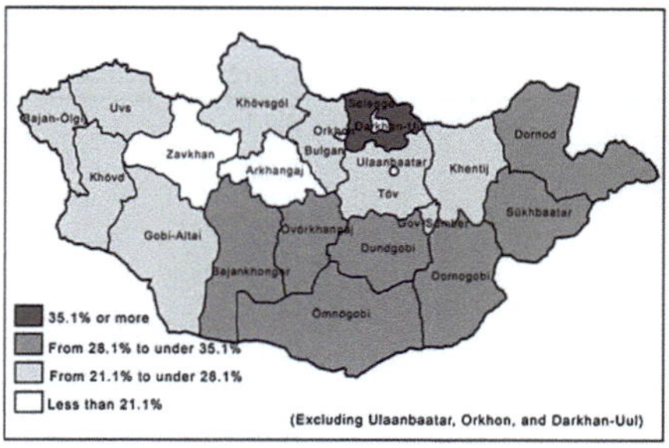

Map 1 Proportion of Adult Female Camels

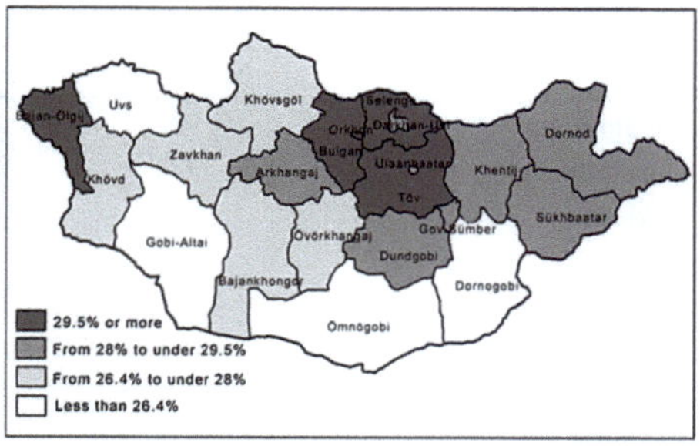

Map 2 Proportion of Adult Female Horses

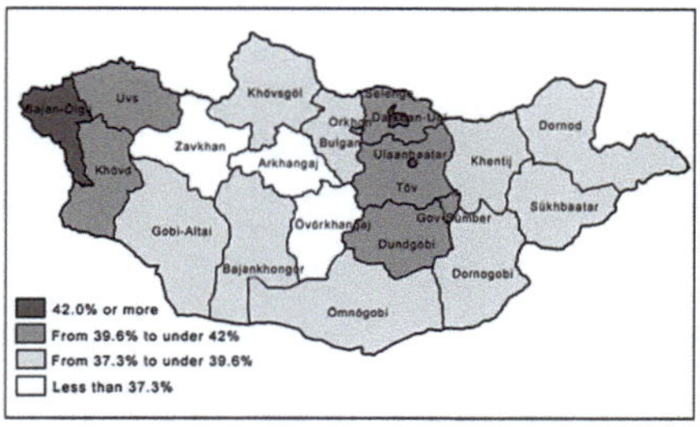

Map 3 Proportion of Adult Female Cattle

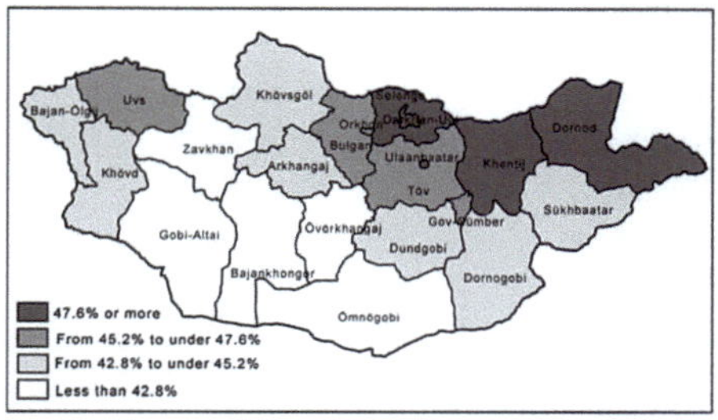

Map 4 Proportion of Adult Female Sheep

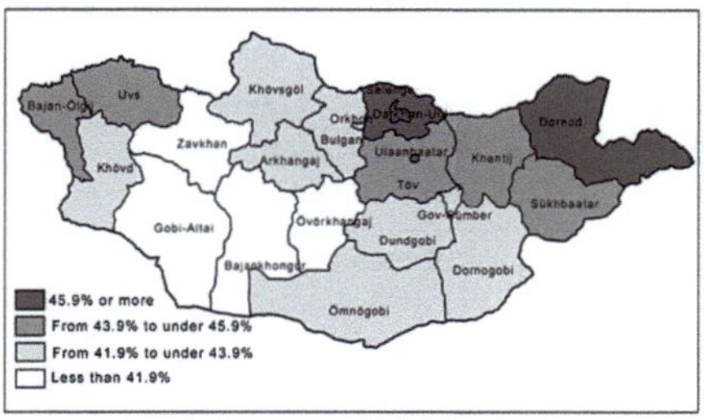

Map 5 Proportion of Adult Female Goats

Fig. 2.1 Regional disparities are discernible in regional differences in the ratio of adult female livestock of five species (*maps 1–5*)
Source: National Statistical Office (2006) *Mongolian Statistical Yearbook 2005*

supply. In other regions, cattle are a main milk source, and cattle are less suited to a dry climate than are camels.

To mitigate the economic difficulties associated with marketization, the nomadic inhabitants of the regions far from the markets have largely migrated to the capital region since the 1990s. Table 2.2 shows that only three industrial cities in central Mongolia gained a positive net migration rate. Chapter 15 by Yamamura et al. provides further understanding of the population shift between urban and rural areas.

The migrants originate from provinces with conspicuously declining populations: Uvs, Zavkhan, Govi-Altai, Bayankhongor, Arkhangai, and Khovd. They are concentrated in the central west, and the proportion of adult female animals is comparatively low, which suggests that castrated males are dominant as a result of the lack of marketability.

Table 2.2 Migration rate in 2002

	Net numbers of migrations[a]	Net migration ratio[b] (%)
Ulaanbaatar	26,344	3.11
Darkhan-Uul	908	1.03
Orkhon	156	0.20
Dornogovi	−165	−0.32
Bayan-Olgii	−371	−0.38
Gobisumber	−51	−0.41
Khovsgol	−550	−0.45
Selenge	−664	−0.65
Bulgan	−556	−0.88
Dornod	−731	−0.98
Tov	−1,026	−1.06
Sukhbaatar	−675	−1.20
Khentii	−933	−1.30
Omnogovi	−718	−1.52
Ovorkhangai	−1,880	−1.65
Dundgovi	−919	−1.80
Khovd	−1,771	−2.00
Arkhangai	−2,372	−2.43
Bayankhongor	−2,499	−2.96
Govi-Altai	−2,476	−3.95
Zavkhan	−3,456	−4.09
Uvs	−3,760	−4.50
National average	−1,835	−28.17

[a]Population increase from 2001 to 2002 minus the natural number of increases
[b]Net number of migrations divided by population
Source: National Statistical Office (2003) *Mongolian Statistical Yearbook 2002*

There are other specific phenomena related to regional differences. Goat breeding is widespread in the Gobi region because of the profits generated by cashmere (see Chaps. 16 and 17). The high ratio of castrated male goats in herds has arisen because male goats can also produce cashmere. Particularly in Dornogovi, where the border trade with China is flourishing, the decline in population is small. These examples show that regional disparities occur because of differing approaches to herd management and the physical distances from the capital.

2.6.2 Appearance of Disparities Among Households

Because of the abolition of the balancing system implemented under the socialist regime, which sought economic equality, economic disparities between families have become apparent. They have also manifested at the interregional level.

According to the national statistics for 2005 (National Statistical Office 2006), nomads who own more than 1,000 head of livestock account for 0.70% of the total.

These wealthy herders are able to invest in fixed animal sheds or wells. Their poorer counterparts, who have no more than 50 head of livestock, comprise 37.00% of the total, while those with 10 head or less constitute as much as 10.77%. These figures tell us that only a minority of herders may be considered wealthy while nearly half of the total number of herders can be classed as poor.

As Chap. 17 shows, there is a tendency for households owning more head of livestock to sell more diversified livestock products. Consequently, they have a steadier cash income. In contrast, the fewer animals a household owns, the more dependent it will be on cashmere from goats, making management less secure. Their management strategies based on the difference of possessing livestock is more and more distinct. Consistent with these regional and household disparities, Mongolian herders are opting for different managerial strategies based on their relative mobility.

2.7 Remaining Issues

This chapter has mainly discussed how each nomad adapts to natural and social environments through a management strategy of migration. Although nomads approach the diverse transitions flexibly, the most problematic condition is the large-scale deterioration of the natural environment. Human activities have caused the following five major problems.

First, the majority of the state farms built during the socialist period have fallen into ruin. The land that was used for those farms generally cannot be again utilized as pasture and farms: they are simply abandoned. Environmental restoration is necessary in such areas. Tamura et al. provide further information on this issue in Chap. 8.

Second is the widespread nonpastoral usage of pastureland. For example, the development of large-scale mining resources and small-scale tourism is proceeding rapidly. Suzuki discusses the conflicts over land use between the mining industry and herding in Chap. 20.

Third, the forests that help to sustain the grasslands are dwindling consequent to fires and insect damage. Tsogtbaatar discusses the problem of deforestation in Chap. 7.

Fourth, the majority of the wells excavated during the socialist period have been allowed to deteriorate because of disuse. Utilization of the grasslands in sustainable combination with underground water resources is necessary. Chapters 5 and 6 detail the issue of water resources in the pasturelands.

Fifth, the Mongolian government has discussed the privatization of pasturelands in recent years. It has not yet reached a conclusion, but this will obviously restrict the spontaneous mobility herders have used to adjust to social and natural environmental changes. The privatization of pastureland might cause further land degradation. Kamimura takes up this issue in Chap. 14.

In the past, nomadic pastoralism has meant access to a changeable natural environment. However, nomads nowadays principally rely on the social environment, such as markets, which are sedentary. It is inevitable that changes to the social environment will prompt additional responses to that of the natural environment, especially when survival in a market-oriented economy is at stake: this remains the case, irrespective of the damage that may eventually be inflicted on the natural environment. Therefore, it is necessary to address each of these issues associated with pastoral nomadism from the perspectives of both the natural sciences and the humanities.

References

Bazargur D. Chinbat B. & Shiirev-Aduya, S (1989) *Bugd nairamdakh mongol ard ulsiin malchidiin nuudel (Nomadic movements in the MPR).* Ulaanbaatar.

Dahl G, Hjort D (1976) *Having herds*. Stockholm: University of Stockholm Press.

Erdenebaatar, B. (1996). Socio-economic aspects of the pastoral movement patterns of Mongolian herders. In Humphrey C, Sneath D (eds) *Culture and environment in Inner Asia*, vol. 1. Cambridge: White Horse Press, pp. 56–110.

Fujita, N. (2003). Sougen shokubutsu no seitai to yuubokuchi no jizokuteki riyou–shokubutsugaku kara mita mongoru kougen (Plant ecology of grasslands and sustainable use of nomadic pasturage–a botanical view of the Mongolian plateau). *Kagaku* 73(5):563–569.

Goto, T. (1968). *Nairiku ajia yuubokumin shakai no kenkyuu (Research on the Nomadic Societies of Inner Asia).* Tokyo: Yoshikawa Kobunkan.

Humphrey, C. & Sneath, D. (1999). *The End of Nomadism?* Durham, NC: Duke University Press.

Konagaya, Y. (1999). Sougen no kuni wo kaetta joseitachi (Women who changed the country of grasslands). In K. Sachiko & Y. Yuko (Eds.) *Shakai henyou to josei–jenda no bunka jinruigaku (Social Change and Women– the anthropology of gender).* (pp. 4–35). Kyoto: Nakanishiya Shuppan.

Konagaya, Y. (2003). Umarekawaru yuubokuron– hito to shizen no aratana kankei wo motomete (The study of nomadism reborn– searching for a new relationship between people and nature). *Kagaku* 73(5):520–524.

Konagaya, Y. (2004). *Mongoru no nijuuseiki– shakai shugi wo ikita hitobito no shougen (Mongolia's 20th Century– testimony of the people who lived under socialism).* Tokyo: Chuo Koron Shinsha.

Konagaya Y. (2005). *Sekai no shokubunka mongoru (World Food Cultures: Mongolia).* Tokyo:Nosangyoson Bunka Kyokai.

Maiskii IM (1921) Sovremennaya Mongoliya *(Modern Mongolia).* Irkutskoe otd-nie, Irkutsk.

Matsubara, M. (1983) *Yuboku no sekai (The world of nomadism).* Tokyo: Chuo koron sha.

Ministry of Foreign Affairs (Japan) (1962) Asia Bureau China Section. *Mongoru jinminkyowakoku (sendenin hikkei) Vol. 1. (Mongolian People's Republic*, public relations officer handbook). Ministry of Foreign Affairs.

National Statistical Office. (2003). *Mongolian Statistical Yearbook 2002*. Ulaanbaatar: National Statistical Office.

National Statistical Office. (2006). *Mongolian Statistical Yearbook 2005*. Ulaanbaatar: National Statistical Office.

Shinoda, M. & Morinaga, Y. (2005). Mongorukoku ni okeru souki saigai keikai sisutemu no kouchiku ni mukete (Towards construction of a disaster early warning system in Mongolia), *Chirigaku Hyouron* 78(11):928–950.

Simukov, A.D. (1935). Pastbisha Mongoliskoi Narodnoi Respubliki (Pastoral Movement in Mongolia), *Sovremennaya Mongoliya* 1935:2(9),76–89.

Tani, Y. (1984). Hitsuji ni totte bokufu to ha nanika (What is the shepherd for sheep?). In Y. Nagata & M. Matsubara (Eds.), Isuramu sekai no hitobito - 3 Bokuchikumin (People of the Islamic world - 3 Pastoralists), Toyo Keizai. (pp. 70–108).

Toshimitsu (Konagaya), Y. (1983). Otoru nouto– mongoru no idou bokuchiku wo megutte ("Otor notes– concerning Mongolia's mobile animal husbandry"). *Jinbun Chiri*, 35(6):548–559.

Toshimitsu (Konagaya), Y. (1986). Mongoru ni okeru kachiku yotaku no kankou (Livestock banking customs in Mongolia). *Shirin* 69:770–794.

Umesao, T. (1976). *Shuryou to yuuboku no sekai* (*The World of Hunting and Nomadism*). Tokyo: Kodansha.

Vreeland, H.H. (1957). *Mongol Community and Kinship Structure*. New Haven, CT: Human Relations Area Files.

Chapter 3
Distribution Patterns of Vegetation as a Fundamental Factor in Mongolian Ecosystems

Noboru Fujita and Narantsetsegiin Amartuvshin

Keywords Desertification • Dry steppe • Forest steppe • Steppe • Vegetation degradation

3.1 Introduction

An ecosystem network is composed of inorganic environments, organisms, and human societies. The vegetation that interacts with these is an important node in the network because it adapts and reacts to both the inorganic environment and human activities and sustains other organisms and human societies through primary production. We first describe general patterns and processes for the pasture vegetation in Central Mongolia to help understand interactions between nodes in the network. The nomenclature used for plant species follows Grubov (2001), except for *Leymus chinensis* (Trin.) Tzvel., which is based on Gubanov (1996). Based on information in this chapter, in Part II: Natural Environment and the Ecosystem Network, future precipitation change and the long-term relationship between precipitation and pasture productivity are addressed. Additionally, within grassland pastures, climate change effects on water dynamics in the soil–vegetation–atmosphere system, interactions between forest and grassland, dynamic soil changes in response to grazing, vegetation dynamics, effects of grazing on productivity and species diversity, and utilization of pasture by pastoral nomads are addressed.

N. Fujita (✉)
Research Institute for Humanity and Nature, Kyoto, Japan
e-mail: fujita@chikyu.ac.jp

N. Amartuvshin
Institute of Botany, Mongolian Academy of Sciences, Ulaanbaatar, Mongolia

N. Yamamura et al. (eds.), *The Mongolian Ecosystem Network: Environmental Issues Under Climate and Social Changes*, Ecological Research Monographs,
DOI 10.1007/978-4-431-54052-6_3,

3.2 Vegetation Patterns

From north to south, Mongolian vegetation is roughly classified into taiga forest, forest steppe, steppe, dry steppe, and desert (Finch 1999; Fig. 3.1). Our research areas are mainly located in the forest to dry steppe in Central Mongolia. Ulan Bator, Mandalgobi, and Dalanzadgad, the main cities, are located in the forest steppe, steppe, and dry steppe, respectively (Fig. 3.1). The average monthly precipitation and air temperature of Ulan Bator (1969–2007), Mandalgobi (1944–2007), and Dalanzadgad (1937–2007) are shown in Fig. 3.2. Average annual air temperatures are –0.6, 0.7, and 3.8°C, and average annual precipitation is 293.2, 170.6, and 130.8 mm, at Ulan Bator, Mandalgobi, and Dalanzadgad, respectively. The transition from forest to dry steppe via steppe vegetation corresponds to a reduction in precipitation and a rise in air temperature from north to south. Along the humidity gradient, the dominant tree species change from tall trees in the forest-steppe zone to shrubs in the steppe and dry steppe zones. In the forest-steppe zone, although the annual precipitation is less than 300 mm, taiga forest is widespread near Ulan Bator because most precipitation occurs in the growing season from June to August and both precipitation and temperatures are low during winter. In this seasonal climate, the soil in the forest-steppe zone does not dry out, and a balance exists between precipitation and evapotranspiration. Therefore, forest can grow mainly on north-facing slopes in the forest-steppe zone. Forests may have covered all areas of the

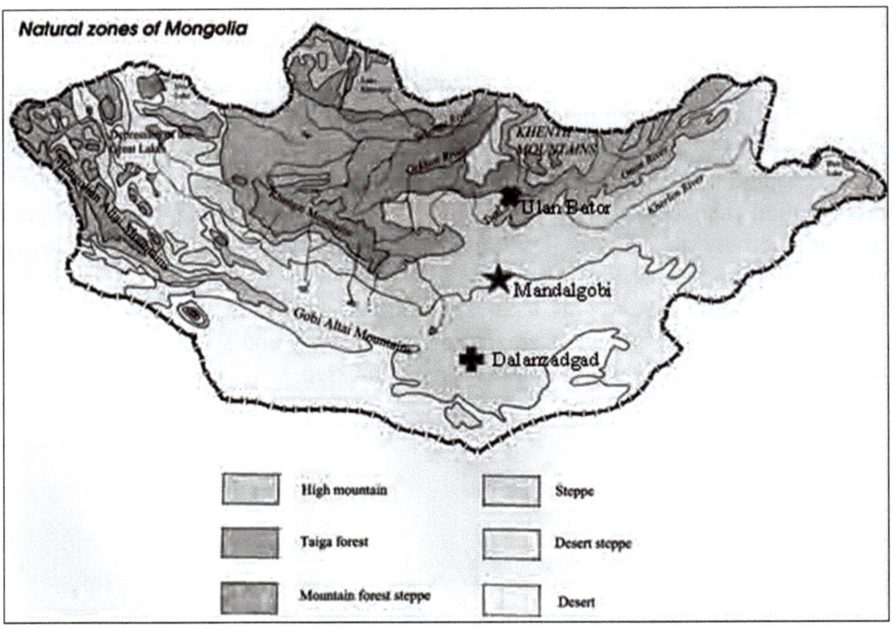

Fig. 3.1 Vegetation zones in Mongolia with city names. (After Finch 1999)

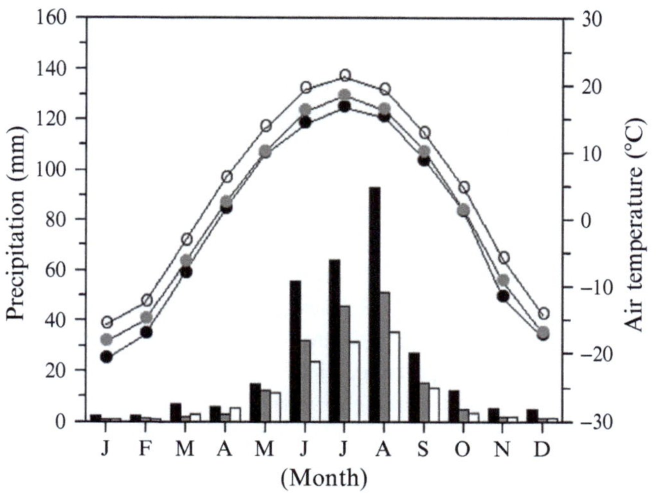

Fig. 3.2 Monthly mean precipitation and air temperature. *Bar and line graphs* show precipitation and air temperature, respectively. *Black, dark, and solid white bars* represent Ulan Bator, Mandalgobi, and Dalanzadgad, respectively

Table 3.1 Dominant species of shrubs, herbs, and grazing-tolerant plants according to vegetation zone

Zone	Shrub	Herb	Grazing-tolerant
Forest steppe			*Artemisia dracunculus*
			Iris lactea
		Stipa krylovii	*Urtica cannabina*
Steppe	*Caragana microphylla*		*Artemisia adamsii*
			Achnatherum splendens
Dry steppe	*Caragana stenophylla*	*Stipa glareosa*	Annuals
		Allium mongolicum	
	Kalidium foliatum	*Stipa gobica*	*Artemisia pectinata*
	Reaumuria soongorica	*Allium polyrrhizum*	*Peganum nigellastrum*

forest-steppe zone in the past (see Fujita et al., this volume, Chap. 13), but at present pastures are widely distributed in flat valleys, on lower slopes, and on south-facing slopes. Primary production of pastures is high in the forest-steppe zone because of the good soil moisture, which allows high livestock production.

In the steppe and dry steppe zones, typical vegetation is composed of shrub species (Table 3.1). Between Ulan Bator and Mandalgobi in the steppe zone, *Caragana microphylla* (Pall.) Lam. widely dominates the pasture. Between Mandalgobi and Dalanzadgad in the dry steppe zone, two main vegetation types exist: *Caragana stenophylla* Pojark. and a mixture of *Kalidium foliatum* (Pall.) Moq. and *Reaumuria soongorica* (Pall.) Maxim. Soil moisture in the steppe zone is generally better than in the dry steppe zone, and *C. microphylla* prefers wetter soils, similar to its habitat in China (Ma et al. 2004). In the dry steppe zone, soil moisture differs between the habitats of *C. stenophylla*, which grows in moister, sandy soil, and *K. foliatum* and *R. soongorica*, which are found on drier, silty soils. This variation in soil moisture results from differences in soil-particle distribution, as shown in a field survey (Fujita et al., this volume, Chap. 13).

Herb species change from north to south, roughly corresponding to shrub distribution zones. Many herb species occur in the forest-steppe zone (Fujita et al. 2009), whereas fewer herb species are found in the steppe and dry steppe zones. In the northern steppe zone, *Stipa krylovii* Roshev., *Agropyron cristatum* (L.) P.B., *Carex duriuscula* C.A. Mey., and *L. chinensis* are representative species. From the southern steppe zone to the dry steppe zone, *Allium* and *Stipa* species become dominant. *Allium mongolicum* Rgl. and *Stipa glareosa* P. Smirn. are dominant around the shrubs *C. microphylla* and *C. stenophylla*, and *Allium polyrrhizum* Turcz. ex Rgl. and *Stipa gobica* Roshev. are distributed with *K. foliatum* and *R. soongorica* (Table 3.1).

3.3 Vegetation Degradation

Overgrazing by livestock disturbs pasture, and the pasture begins to degrade when subjected to overgrazing. When the overgrazing continues over a long time and is repeated yearly, the physiognomy and species composition of pasture vegetation are changed by degradation. Pasture degradation by overgrazing was found especially around sites where livestock gathered, such as water sources (Stumpp et al. 2005), livestock camps (Sasaki et al. 2008), and Sum (village) centers.

Herbs are normally grazed by livestock in pastures, and in many taxa, the pasture species and the forest floor species are in the same genus. The pasture species must have some grazing tolerance to large herbivores. Poaceae, *Carex*, and *Allium* that are good feeds for livestock have a meristem at the leaf base and can continue to grow even if the top of the plant is grazed. *Urtica* has irritating trichomes on its stems and leaves, and the leaves of some *Artemisia* and *Iris* species and the flowers of *Taraxacum* have a bitter taste. Old culms of *Achnatherum splendens* (Trin,) Nevski become hard to protect soft new shoots. These plants that are resistant to livestock grazing by chemical and physical protection are avoided by livestock in grazed pasture and become taller than grazed plants. *Taraxacum* and *Potentilla* have flat leaf rosettes to escape livestock grazing and can be dominant in grazed pasture. However, this reduces their competitive ability for light. In Poaceae, low grazing pressure leads to a reduction in *L. chinensis* with flat leaves, while *S. krylovii*, with upright leaves, increases (van Staalduinen et al. 2007). Species diversity of pasture plants can be enhanced by livestock grazing in habitats with good soil moisture in the forest-steppe zone as a result of relaxation of competition for light (Fujita et al. 2009). Intermediate grazing pressure by livestock enhances species diversity and production of pasture compared with no or overly strong grazing pressure

(Fujita et al., this volume, Chap. 11). Too much grazing, however, can damage pasture vegetation, reducing species diversity and the biomass production of pasture plants (Zhao et al. 2004), and overgrazing eventually denudes the soil surface of a pasture. On bare ground in all vegetation zones, annual plants dominate after precipitation, for example, *Chenopodium album* L., *Salsola pestifera* Nels., *Bassia dasyphylla* (Fisch. et Mey.) Ktze., and *Kochia densiflora* Turcz. Long-term overgrazing leads to domination by perennial tall, grazing-tolerant plants in the pasture. Distribution of most grazing-tolerant perennials changes among vegetation zones (Table 3.1). In the forest-steppe, steppe, and dry steppe zones, different degradation transitions of pasture are found, depending on the duration and strength of the overgrazing, and grazing-tolerant plants (Table 3.1) occur in degraded pasture. Annuals listed in grazing-tolerant plants in Table 3.1 grow in degraded pasture but differ in grazing tolerance among species.

In the forest-steppe zone, common tall, grazing-tolerant herbs are *Artemisia dracunculus* L., *Iris lactea* Pall., *A. splendens*, *Urtica cannabina* L., and *Stelleria chamaejasme* L. In the northern steppe zone, *Artemisia adamsii* Bess. is typical. From the southern steppe to the dry steppe zone, *Artemisia pectinata* Pall. and *Peganum nigellastrum* Bge. are found in strongly degraded sites. In *C. microphylla* and *C. stenophylla* sites, *Allium* and *Stipa* species change from *A. mongolicum* and *S. glareosa* to *A. polyrrhizum* and *S. gobica* in the process of shrub degradation. The occurrence of *Allium* and *Stipa* supports pasture production after sufficient precipitation on degraded shrub vegetation (Fujita et al., this volume, Chap. 13). *A. splendens*, a grazing-tolerant species of the forest-steppe zone with hard old culms, is, however, grazed by livestock in degraded sites in the steppe and dry steppe zones. Livestock that are large herbivores generally have high tolerance to the hardness and toxicity of plants and will graze most plants when other food is short (Newman et al. 1994). *A. splendens* is said to grow in habitats with a shallow groundwater table, although its habitat is mainly limited to degraded pasture. After the growing season, in autumn, the aboveground parts and leaves, respectively, of herbs and deciduous shrubs wither. After withering, the aboveground biomass and nutrient quality of pasture plants decrease, but these plants still provide essential grazing for livestock up to the next spring. Grazing-tolerant tall plants that are not grazed when live are fully grazed during winter. Pasture degradation leads to desertification, which is a global environmental problem, and it is also a local problem for nomads because it increases feed cost for livestock (Yoshihara et al. 2009).

Herbs die back in winter and in drought, although the underground parts of perennial herbs remain alive. Trees and shrubs have live above- and belowground parts throughout the year. Deer are known to kill big trees by bark stripping (Gill 1992), but livestock strip bark very rarely in the forest-steppe zone of Mongolia. Tree seedlings are grazed and browsed by livestock, which can reduce forest regeneration, and leaves and branches of all shrubs are also grazed. Therefore, rapid shrub degradation by direct livestock grazing is more common than for trees. Shrub encroachment under grazing systems is reported to be widespread (Asner et al. 2004). On the other hand, shrub elimination under overgrazing was also reported (Adámoli et al. 1990; Sasaki et al. 2008). Heavy grazing resulted in reduced size of palatable shrubs but increased the density of an unpalatable shrub (Todd and Hoffman 1999). It is reasonable to conclude that the palatability of shrubs affects whether shrubs encroach or are eliminated under various grazing systems. In the steppe and dry steppe zones from Ulan Bator to Dalandadgad in Mongolia, all shrubs are grazed by livestock to varying degrees. Then, when shrubs are grazed frequently, their growth is prevented, which can lead to reduction in both shrub size and patch density. Finally, shrubs might disappear completely from pastures. A lack of shrub patches in pastures at present may be a result of strong and continuous grazing by livestock. Shrub degradation has occurred very widely up to now (Fujita et al., this volume, Chap. 13), so shrub degradation might have begun long ago.

3.4 Desertification

Desertification is said to be increasing in Mongolia, but it does not advance in the sense of the complete loss of vegetation. In the sense of vegetation and soil degradation, it is widespread through overgrazing. Vegetation degradation is accompanied by soil degradation (Villamil et al. 2001), erosion, blowing sand, and alkalization (Fujita et al., this volume, Chap. 13), which can also lead to further desertification. Distinguishing between vegetation degradation and desertification is difficult because the two are interrelated. Healthy vegetation in pastures will take up more water from the ground and release it to the atmosphere through evapotranspiration compared to poor vegetation (Baldocchi et al. 2004). However, soil water is depleted in degraded pasture in spite of reduced evapotranspiration, and pasture alkalization increases with degradation (Bryant et al. 1990). Other factors such as reduced water-holding capacity in the soil as a result of reduced underground biomass of pasture plants and increased soil surface temperature caused by a lack of vegetation shading may lead to dry soils in the degraded pastures. Shrub degradation is accompanied with pasture alkalization in the steppe and dry steppe zones in Mongolia.

3.5 Conclusions

By observing the present vegetation, we can obtain information on the status of the inorganic environment, livestock, pastoral nomadism, and the social system. Degrees of domination of trees and shrubs and of grazing-tolerant plants are, respectively, positive and negative indicators of pasture health. Therefore, vegetation is a fundamental node of the ecosystem network in Mongolia.

Acknowledgments This work was financially supported in part by the RIHN research project "Collapse and Restoration of Ecosystem Networks through Human Activity." We are grateful to Professor N. Yamamura for his helpful suggestions on the draft of this manuscript.

References

Adámoli J, Sennhauser E, Acero JM, Rescia A (1990) Stress and disturbance: vegetation dynamics in the dry Chaco region of Argentina. J Biogeogr 17: 491–500.

Asner GP, Elmore AJ, Olander LP, Martin RE, Harris AT (2004) Grazing systems, ecosystem responses, and global change. Annu Rev Environ Resour 29: 261–299.

Baldocchi, DD, Xu, L, Kiang, N (2004) How plant functional type, weather, seasonal drought, and soil physical properties alter water and energy fluxes of an oak–grass savanna and an annual grassland. Agric For Meteorol 123: 13–39.

Bryant, NA, Johnson, LF, Brazel, AJ, Balling, RC, Hutchinson, CF, Beck, LR (1990) Measuring the effect of overgrazing in the Sonoran Desert. Climatic Change 17: 243–264.

Finch, C (1999) *Mongolia's Wild Heritage*. Avery Press, Boulder, CO.

Fujita, N, Amartuvshin, N, Yamada, Y., Matsui, K., Sakai, S, Yamamura, N (2009) Positive and negative effects of livestock grazing on plant diversity of Mongolian nomadic pasturelands along a slope with soil moisture gradient. Grassland Sci 55: 126–134.

Gill, RMA (1992) A review of damage by mammals in north temperate forests. 1. Deer. Forestry 65, 145–169.

Grubov, VL (2001) Key to the Vascular Plants of Mongolia, Vols. I, II. Science Publishers, Enfield, UK, pp 1–817.

Gubanov, IA (1996) *Conspectus of Flora of Outer Mongolia*. Valang Publishers, Moscow, pp 1–132.

Ma, CC, Gao, YB, Guo, HY, Wang, JL (2004) Photosynthesis, transpiration, and water use efficiency of *Caragana microphylla*, *C. intermedia*, and *C. korshinskii*. Photosynthetica 42: 65–70.

Newman, JA, Penning, PD, Parsons, AD, Harvey, A, Orr, RJ (1994) Fasting affects intake behavior and diet preference of grazing sheep. Anim Behav 47: 185–193.

Sasaki T, Okayasu T, Jamsran U, Takeuchi K (2008) Threshold changes in vegetation along a grazing gradient in Mongolian rangelands. J Ecol 96: 145–154.

Stumpp M, Wesche K, Retzer V, Miehe G (2005) Impact of grazing livestock and distance from water source on soil fertility in southern Mongolia. Mountain Res Dev 25: 244–251.

van Staalduinen, MA, Heinjo, D, Werger, MJA (2007) Impact of grazing regime on a Mongolian forest steppe. Appl Veget Sci 10: 299–306.

Todd SW, Hoffman MT (1999) A fence-line contrast reveals effects of heavy grazing on plant diversity and community composition in Namaqualand, South Africa. Plant Ecol 142: 169–178.

Villamil, MB, Amiotti, NM, Peinemann, N (2001) Soil degradation related to overgrazing in the semi-arid southern Caldenal area of Argentina. Soil Sci 166: 441–452.

Yoshihara, Y, Chimeddorj, B, Buuveibaatar, B (2009) Heavy grazing constraints on foraging behavior of Mongolian livestock. Grassland Sci 55: 29–35.

Zhao, HL, Li, SG, Zhang, TH, Ohkuro, T, Zhou, RL (2004) Sheep gain and species diversity in sandy grassland, Inner Mongolia. Rangeland Ecol Manag 57: 187–190.

Part II
Natural Environment and the Ecosystem Network

Chapter 4
Long-Term Study of the Relationship Between Precipitation and Productivity in the Main Pasture Vegetation of a Steppe Ecosystem in Eastern Mongolia

Luvsandorjiin Jargalsaikhan

Keywords Aboveground phytomass • Eastern Mongolia • Plant community • Steppe

4.1 Introduction

Large-scale global changes in the natural environment have become more apparent in recent years. These changes are tracked on different levels in all geospheres of the earth and are having an increasing influence on human society. One indication of global change is increased air temperature, which began approximately 100 years ago (Kasimov and Klige 2006). The consequences of this warming, which have become apparent during the past few decades, include soil erosion, decreased vegetation cover, and alterations in the functioning of arid and semiarid ecosystems of which primary production is determined by annual temperature and precipitation. Understanding the consequences of climate change to plants and ecosystems is important for Mongolia because its economic condition depends on livestock production. We have been studying the dynamics of aboveground phytomass in three dominant vegetation communities in Tumentsogt since 1982. The objective of this study was to reveal how the dynamics of aboveground phytomass relate to changes in climatic factors.

L. Jargalsaikhan (✉)
Institute of Botany, Mongolian Academy of Sciences, Ulaanbaatar, Mongolia
e-mail: l_jaga_cj@mail.ru

N. Yamamura et al. (eds.), *The Mongolian Ecosystem Network: Environmental Issues Under Climate and Social Changes*, Ecological Research Monographs,
DOI 10.1007/978-4-431-54052-6_4,

4.2 Study Site and Materials

4.2.1 Location

The eastern steppe permanent study site of the Joint Russian–Mongolian Complex Biological Expedition is located in the territory of Tumentsogt sum of Suhbaatar aimag between 47°15′ and 47°38′ N and 112°02′ and 112°43′ E.

In 1960, Tumentsogt sum was created as an administrative unit, simultaneously with a state farm of the same name. The main economic activity has been animal husbandry, but since 1966 pasture stock-breeding has also been conducted. Before 1998 the general area of agricultural fodder comprised 212,350 ga, of which 193,198 ga was used as pasture, 9,724 ga was hayfields, and 9,428 ga was ploughed fields.

4.2.2 Landscape

The territory of Tumentsogt sum lies in Dundad-Halha county of the eastern Mongolian flat area (Tsegmid 1968). Based on a geomorphological map (Timofeev 1984), the dominant landform types include high (1,100–1,260 m), medium (1,050–1,150 m), and low (900–1,080 m) plains, as well as mountains of two morphogenetic groups: (1) more recently formed low mountains and plateaus, located primarily in the west and to some degree east of the center of the sum, and (2) older formations, including denudational and lithomorphic mountains, which are widespread in the south and east of the sum.

4.2.3 Climate

The most important factors affecting the eastern Mongolian steppe ecosystem are the specific temperature and moisture conditions during the vegetative growth period. These factors are more important than the dry conditions during the winter period. This region is subjected to a harsh continental climate, which reduces the period of vegetative growth.

Precipitation falls mainly in summer, renewing the soil moisture supply. Based on its climate type, the territory of Tumentsogt sum is a part of the comparatively moist Kherlen area (Badarch 1971).

The average annual air temperature is +1.5°C. The length of the frost-free period is 198 days, annual average wind velocity is 3.9 m/s, and annual precipitation is 249.7 mm (Fig. 4.1). The majority of the precipitation is associated with the summer monsoon (May–September), when an average of 239.2 mm falls (95.8% of annual total).

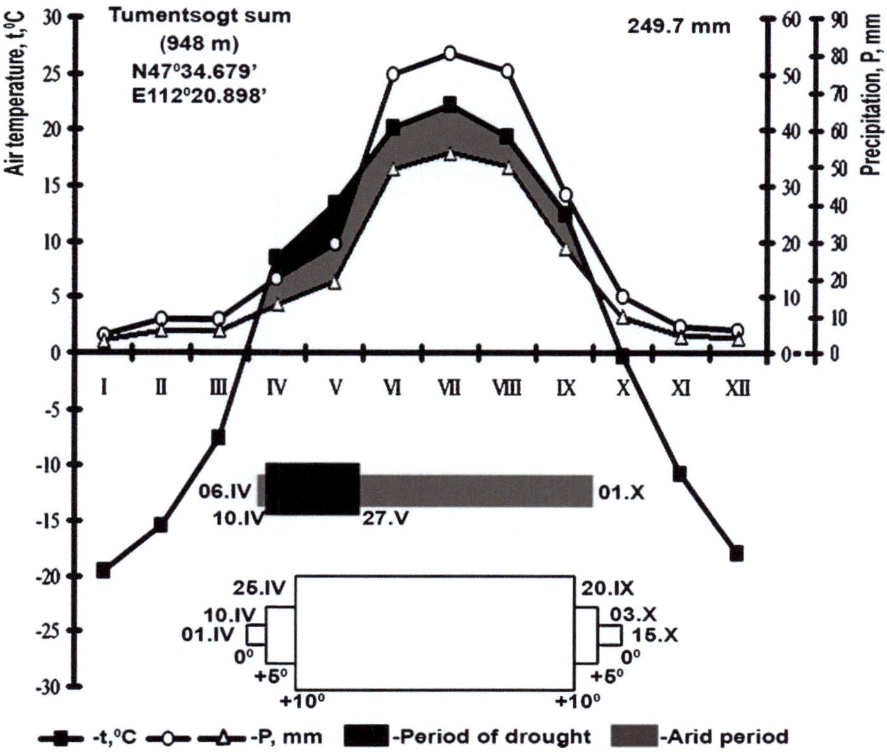

Fig. 4.1 Climate diagram of the meteorological stations at Tumentsogt sum

During this long-term study, 38.5% of years had drought, 34.6% were normal, and 26.9% were humid. The long-term trends in precipitation and average air temperatures are shown in Fig. 4.2.

Trends in mean annual precipitation and air temperature (Fig. 4.2) during the research years show that mean annual precipitation decreased by 34 mm and mean annual air temperature increased by +1.3°C, indicating climate change in the eastern steppe of Mongolia.

We conducted long-term observations on the dynamics of aboveground biomass in three plant communities representative of the eastern Mongolian steppe.

1. *Filifolium sibiricum* community. The study site representing the *F. sibiricum* community (*F. sibiricum* + *Bupleurum scorzonerifolium* + *Ptilotrichum dahuricum* + *Cymbaria dahurica* + *Stipa grandis* + *S. sibirica*) was located on the flat top of a small ridge formed on eluvial-diluvial granite and sandstone at a height of 1,100 m, 12 km northwest of the central Tumentsogt sum homestead. In an area of 60 × 50 m, 107 species of vascular plants belonging to 29 families and 69 genera were recorded. The most representative families were Leguminosae

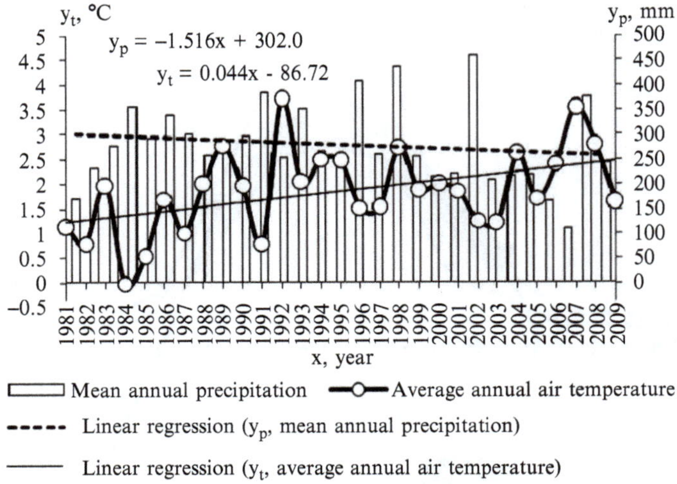

Fig. 4.2 Trend in mean annual precipitation (mm) and mean annual air temperature (°C) at Tumentsogt meteorological station

(16 genera), Compositae (15 genera), and Gramineae (11 genera). Of the vascular plant species, 82.9% were perennials, 13.5% were annual-biennials, and 3.6% were shrubs and semi-shrubs.

2. *Elymus chinensis* community. The *E. chinensis* community (*E. chinensis* + *Poa attenuata* + *Potentilla bifurca* + *Iris dichotoma* + *Saposhnikovia divaricata* + *Stipa krylovii*) was located in the higher part of a valley at a height of 1,070 m, 15 km west of sum center. In a 60×50 m area, 79 species of vascular plants, belonging to 22 families and 53 genera, were recorded. The most dominant families were Compositae (14 genera), Leguminosae (11 genera), and Gramineae (8 genera). Of the vascular plant species, 78.1% were perennials, 17.1% were annual-biennials, and 4.8% were shrubs and semi-shrubs. Canopy cover varied from 40% to 65%. An indicator of this community is the Chinese-Manchurian–Daurian-Mongolian species *E. chinensis* (Lavrenko 1978). It is one of the most widespread species throughout Mongolia and is very abundant in the eastern part of the country (Yunatov 1950; Dashnyam 1974).
3. *Stipa grandis* community. The *S. grandis* community (*S. grandis* + *S. sibirica* + *Serratula centauroides* + *Potentilla tanacetifolia* + *Polygonum divaricatum* + *Caragana microphylla*) was located in the center of a plain between hills, at an elevation of 925 m above sea level. A 100×100 m area was enclosed in 1982 to study the influence of pasture livestock on the steppe ecosystem. Plants belonging to 26 families and 59 genera were recorded in this enclosed area. The most representative families were Compositae (12 genera), Gramineae (12 genera),

Rosaceae (9 genera), and Leguminosae (9 genera). There were 17 species of annuals and 14 species of biennials. Canopy cover was 45–65%. Edificators of the community are *S. grandis* and *S. sibirica.*

4.3 Methods

Aboveground biomass dynamics studies were conducted in the *F. sibiricum* and *E. chinensis* communities from 1982 to 1997, and from 1982 to 2009 in the *S. grandis* plot. Phytomass calculations were conducted for 1-m^2 plots during the spring–summer–autumn period (May–September) on the 10th day of every month. The vegetation was cut at ground level and separated by species.

The total plant canopy cover was determined, and the average height and phenological condition of each species was noted. Cut grasses were air-dried and weighed separately by species.

Meteorological data for 1982–2009 from the weather station in Tumentsogt were used. Data were processed by means of standard statistical methods.

4.4 Results and Discussion

4.4.1 Productivity Dynamics of Aboveground Green Phytomass

The general performance of perennials in the *Filifolium* community in terms of aboveground phytomass is shown in Fig. 4.3.

The maximum aboveground living phytomass in the *F. sibiricum* community ranged from 11.8 (1986) to 23.5 centner/ha (1985), and averaged 15.9 centner/ha. The general trend in perennial green phytomass was described by a polynomial equation ($R^2=0.98$).

The general performance of perennials in the *Elymus* community in terms of aboveground phytomass is shown in Fig. 4.4.

The maximum aboveground living phytomass in the *E. chinensis* community ranged from 8.3 (1986) to 28.4 centner/ha (1985) and averaged 14.7 centner/ha. The general trend in perennial green phytomass was expressed as a polynomial equation ($R^2=0.994$).

The general performance of perennials in the *Stipa* community in terms of aboveground phytomass is shown in Fig. 4.5.

The maximum aboveground living phytomass in the *S. grandis* community ranged from 4.7 (2007) to 21.2 centner/ha (1985) and averaged 13.0 centner/ha. The general trend in perennial green phytomass was expressed as a polynomial equation ($R^2=0.90$).

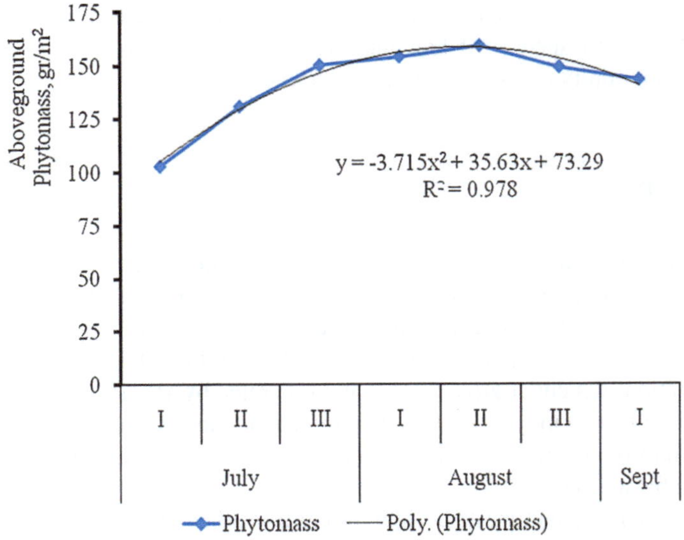

Fig. 4.3 Average 10-day accumulation of green phytomass in the *Filifolium sibiricum* community for the period of vegetation growth during the study period (1982–1997)

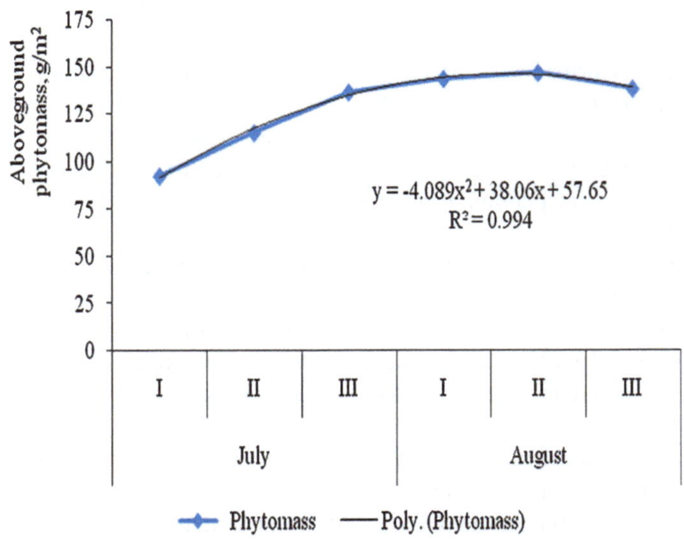

Fig. 4.4 Average 10-day accumulation of green phytomass in the *Elymus chinensis* community for the period of vegetative growth during the study period (1982–1997)

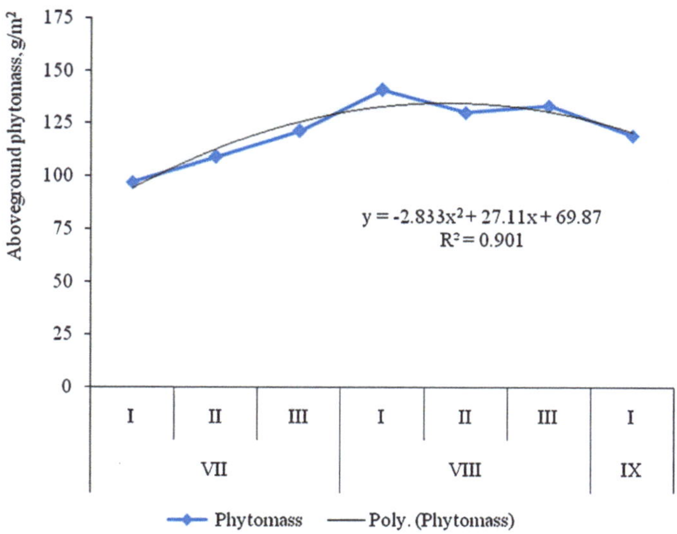

Fig. 4.5 Average 10-day accumulation of green phytomass in the *Stipa grandis* community for the period of vegetative growth during the study period (1982–2007)

4.4.2 Dependency of Green Phytomass Dynamics on Weather Conditions

The main factors defining plant communities and the formation of green phytomass in the dry steppes of eastern Mongolia were rainfall and the pattern of increasing temperature in the spring. Table 4.1 shows that a correlation exists between years of maximum phytomass and spring precipitation and temperature. Maximum phytomass depended on precipitation in April, May, and June (r=0.56, 0.82, 0.65), but differed among communities, which was caused by differences in relief, soil conditions, and, most significantly, plant community structure. The negative correlation with temperature (r=–0.63, –0.55, –0.50) can be explained by the association between lower temperatures and precipitation. That is to say, during hot years, dry springtime maximum phytomass was low, but during wet years with low springtime temperatures, phytomass was greater. There was a negative relationship between the amount of spring (IV+V+VI) precipitation and the time required to reach maximum phytomass in all three communities (r=–0.64, –0.75, and –0.63, for the *Filiformis*, *Elymus*, and *Stipa* communities, respectively). Thus, in years with moist springs, maximum green biomass was reached more quickly than in years with drier springs.

Precipitation and temperature did not produce immediate effects on phytomass; rather, time was required for these abiotic factors to influence the vegetation. Table 4.2 shows the correlation between the weather and biomass at 10-day intervals. Phytomass depended significantly on the rainfall (r=0.98, 0.92, 0.93) of previous

Table 4.1 Correlation matrix between spring weather factors and maximum amount of green phytomass and the period required to attain maximum phytomass

Factor	Sampling interval (month)	Maximum phytomass 1	2	3	Period of maximum phytomass 1	2	3
Precipitation	IV	0.45	0.66	0.43	–0.38	–0.11	–0.22
	V	0.27	0.46	0.22	–0.33	–0.52	–0.40
	VI	0.43	0.72	0.74	0.01	–0.68	–0.33
	IV+V	0.39	0.63	0.07	–0.49	–0.51	–0.44
	IV+V+VI	0.56	0.82	0.65	–0.64	–0.75	–0.63
	V+VI	0.46	0.76	0.57	–0.09	–0.56	–0.41
Temperature	IV	–0.16	–0.02	–0.43	–0.12	–0.08	–0.02
	V	–0.13	–0.06	–0.33	–0.05	–0.53	–0.38
	VI	–0.51	–0.59	–0.30	–0.12	–0.30	–0.18
	IV+V	–0.32	–0.04	–0.32	–0.18	–0.56	–0.35
	IV+V+VI	–0.63	–0.55	–0.50	–0.23	–0.01	–0.35

1, *Filifolium sibiricum* community; 2, *Elymus chinensis* community; 3, *Stipa grandis* community

Table 4.2 Correlation matrix for 10 days of weather factors and phytomass amount

Factor		Phytomass 1	2	3
Precipitation	P_i	0.05	0.15	0.01
	P_{i-1}	0.38	0,41	0,57
	P_{i-2}	0.83	0.85	0.88
	P_{i-3}	0.98	0.92	0.93
Temperature	T_i	0.58	0.59	0.61
	T_{i-1}	0.65	0.80	0.78
	T_{i-2}	0.77	0.81	0.76
	T_{i-3}	0.67	0.77	0.74

i, 10-day period (commencing on 30 April); 1, *Filifolium sibiricum* community; 2, *Elymus chinensis* community; 3, *Stipa grandis* community.

periods (particularly 20–30 days prior), and on the temperature (r=0.77, 0.81, 0.76) of the preceding 10 days (10–20 days). Table 4.3 shows the various meteorological factors that directly influenced green phytomass. This relationship was described by the following multivariate regression equations.

For the *F. sibiricum* community:

$$Y = 0.9638 + 0.3313X_3 + 0.0594X_2 - 0.0179X_1$$

The multivariate correlation coefficient was R=0.956, and the determination coefficient was D=0.92. The shares of the influence variables, obtained using the calculation method of Liepa (1973), were as follows: X_1=36.4%, X_2=18.2%, X_3=37.3%.

Table 4.3 Correlation coefficients between integrated (total) meteorological factors and amount of green phytomass

Community	Factor	X_1	X_2	X_3
Filifolium sibiricum	*Y*	0.949	0.958	0.948
Elymus chinensis	*Y*	0.932	0.934	0.936
Stipa grandis	*Y*	0.870	0.904	0.878

Y, phytomass; X_1, temporal factor (before cropping), commencing on 1 May; X_2, volume of rainfall before cropping (sample collection), since 1 May; X_3, mean day temperature since 1 May

For the *E. chinensis* community:

$$Y = 8.5698 - 0.49497X_3 + 0.025X_2 + 0.0301X_1$$

Here, the multivariate correlation coefficient was $R=0.89$, the determination coefficient was $D=0.79$, and the shares of the influence variables were $X_1=30.2\%$, $X_2=13.4\%$, $X_3=35.7\%$. These influencing factors had a combined influence of 79.3%. The average percentage error in the calculations was 0.13.

For the *S. grandis* community:

$$Y = 17.83 - 0.928X_3 + 0.071X_2 - 0.0444X_1$$

In this community, the multivariate correlation coefficient was $R=0.93$, the determination coefficient was $D=0.87$, and the shares of the influence variables were $X_1=41.0\%$, $X_2=8.7\%$, $X_3=37.8\%$. These influencing factors had a combined influence of 87.5%, but the average percentage error in the calculations was 0.1.

4.5 Conclusion

Observations of phytomass dynamics in a steppe ecosystem in eastern Mongolia confirmed high variability in phytomass resulting from the effects of meteorological factors on productivity. An interpretation of 27 years of field data has allowed us to achieve the main goals of the study: that is, to define the phytocenotic features of the three main plant communities in the eastern Mongolian steppe, to study the dynamics of seasonal plant development in typical steppe communities, to relate phytomass productivity and the structures of the main plant communities in the steppe ecosystem to changes in climatic factors, and to reveal the main trends in plant cover for the observation period in relationship with natural and anthropogenic factors.

Our results suggest the following findings:

1. In all communities, the phytomass accumulation process was described by a polynomial equation with high determination factors (0.98, 0.994, and 0.90, respectively).

2. There was a negative relationship between spring (IV + V + VI) precipitation and the period required to reach maximum phytomass in all three communities ($r = -0.64$, -0.75, and -0.63, for the *Filiformis*, *Elymus*, and *Stipa* communities, respectively). Thus, in years with wet springs, maximum green biomass was attained more quickly than in years with drier springs.
3. Precipitation and temperature did not have immediate effects on phytomass; rather, 10–30 days were required for these factors to impact growth.
4. The green phytomass seasonal dynamics were defined by the integral amount of precipitation and temperature during the vegetative growth period and were described by multivariate linear equations ($R^2 = 0.92$, 0.79, and 0.87).

References

Badarch N. Climate of Mongolia. UB. 1971. 205 pp.

Dashnyam B. Flora and vegetation of Eastern Mongolia. UB. 1974. 146 pp.

Kasimov N.S., Klige R.K. The problems global change. Modern global change the natural ambience. In 2 vol. Vol. 1. -Moscow. Scientific world. 2006. P. 19–34

Lavrenko E.M. About vegetation of the steppes and desert MPR // Problems mastering desert. 1978, #1. P. 3–18.

Liepa I. Ya. To estimation mathematical model. Uch. Zap. Latv. Gos. Univ. 1973. Vol. 181. P. 70–76.

Timofeev D.A. Geomorphological map of Tumentsogt sum., 1984.

Tsegmid Sh. Physical geography of Mongolia. UB. 1968. 405 pp.

Yunatov A.A. The main line of the vegetable cover of the Mongolian Public Republic. Works Mongolian Commissions AS SSSR. Issue 39. 1950. 223 pp.

Chapter 5
Water Dynamics Within the Soil–Vegetation–Atmosphere System in a Steppe Region Covered by Shrubs and Herbaceous Plants

Takashi Satoh, Michiaki Sugita, Tsutomu Yamanaka, Maki Tsujimura, and Reiichiro Ishii

Keywords Arid region • Evaporation • Mongolia • Oxygen and hydrogen stable isotopes • Transpiration • Water use

5.1 Introduction

In many parts of arid regions, the growth, survival, and sustainability of vegetation depend directly on the availability of water. Because the only source of water is usually precipitation, its fate is of great concern. It has been reported that evapotranspiration (ET) occupies the greater portion of water consumption in an arid region (e.g., Kurc and Small 2004; Yamanaka et al. 2007a). Thus, the amounts and time variation of ET are relevant information for understanding the role of water in an ecosystem. However, the knowledge of ET alone is not sufficient. Such information as the water

T. Satoh
Graduate School of Life and Environmental Sciences, University of Tsukuba, Tsukuba, Japan

(Present affiliation: Kanematsu Corporation, Tokyo, Japan)

M. Sugita (✉) • M. Tsujimura
Faculty of Life and Environmental Sciences, University of Tsukuba, Tsukuba, Japan
e-mail: sugita@geoenv.tsukuba.ac.jp

T. Yamanaka
Faculty of Life and Environmental Sciences, University of Tsukuba, Tsukuba, Japan

Terrestrial Environment Research Center, University of Tsukuba, Tsukuba, Japan

R. Ishii
Research Institute for Global Change, Japan Agency for Marine-Earth Science and Technology, Yokosuka, Japan

N. Yamamura et al. (eds.), *The Mongolian Ecosystem Network: Environmental Issues Under Climate and Social Changes*, Ecological Research Monographs, DOI 10.1007/978-4-431-54052-6_5, © Springer Japan 2013 dynamics below the ground surface and the source of water used by the vegetation are also essential to fully understand dryland ecohydrology. This information is important because water is not uniformly available at all surfaces and through all depths of the soil column, and different plant types or species may use different water sources to make the best use of the small amounts of water in an arid area. As such, the aboveground processes of ET, bare ground evaporation (EV), transpiration (TR), and the belowground processes of soil water movement all need to be evaluated in many cases.

For the evaluation of ET, many methodologies have been developed. For example, the eddy correlation method (e.g., Li et al. 2007) is now considered to be the standard to estimate ET over a horizontal scale of 10^2 m. However, this approach does not allow separate evaluation of EV and TR. In contrast, the chamber method (e.g., Heijmans et al. 2004; Yepez et al. 2005; Stannard and Weltz 2006; Nakano et al. 2007) can be used for independent measurements of ET and EV by applying the same procedure for surfaces with or without vegetation. With additional information about vegetation cover fraction, TR can also be inferred. However, what is measured is the flux representing a small site with a horizontal scale of the order of 10^{-1} m. Because site-to-site variations can be large, great care is needed to obtain the average condition of an area, for example, by making measurements at a sufficient number of sites. Also, the chamber measurement has often been criticized for its potential to change the environment by covering the surface with a chamber, which could disturb the flux to be measured (Heijmans et al. 2004). This problem has been minimized by making measurements within a short time period in which temporal environmental changes could be limited (Stannard and Weltz 2006). Thus, depending on the scale and target to be measured, a different approach could and should be utilized. In the present study, the target is a sparse vegetation cover in Mongolia, as explained below, where the principal vegetation consists of a shrub and a herbaceous plant. Thus, the chamber method is probably a better choice to investigate the difference of TR between the plant types and also that between TR and EV.

For the determination of sources of water used by plants, the stable isotope ratio of oxygen and hydrogen has been a powerful tool because the isotopic composition in xylem sap corresponds to the average isotopic composition in the soil water utilized (Ehleringer et al. 1991; Smith et al. 1997; Weltzin and McPherson 1997; Zenchich et al. 2002; Romero-Saltos et al. 2005). Thus, comparison of the isotopic ratio of the plant body and of the soil column at different depths should allow identification of the source(s) of plant water. Moreover, with additional measurements of the profile of soil water content (SWC), depth to the water table, root density, and so on, not only the source of water for plants but also the belowground water dynamics can be evaluated.

The foregoing is a brief background of the methodologies and tools that can be used to measure, analyze, and understand dryland ecohydrology, in particular, for application in the dry steppe region in Mongolia. As indicated, relevant methods have been established. However, these are all quite labor intensive, and it is not easy to apply them together in a remote area. Also, investigators who work with isotopes and those taking flux measurements generally belong to different groups. Thus, a thorough and integrated application of these methods in the same area or the site has been limited, except perhaps in a large-scale field experiment (Sugita et al. 2007) where relevant measurements can be made by multiple groups of investigators. In this study, the chamber method, the isotopic approach, and a soil/vegetation survey were all carried out simultaneously at the same site by the same investigators for two summers, permitting efficient analysis. The main purpose of the study is to investigate the relationship between the ecosystem and hydrology in a dry area, and, more specifically, the relative strength of ET, EV, and TR was determined, and then the role of soil water and vegetation contributing to the flux differences was studied.

5.2 Materials and Methods

5.2.1 Study Area

Intensive observations were carried out in 2008 (5–11 July, 12–17 August) and in 2009 (21 May–11 June, 9 July–2 August), mainly in a small watershed (W1) in a gentle hilly terrain, located in Saintsagaan *soum* (county) of Dundgobi *aimag* (prefecture) at some 4 km northeast of Mandalgobi city in Mongolia (Fig. 5.1). Observations in the adjacent watershed (W2) were also used to supplement those in the W1 watershed. The climate in this area is classified as arid, with the mean annual precipitation of 153 mm (for 1944–2007) measured at the nearby meteorological station (about 5.5 km southeast from the study area) of the Institute of Meteorology and Hydrology (hereafter referred to as the IMH station). About 78% of the annual precipitation falls during the summer season of June–August. The mean annual air temperature is 1.5°C; the annual range of the monthly means is as great as 37°C, with the lowest of –17.7°C occurring in January and the highest of 19.3°C in July.

The dominant vegetation consists of a shrub (*Caragana microphylla*, a C_3 plant) and a herbaceous plant (*Allium polyrrhizum*, a C_3 plant) in the W1 watershed, and it is *Allium* without the shrub in W2. However, even during the typical growing season from April through August, only intermittent appearances of *Allium* after rainfall events were observed on the surface. *Caragana* forms a mound of soil particles at its basal area with a height of about 0.3 m by trapping windblown topsoil particles. The soil in this area is classified as semidesert brown soil or semidesert brown friable sandy soil (Dorzhgotov 2003), and the morphology of the A horizon can be characterized by high compactness, brown color, and weak development of soil structure. High sand and gravel contents can also be observed in the surface soil (Asano 2010; personal communication). Plants in this area, especially *Allium*, are under grazing pressure, mainly from the sheep and goats of local nomadic herdsmen. The mean animal number per unit area was 0.77 SEU ha^{-1} for Saintsagaan *soum* and approximately 0.56 SEU ha^{-1} in the two watersheds, W1 and W2 [here, SEU represents the sheep equivalent unit, and the numbers of camel, horses, cows, and goats are converted to that of sheep by common conversion factors (Pratt and Rasmussen 2001)].

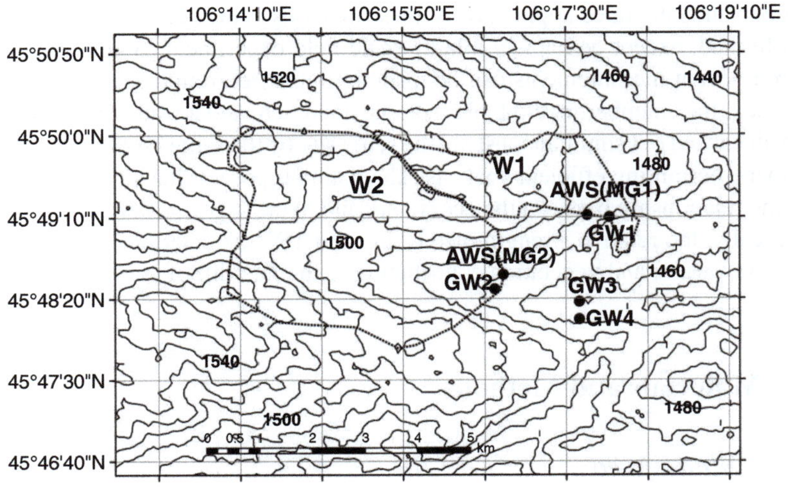

Fig. 5.1 Topographic map of the study area created from NASA Shuttle Radar Topographic Mission (SRTM) 90-m DEM data set [Consultative Group for International Agriculture Research–Consortium for Spatial Information (CGIAR-CSI) 2008; Yoshizawa 2010]. *Closed circles* indicate the locations of automatic weathering station (AWS) and wells (GW1–GW4). Watershed boundaries are shown with *dotted lines*; contours are given by *continuous lines* at 10-m intervals. The smaller watershed located to the east is W1; the larger watershed to the west is W2

The animal numbers were obtained from the annual official statistics in 2007 compiled by the National Statistical Office and from information provided by the local herdsmen in 2009.

5.2.2 *Observations and Measurements*

Two sets of automatic weather stations (AWS) called MG1 (45°49′11.30″ N, 106°17′43.00″ E) and MG2 (45°48′34.80″ N, 106°16′51.40″ E) were installed on 7 July 2008 and 3 May 2009, respectively, in each of the watersheds to obtain basic meteorological and hydrological variables (Fig. 5.1; Table 5.1). As the downward components of radiation are assumed to be the same at the two sites, only the upward components were measured at MG2. Because the volumetric SWC is known to vary both horizontally and vertically, the horizontal distribution of SWC in the top 0.1-m layer was estimated 11 times in the summers of 2008 and 2009 by measuring SWC at 160 points along eight transect lines to investigate how representative are these SWC measurements at AWSs. The horizontal difference was found to be generally small with a standard deviation of the order of 0.3–1.2%. The TDR (time domain reflectometry) sensors for the SWC measurements were intercalibrated before the installation, and the outputs were corrected with the soil temperature data to minimize the temperature influence (see Yamanaka et al. 2003).

Table 5.1 List of variables measured at two stations

Variable	Station	Height (m)	Sensor	Remarks
Wind speed	MG1	3.0	R.M. Young, 3101	
Temperature and relative humidity	MG1	2.65	Campbell Scientific, 41003-5; Vaisala, HMP45D	The probe is housed in a radiation shield, with natural ventilation
Radiation	MG1	2.5	Hukseflux Thermal Sensors, NR01	Four radiation components (up and down, short- and long-wave)
	MG2	2.5	Hukseflux Thermal Sensors, RA01	Two components (upward only)
Soil heat flux	MG1	−0.01	Hukseflux Thermal Sensors, HFP01	
Rainfall	MG1	0.6	Texas Electronics, TR-525M	Tipping-bucket rain gauge
Soil moisture	MG1	−0.05, −0.1, −0.2, −0.3, −0.7, −1.1	Campbell Scientific, CS-616	Time domain reflectometry (TDR) sensor
	MG2	−0.1, −0.2, −0.3, −0.5, −0.7		
Soil temperature	MG1	−0.1, −0.3, −0.7		Platinum resistance thermometer
	MG2	−0.1, −0.2, −0.3, −0.5, −0.7		

Vegetation coverage, *f*, was determined from photographic images of the vegetated surface by a nadir-looking camera (EX-Z600; Casio Computer) taken on cloudy days on 2 July, 11 July, 19 July, 25 July, and 29 July 2009. On each day, 160 images were obtained at 1.5-m intervals along the eight transect lines; they were transformed from RGB to IHS color, and then the supervised classification algorithm built within the ERDAS IMAGINE software (Leica Geosystems, ver. 9.1) was applied to distinguish bare soil from vegetation to calculate the plant coverage, *f*. The mean *f* value over the area was estimated as the average *f* values of the 160 images. The same procedure was also applied to the image taken over the chamber sites (see below) to determine *f* of each chamber site. Details on this procedure can be found in Byambakhuu et al. (2010) and Satoh (2010).

Vertical profiles of fine and thick roots were estimated by determining dry weight of belowground biomass. The fine roots were separated by using 500-μm- and 211-μm-mesh sieves from a 10^{-3}-m^3 soil cube, which was sampled from the side wall of a trench at a 0.1-m-depth interval from surface down to $z=-1$ m on 7 July 2008 and 13 July 2009. The large-sized roots were sampled on 28 July 2009 by digging into the soil at a 0.1-m interval under a semicircle of a *Caragana* mound about 1.2 m in diameter. These samples were weighed to determine the amount of dry biomass. Soil samples were obtained for the determination of the physical properties with a 100-ml stainless steel sampling tube from surface down to $z=-1$ m at 0.1-m intervals on 7 July 2008 and 2 June and 31 July 2009.

The chamber measurements were carried out from 11 to 15 MST (Mongolian Standard Time) for 3 days in early June (1, 3, and 4 June) and 22 days in July (10–14 and 16–31 July) in 2009 to determine the amount of ET and its components, EV and TR. Among them, the measurements on 4 June, 14 July, 18 July, 21 July, and 25 July extended longer than the others, from sunrise through sunset, to allow estimation of the diurnal change and the daily mean. The chamber was of cylindrical shape with a height of 0.3 m and 0.4 m in diameter, made of acrylic plastic (5 mm thick). Instead of a collar inserted into the soil to fix the chamber, a polyvinylchloride film extension was attached to the bottom edge of the chamber to minimize the effect of air exchange between the inside and outside of the chamber. This design has the advantage of not damaging plant roots and of minimizing the impact to the soil surface environment. A preliminary test to compare the performance of the two types of chamber, one with a collar at its base and the current design, has indicated that the difference of the resulting fluxes from them is small. A fan was installed at the center of the top plate inside the chamber to circulate and mix the air and to simulate natural conditions (Heijmans et al. 2004). The vertical wind speed in the chamber was 1.8 $m\,s^{-1}$ just under the fan, and 0.7–0.9 $m\,s^{-1}$ at $z=0.1$, whereas horizontal wind speed was 0.2 $m\,s^{-1}$, also at $z=0.1$ m. The chamber was connected to an air pump that provides a 1 $l\,min^{-1}$ steady flow to and from a CO_2/H_2O gas analyzer (LI-840; LI-COR Biosciences) with a vinyl tube. Temperature of the air inside the chamber and that of the top soil ($z=-5$ cm), both inside and outside the chamber, were also measured. The outside sensor readings were used as a reference. Incident photosynthetically active radiation in wavelengths of 400–700 nm was also measured at $z=1.35$ m near the chamber measurement with a quantum sensor (LI-190; LI-COR Biosciences) to provide photosynthetic photon flux density (PPFD). Because the acrylic material of the chamber has a transmissivity of 83% in photosynthetically active radiation, those PPFD measurements outside the chamber were corrected by the transmissivity to derive PPFD inside the chamber. All data were recorded at 1-s intervals. For the analysis of diurnal changes and daytime mean, it is preferable to use continuous AWS outputs rather than the intermittent chamber measurements of PPFD. In such cases, PPFD values were estimated from the solar radiation measurements of the AWS. Both are related by a single reduction factor (Udo and Aro 1999; Satoh 2010).

Five sites covered partially with shrubs (hereafter referred to as sites CA1 through CA5) were chosen for the ET measurement and five sites of bare ground were chosen for the EV measurement. After rainfall of 5.9 mm on 16–17 July 2009, five additional sites (hereafter referred to as sites AL1 through AL5), whose surface was covered with newly grown herbaceous plants, were added for the measurements. Details of the plants at each site are listed in Table 5.2. Note that these sites for the chamber measurement were not affected directly by grazing during the measuring periods.

For each measurement, the surface was covered with the chamber for about 2 min. The measurement was repeated at least three times for each site. Water vapor flux E ($L\,T^{-1}$) was derived from the time variation of H_2O gas density within the chamber by

Table 5.2 List of the chamber measurement sites

Site	Biomass (g)	LAI	Vegetation coverage, *f* (%)
CA1	64.76	1.28	41.9
CA2	56.93	1.16	44.6
CA3	19.74	0.41	21.5
CA4	19.92	0.31	13.6
CA5	16.85	0.30	15.0
AL1	0.58	0.04	1.2
AL2	1.09	0.09	4.3
AL3	0.56	0.04	2.6
AL4	0.93	0.08	2.4
AL5	0.6	0.03	2.0

LAI, leaf area index, defined here as total one-sided leaf area over the basal area of the chamber; LAI, biomass, and *f* were measured over 3 days on 2009/7/30–8/1

CA represents a site with *Caragana microphylla*; AL represents a site with *Allium polyrrhizum*

$$E = \frac{v}{a}\frac{\rho_a}{\rho_w}\frac{dq}{dt} \tag{5.1}$$

where d*q*/d*t* is the time variation of the specific humidity ($M^3\ M^3 T^{-1}$), *v* is the volume of the chamber (L^3), *a* is the basal area of the chamber (L^2), and ρ_a and ρ_w are the density of air and water ($M^3\ L^{-3}$), respectively. The air pressure needed for ρ_a was obtained from the IMH station. To calculate mean d*q*/d*t*, data over 30 s were used. This period was selected from those data taken after the surface had been covered with the chamber and after the steady mixing of the air inside the chamber had been established to indicate a more or less constant slope of d*q*/d*t*. The latter condition was usually established at about 15 s after the chamber closure.

The derived *E* from the measurements at the CA sites and AL sites represents ET. Thus, TR flux was evaluated as $E_{TR} = [E_{ET} - (1-f)E_{EV})]/f$ in which *f* is the fractional plant coverage of each site derived by the procedures described above, and E_{EV} and E_{ET} are the flux of EV and ET, respectively. However, this would result in a large error in E_{TR} when *f* is small and when ET and EV are of the same order of magnitude. As this was the case for the AL sites after rainfall events, these cases were excluded from the following analysis.

5.2.3 Sampling of Water and Stable Isotope Analysis

For the isotopic analysis, precipitation was accumulated within a polyethylene tank, placed at the IMH station, to which a funnel (120 mm in diameter) containing a ping-pong ball to prevent evaporation loss (Yamanaka et al. 2007b) was attached; the accumulated water was collected at the end of each month from April through September, 2009. Groundwater was collected manually on 1 June 2009 and 20 June from GW1 and on 1 June, 11 July, 18 July, and 1 August 2009 from GW2.

Plant and soil samples were collected on 7 July 2008 and 2 June, 13 July, 19 July, 25 July, and 28 July 2009, except for 13 July (shrubs only) and 22 July (plant samples only). For plants, stems and roots fully covered with periderm were randomly selected, the periderm cover was peeled off, and the remaining parts were stored in a 100-ml fully sealed test tube and kept refrigerated. Three soil samples were collected at each depth and placed into 200-ml glass vials double sealed with polyethylene and polypropylene caps. The target depths were $z=-0.05$, and from $z=-0.1$ to −1.0 m at 0.1-m intervals.

Water in the plant and the soil samples was later extracted in a laboratory using the cryogenic vacuum distillation method (Iizuka et al. 2004) and the ordinary temperature distillation (OTD) method (Yamanaka and Shimada 1997), respectively. The OTD method is better suited for a dry soil sample in that it will not extract water in hydrated minerals, which cannot be utilized by plants.

The hydrogen stable isotope ratio was determined with the gaseous hydrogen equilibration method (Coplen et al. 1991) using a hydrophobic platinum catalyst, whereas the oxygen isotope ratio was determined with the carbon dioxide equilibration method (Socki et al. 1999), both by an isotope ratio mass spectrometer (Finnigan MAT252; Thermo Fisher Scientific). The precision of the analysis by the spectrometer is approximately ±1‰ for δD and ±0.1‰ for $\delta^{18}O$.

5.3 Results and Discussion

5.3.1 Evaporation and Transpiration

Figure 5.2 represents the relationship between solar radiation and EV for 14–21 July and 25–31 July 2009, during which the main rainfall events took place on 16 July (2250–0050, 1.5 mm), 17 July (0440–0600, 4.4 mm), 26 July (0700–0940, 4.4 mm), and 27 July (0150–0220, 3.5 mm). First, the general increase of EV against the solar radiation on each day can be observed. However, the slope is different from one day to another. Clearly, EV increased rapidly with solar radiation on the days of rainfall, which was followed by a rapid decline of the slope and magnitude of EV. In both cases, EV came back to the pre-rainfall magnitude only in several days. Figure 5.3 shows the corresponding relationship between PPFD and TR_{CA} for the same periods. The general trend of TR_{CA} increase with solar radiation (or more exactly with PPFD in the case of plants) on each day is the same as EV. However, in contrast to EV, TR_{CA} did not change much in response to the rainfall events.

The behavior of *Allium* transpiration TR_{AL} was different from and less obvious than that of TR_{CA}. As shown in Fig. 5.4, TR_{AL} varied largely from one day to another. However, in response to the rainfall events, TR_{AL} increased for 20–22 July and 30–31 July, and decreased for the drier period of 24–25 July; this appears to be because TR_{AL} depends more strongly than TR_{CA} on the surface SWC, just as the case of EV. Time variation of SWC near the surface at $z=-0.1$ m and the daily precipitation are shown in Fig. 5.5. Clearly, those 2 days of 24–25 July were with

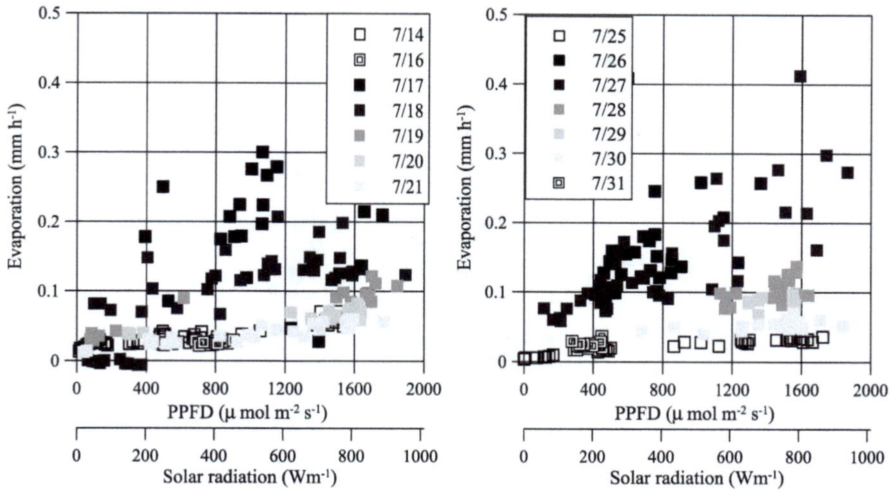

Fig. 5.2 Relationship between the solar radiation and evaporation (EV) before and after the rainfall event on 16–17 July and 26–27 July 2009. *PPFD*, photosynthetic photon flux density

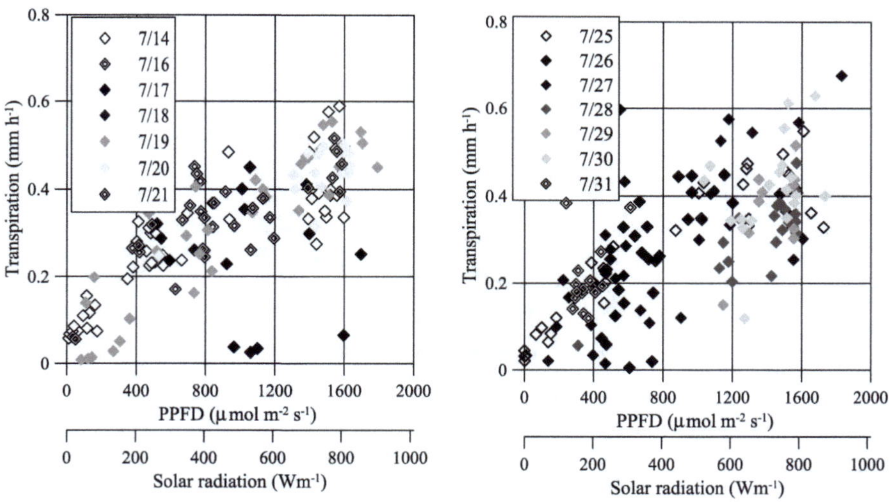

Fig. 5.3 Relationship between photosynthetic photon flux density (*PPFD*) and transpiration (TR) from shrubs before and after the rainfall event on 16–17 July and 26–27 July 2009

minimum SWC. Also important is the plant growth of *Allium* in response to the rainfall. Figure 5.5 gives the time variation of the midday mean TR_{AL} for PPFD > 1,400 μmol m^{-2} s^{-1} and the maximum plant height of *Allium*. After the rainfall event on 16–17 July, SWC and TR_{AL} started to decrease on 20 July. Three days later, *Allium* started to die in part, as indicated by the slight decrease of plant height as a result of withering at the tip of leaves. However, after the rainfall events on 26

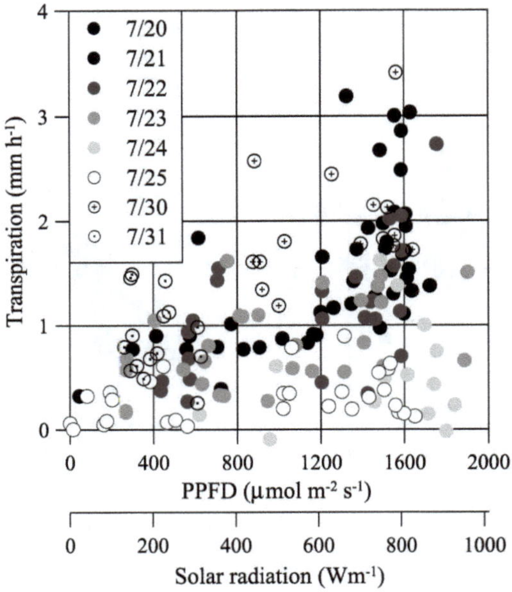

Fig. 5.4 Relationship between photosynthetic photon flux density (*PPFD*) and transpiration (TR) from herbaceous plants

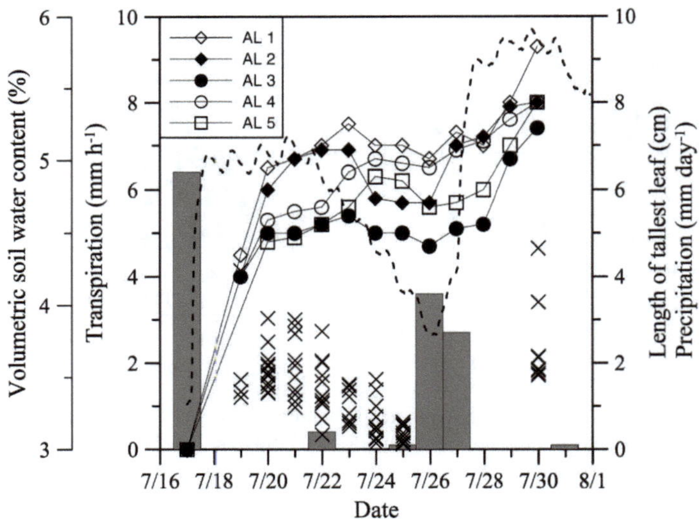

Fig. 5.5 Relationship between daily precipitation (*bar chart*), soil water content (SWC) at $z=-0.1$-m depth measured on the MG2 station (*broken line*), TR_{AC} for $R_p>1{,}400$ (μmol m^{-2} s^{-1}) (*crosses*), and length of tallest leaf in each *AL* site. (AL represents a site with *Allium polyrrhizum*)

and 27 July, regeneration of these plants was observed in response to the increase of SWC and TR_{AL}. This observation agrees with a detailed assessment of the factors influencing the plant growth made near Mandalgobi by Hirata et al. (2009). Physiologically, this can be explained as follows. After the rainfall, herbaceous plants start to produce sugar to encourage growth. As a result, TR_{AL} increases shortly after the rainfall. When SWC becomes depleted after several days from the event, plants experience water stresses. Under this circumstance, plants try to close their stoma to keep the water inside the plants. However, vapor would be lost through the cuticle (Chapin et al. 2002; Larcher 2003), which should partly be reflected to TR_{AL} in the dry period.

Also different between the shrubs and herbaceous plants is the magnitude of TR. Clearly, TR_{AL} is much larger than TR_{CA} on a unit vegetation cover basis. However, because coverage f is also different between them, the area-averaged TR, that is, the landscape TR, should be different from TR_{AL} and TR_{CA} on a unit vegetation cover basis. The landscape TR can be estimated by combining the area average f and the chamber measurements. Because the diurnal variation of the vapor fluxes is quite large, it is desirable to use averages to make a simple comparison. In the present analysis, a typical 3 days were chosen from those days when the diurnal changes were observed by the chamber measurements and the averages for each day were derived. Selected dates are (1) dry day on 15 July 2009 before the rainfall event on 16–17 July without herbaceous plants, (2) wet day on 21 July 2009 shortly after the rainfall event, and (3) dry day on 25 July 2009 after a week from the rainfall event.

For the estimation of the daily mean TR, EV, and ET, first, empirical equations that relate vapor fluxes E with meteorological factor(s) were derived. A stepwise approach to select relevant variables in a multiple regression analysis (Draper and Smith 1981) to consider all factors that could possibly affect E was applied to all chamber measurements in each of the selected 3 days. However, the only variable chosen for the regression was radiation for all cases, probably because the diurnal ranges of other meteorological factors were not large enough to explain the E variation within 1 day. Thus, it was decided to use PPFD to estimate TR and the solar radiation for EV. The selected equations on the basis of R^2 value (which ranged from 0.65 to 0.97, except for TR_{AL} on 25 July, for which it is 0.30) are listed in Table 5.3. Note that a single equation for these 3 days was derived for TR_{CA} because TR_{CA} behavior on these 3 days was quite similar. The nighttime TR was assumed as zero because C_3 plants in general close their stomata during the night time to stop TR. The EV during night was estimated as 4.23×10^{-3}–1.83×10^{-2} mm h^{-1} from the derived equations. Daily totals of TR_{CA}, TR_{AL}, and EV were thus obtained by integrating 30-min values estimated from equations in Table 5.3, and then area averages were derived for each component, and also for ET as a weighted means of TR and EV (Table 5.4). The weighting factors are the fractional coverage of *Caragana*, *Allium*, and bare soil, also listed in Table 5.4.

Table 5.4 shows that summertime landscape ET was of the order of 0.6–1.0 mm day^{-1} in this area. EV is clearly the largest (64–85%) consumer of water, followed by TR_{CA} (15–31%) and TR_{AL} (<13%). The TR to ET ratio is smaller than that reported for similar climate and surface conditions in the literature. For example,

Table 5.3 Empirical equations to estimate TR_{CA}, TR_{AL}, and EV

Date	TR_{CA} $E = a_1 R_P^{b_1}$		TR_{AL} $E = a_1 R_P^2 + a_2 R_P^1 + b_1$			EV $E = a_1 R_s^2 + a_2 R_s^1 + b_1$		
	a_1	b_1	a_1	a_2	b_1	a_1	a_2	b_1
14 July	1.47×10^{-2}	4.54×10^{-1}	N/A			0	5.39×10^{-5}	1.83×10^{-2}
21 July			7.09×10^{-7}	-7.44×10^{-4}	9.36×10^{-1}	2.02×10^{-9}	4.73×10^{-5}	1.46×10^{-2}
25 July			-3.31×10^{-7}	7.04×10^{-4}	4.47×10^{-2}	2.76×10^{-8}	5.78×10^{-5}	4.23×10^{-3}

E, vapor flux (mm h^{-1}); TR_{CA}, transpiration of *Caragana*, TR_{AL}, transpiration of *Allium*, EV, evaporation; R_p, photosynthetic photon flux density (PPFD) (μmol m^{-2} s^{-1}); R_s, solar radiation (Wm^{-1}); a_1, a_2, a_3, b_1, regression coefficients

Table 5.4 Estimates of landscape mean transpiration (TR), evaporation (EV), and evapotranspiration (ET)

	TR_{CA}		TR_{AL}		EV		ET	SWC
Date	f_{CA} (%)	E (mm/day)	f_{AL} (%)	E (mm/day)	$1-f_{CA}-f_{AL}$ (%)	E (mm/day)	E (mm/day)	(%)
14 July	3.12 (11 July)	0.150	0 (11 July)	0	96.88 (11 July)	0.846	1.00	3.6
21 July	3.86 (29 July)	0.169	0.69 (29 July)	0.121	95.45 (29 July)	0.653	0.943	5.0
25 July		0.185		0.0279		0.382	0.594	4.0

SWC is at $z = -0.1$ m at MG2 station

Tsujimura et al. (2007b) found the ratio to be 35–59% in semiarid grassland of mainly *Stipa*, *Carex*, and *Artemisia* (f=47–52%, LAI=0.50–0.57, SWC=14–17% at z=−0.05 m) in northeastern Mongolia in the July–August period. In the shrub land that consists mainly of *Zinnia*, *Flourensia*, and *Acacia* (f=26%, SWC=1–4% at z=−0.025 m) of southern Arizona, Stannard and Weltz (2006) determined that TR consists 84% of ET on average for 3 days in August. Thus, even a small difference in vegetation coverage, plant type, and soil moisture appears to be important. However, if TR/ET per 1% of shrub vegetation coverage is considered, the value for TR_{CA} (5–8%) agrees with that determined by Stannard and Weltz (2006) for *Zinnia* (4%). For herbaceous plants, the agreement is less clear with 7–19% for TR_{AL} and 67–117% from Tsujimura et al. (2007a). This finding is not surprising as shrubs are not affected as much by day-to-day variation of SWC as are herbaceous plants.

Also of interest is a thought experiment to consider the two extreme cases, one with no vegetation cover (f=0) and another with full vegetation cover (f=1) in the study area. The ET for f=0 could be estimated by calculating E with f=0. It can easily be shown that E=0.87 mm day^{-1} for 14 July. Similarly, for f=1 covered with shrubs, ET can be estimated as E=4.94 mm day^{-1}. This value is clearly unrealistic, as this level of ET would consume water equivalent to annual precipitation within only 1 month. Clearly, shrubs of f=1 would not survive in this area and eventually come to an equilibrium surface cover, which from the foregoing simple theoretical experiment appears to be more closer to bare ground than the vegetated surface.

5.3.2 Subsurface Water Dynamics

As already shown, shrubs are a more steady consumer of water than the soil surface or grassy vegetation regardless of rainfall events, even though the relative amount of landscape shrub transpiration is much smaller than soil evaporation because of its small fractional coverage. To scrutinize the mechanism causing this difference, water transfer through the soil–plant–atmosphere system was further examined. Figure 5.6 indicates the seasonal change of soil water profile for the period of 1 March–31 July 2009. The groundwater level at GW1 was more or less stable at 4.5–5.1 m from the surface (Sugita et al. 2012). The soil water recharge near the surface (z>−0.2 m) took place first in March when the soil temperature exceeded 0°C by melting frozen soil water and/or accumulated snowfall over the soil surface. According to the IMH station data, snowfall from November 2008 through March 2009 was 2.6 mm; this was followed by the infiltration of the rainwater into the soil column down to −0.7 m by a heavy rainfall event of 4.7 mm per 1.3 h on 19 April.

Also noticed in Fig. 5.6 is the presence of soil layer at around z=−0.2 m in which SWC started decreasing shortly after the heavy rainfall in April, and that layer remained with the least SWC within the soil column. Three-phase analysis has shown that the porosity is 25–30% at 0.2–0.4 m depth in contrast to 30–40% in other layers. In contrast, the soil layer at around z=−0.3 m retained more SWC through the year, probably the result of the presence of Calcic horizon extending

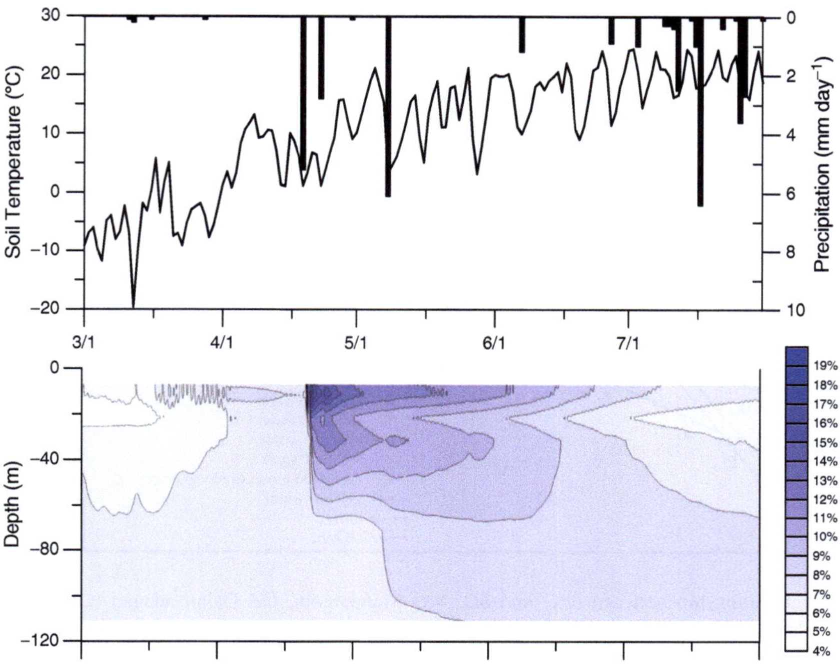

Fig. 5.6 Relationship of daily precipitation and soil temperature from May through July 2009 (*top*) and isopleths of volumetric soil water content (%) (*bottom*)

from at around $z = -0.3$ m downward. Soil permeability tests have indicated that the saturated hydraulic conductivity of the soil in this horizon is of the order of 10^{-7} to 10^{-8} m s^{-1}, as compared to 10^{-5} to 10^{-6} m s^{-1} in other layers. Thus, the water that has penetrated tends to be retained, at least partially, at around $z = -0.3$ m.

Note in Fig. 5.6 that the TDR sensors at the shallower depth of the MG1 station did not respond to the series of rainfall events in the June–July period. The reason for this is not immediately clear. However, the fact that the TDR sensors of the MG1 station were buried under a mound may have been a factor. Vegetation cover, particularly during summer, may have intercepted rainfall. Indeed, SWC data at the MG2 station without shrubs show response to these rainfall events. Thus, whenever necessary, MG2 station data are also used in the analysis.

Figure 5.7 shows the relationship between δ^{18}O and δD values for soil water, groundwater, and precipitation in 2009. The volume-weighted mean value of the rainwater was determined by using the monthly precipitation measured by the MG1 station. A global meteoric water line (GMWL) is a regression line determined from stable isotope ratios of hydrogen and oxygen of monthly precipitation over the globe (Craig 1961). On a local scale, an equivalent local meteoric water line (LMWL) was derived from the precipitation isotopic value of Mandalgobi. The slope of LMWL was $a = 7.36$ with an intercept $b = 6.19$. This finding corresponds well with the result from Yamanaka et al. (2007b) for the monthly rain samples collected over 1 year

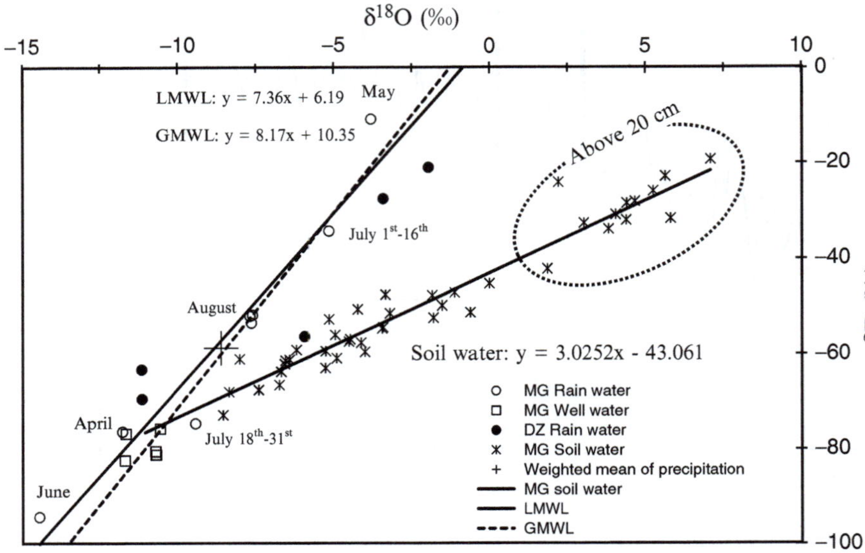

Fig. 5.7 Relationship between δ18O and δD. MG Mandalgobi, DZ Dalanzadagad (Omungobi *aimag*, about 270 km south of MG). DZ data, shown for comparison, were sampled at IMH station there and analyzed in the same manner as in MG. Precipitation values are the monthly means as accumulated rainfall was collected once in a month, except for July when it was sampled twice. LMWL local meteoric water line, GMWL global meteoric water line

from October 2002 to September 2003 at Mandalgobi, with $a=7.75$ (±0.77)‰ and $b=1.47$ (±14.19)‰, which tends to indicate that the observed years were not unusual in terms of the general weather system and water inputs to the area.

All the data from groundwater are clustered along the LMWL, which suggests that groundwater should originate from precipitation, without much influence of evaporation during the recharge process to the groundwater. This in turn implies that groundwater recharge took place within a short time period, by for example rapid infiltration of strong rainfall events, perhaps by means of channel-flows. Tsujimura et al. (2007b) also speculated from isotopic analysis of groundwater and precipitation in the northeastern part of Mongolia that the recharge of groundwater occurs only by relatively large rainfall events. Thus, under the normal situation, rainwater that infiltrates into the soil does not reach groundwater; rather, it remains in the shallow soil layers and returns to the atmosphere only through evaporation and transpiration. This finding can also be confirmed by Fig. 5.5 in which activities of liquid water are limited mainly within the layers of $z>-1$ m, agreeing with the more detailed assessment of the surface water balance in this region by Yamanaka et al. (2007a) and in this site by Sugita et al. (2012), who found only the soil column of uppermost 0.2–0.25 m was actively involved in the hydrological processes of infiltration and evaporation.

Further evidence that the infiltrated water to the shallow soils goes back to the atmosphere can be obtained from Fig. 5.7. The data for the soil water from surface down to $z = -1$ m, particularly those from $z > -0.2$ m, are scattered in the right-hand side of LMWL, which means that soil water was affected by the evaporation process (Schwartz and Zhang 2003; Gat and Airey 2006; Gibson et al. 2008). Moreover, the slope of soil water regression of $a = 2$ (or larger in mid- to higher latitude) indicates that the soil water was vaporized within the soil column and then transported mainly through molecular diffusion (Gibson et al. 2008). For the soil water in this region, $a = 2.97$, on average, and thus this tends to support that vaporization within the soil takes place and water vapor is transported by molecular diffusion through the soil layers from at least $z = -1$ m to the surface. However, the relative role of vapor transportation from deeper layers is probably small when accounting for the overall amount of EV. As shown above by the chamber measurements, EV responded to rainfall events. Vaporization of soil water and molecular diffusion transport within the soil column should not be affected by rainfall events. Therefore, the main source of water for soil evaporation is probably the soil water in the layers just below the surface.

The regression line of soil water isotope ratios intersects with LMWL at around $\delta^{18}O = -11.4$‰ and $\delta D = -77.4$, which is quite close to the point for the April precipitation ($\delta^{18}O = -11.8$‰ and $\delta D = -76.3$). It is also close to a crude estimate of the volume-weighted annual mean precipitation values of $\delta^{18}O = -11$‰ and $\delta D = -77$, made by the current data set, the 2008 winter snowfall amount data at the IMH station, and the values of $\delta^{18}O$ and δD of the snowfalls determined in 2003 by Yamanaka et al. (2007b). Thus, two different interpretations of the source of soil water could be made. The first interpretation would be that the major portion of the source of soil water below at least $z < -0.2$ m in 2009 is the rainfall in April. The second interpretation would be that infiltrated water of a series of rainfalls throughout the year was mixed well within the soil columns, and then subjected to vaporization, forming the observed specific isotopic profile. However, based on the assessment shown by Fig. 5.6, that mainly the rainfall event on April 19 reached and recharged the soil layer down to $z = -1$ m, we tend to support the first interpretation.

Note that the fact that those points of Fig. 5.7 for the shallower layer of $z > -0.2$ m are on the same regression line for the entire group of soil water samples is probably a coincidence, and should not be interpret as the indication that those waters also came from the April rainfall. It is more likely that the source of soil water at shallower layers is changing day to day, and it is probably those rainfall events that took place in close proximity. In fact, Yamanaka et al. (2007a) indicated that the mean summer residence time of water in the upper 0.2-m layer was only 20–25 days in this area.

The vertical profiles of $\delta^{18}O$ in soil water sampled on 13 July, 18 July, 25 July, 28 July, and 1 August 2009 are shown in Fig. 5.8. The $\delta^{18}O$ values in the shallow soil water tend to be higher than those of deeper soil water because of the evaporation effect. Clearly, the rainfall event on 16–17 July did not cause much difference in the soil water isotopic value, which is not surprising as the visual observation of the soil profile shortly after this rainfall event revealed that water penetrated to a depth of only 0.1 m (Satoh 2010). The SWC increase observed only at 0.1 m at MG2 also supports this penetrating depth.

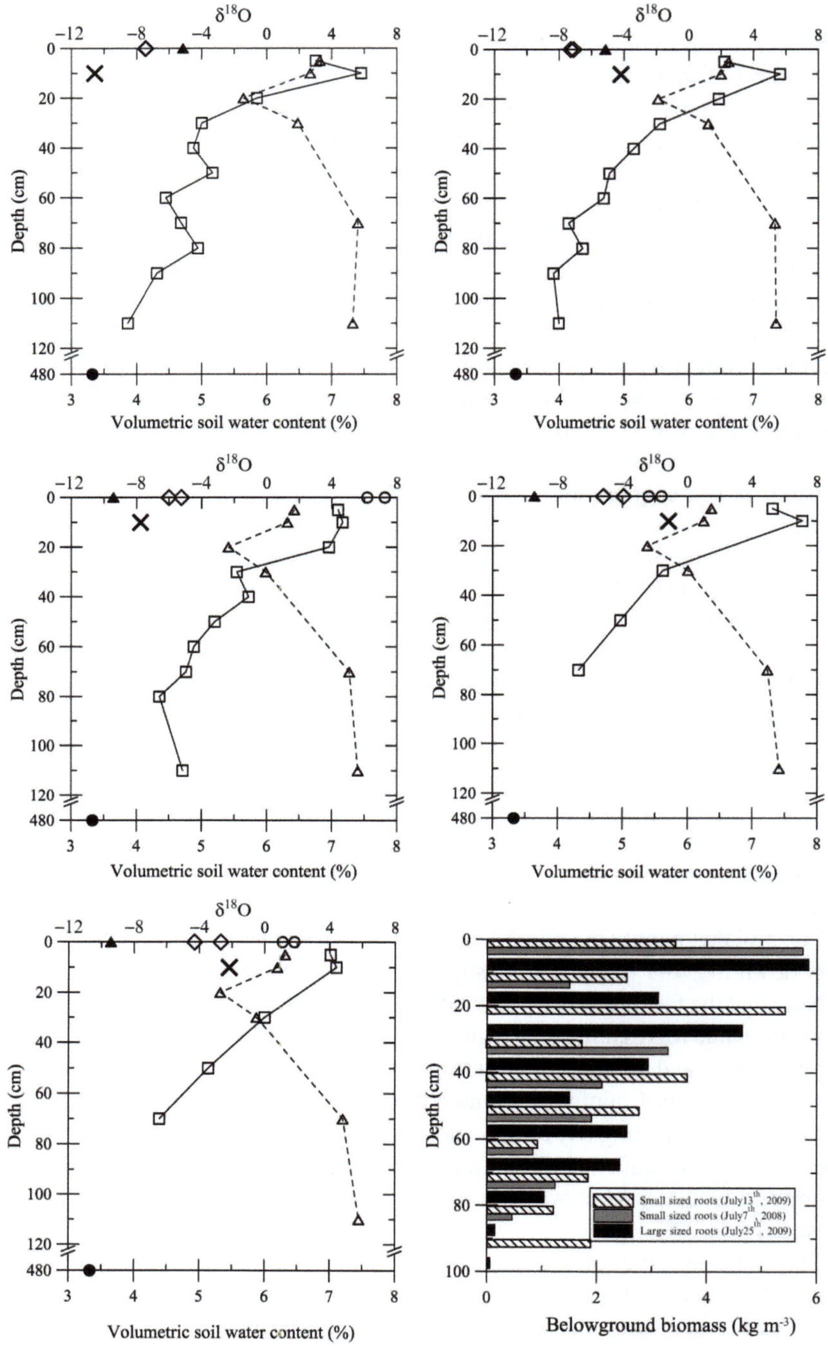

Fig. 5.8 Vertical profiles of the $\delta^{18}O$ in soil water (*open squares*) with the $\delta^{18}O$ of well water (*closed circles*) and in xylem sap of shrubs (*open diamonds*) and herbaceous plants (*open circles*). Also shown are SWC profile at MG1 station (*open triangles*) and surface SWC at MG2 station (*crosses*). *Top left panel*, data from 13 July 2009; *top right panel*, data from 18 July 2009; *middle left panel*, data from 25 July 2009; *middle right panel*, data from 28 July 2009; *lower left panel*, data from 1 August 2009. *Lower right panel* shows belowground biomass. Note that biomass amount for small-size root at $z = -0.2$ m on 7 July is not available

Another important feature one can observe in Fig. 5.8 is the difference of water source for TR and EV. The comparison of $\delta^{18}O$ values indicates that shrubs take water from around $z=-0.7$ to 0.9 m, the layers with higher SWC than the other layers. Indeed, the root system of *Caragana* extended down to 1 m (Fig. 5.8); however, the amount of roots found around $z=-0.7$ to 0.9 m is not necessarily larger as compared to that at shallower layers. Thus, the amount of SWC appears to be more important to determine the source of water that the plants actually utilize. Physiologically, this factor should be the difference of water potential inside and outside the plant roots. Thus, those soil layers that can provide a larger difference which allows easier soil water absorption by plant roots are likely to be the source of water for the plants. For TR_{AL}, the water source appears to be the layer close to the soil surface in most cases, which is in agreement with its root extension from the surface down to 0.1–0.2 m. For EV, as already mentioned, the main water source is the soil near the surface, but, in addition, soil layers from the surface down to at least 1 m also contribute to EV to some extent. As mentioned, soil water evaporates within soils, and water vapor is transported to the surface by means of molecular diffusion, and then by turbulent diffusion to the upper atmosphere.

5.4 Conclusions

The study was conducted to clarify the characteristics of the surface–atmosphere interaction by taking into consideration the soil water dynamics as well as the role of the plants in an arid region of central Mongolia, by combining chamber measurements, the stable isotope analysis of oxygen and hydrogen, and other various in situ measurements. The dominant vegetation consists of a shrub (*C. microphylla*, C_3 plant) and a herbaceous plant (*A. polyrrhizum*, C_3 plant).

Based on the analysis of the data set obtained for the two summer seasons, it was found that there is a clear difference in the response to rainfall, between plant types, and also between vegetation and the soil surface. Soil evaporation increased rapidly shortly after rainfall events. Subsequently, it declined again rapidly during the dry period as a result of the depletion of near-surface soil water. A similar behavior can be noted for the herbaceous plants. In contrast, transpiration of shrubs was found to be more stable and to not change much with rainfall events. This contrast was the result of the difference of the source of water used for transpiration by shrubs and herbaceous plants and for evaporation by the soil surface. The shrubs take water mostly from the deeper soil layers at around $z=-0.7$ to 0.9 m, whereas transpiration of grassy vegetation and soil evaporation tend to use water near the surface, as indicated by isotopic analysis and observation of the vertical distribution of the root systems and SWC.

The relative strength of water vapor transfer from the surface into the atmosphere over a unit cover of vegetation or unit soil surface was most effective with *Allium* (with hourly transpiration of the order of 0–3 mm h^{-1}), followed by *Caragana* (0–0.6 mm h^{-1}) and soil evaporation (0–0.3 mm h^{-1}) during summer months.

However, when the small amount of surface cover of vegetation (<5%) is taken into consideration, the landscape evapotranspiration was found to be dominated mainly by soil evaporation (0.4–0.8 mm day^{-1}), with a smaller contribution by the transpiration of *Caragana* (0.2 mm day^{-1}) and the transpiration of *Allium* (0–0.1 mm day^{-1}).

Acknowledgments The authors thank G. Davaa and D. Oyunbaatar (Institute of Meteorology and Hydrology, IMH) for facilitating local logistics and for arranging maintenance of the stations through Meteorological Centers of IMH at Mandalgobi and Dalanzadgado. They are also grateful for I. Byambakhuu, S. Yoshizawa, and Y. Morinaga (University of Tsukuba) for participating in field observations for this study. Finally, the authors also express their appreciation to N. Fujita and N. Yamamura (Research Institute for Humanity and Nature, RIHN), without whose initiative and leadership the experiments for this study would not have been possible. This research has been supported and financed, in part, through a RIHN project "Collapse and Restoration of Ecosystem Networks with Human Activity."

References

Byambakhuu, I, Sugita M, Matsushima D (2010) Spectral unmixing model to assess land cover fractions in Mongolian steppe regions. Remote Sens Environ 114:2361–2372

CGIAR-CSI (2008): SRTM 90 m digital elevation data, http://srtm.csi.cgiar.org/Index.asp. Accessed 12 April 2009

Chapin FS III, Matson PA, Mooney HA (2002) Principles of terrestrial ecosystem ecology. Springer, Berlin

Coplen TB, Wildman JD, Chen J (1991) Improvements in the gaseous hydrogen-water equilibration technique for hydrogen isotope ratio analysis. Anal Chem 63:910–912

Craig H (1961) Isotopic variations in meteoric waters. Science 133:1702–1703

Dorzhgotov D (2003) Soils of Mongolia. Admon, Ulaanbaatar (in Mongolian with English abstract)

Draper NR, Smith H (1981) Multivariate analysis. Wiley, New York

Ehleringer JR, Philips SL, Schuster WSF, Sandquist DR (1991) Differential utilization of summer rains by desert plants. Oecologia (Berl) 88:430–434

Gat JR, Airey PL (2006) Stable water isotopes in the atmosphere/biosphere/lithosphere interface: scaling-up from the local to continental scale, under humid and dry conditions. Global Planet Change 51:25–33

Gibson JJ, Birks SJ, Edwards TWD (2008) Global prediction of δ_A and δ^2H-δ^{18}O evaporation slopes for lakes and soil water accounting for seasonality. Global Biogeochem Cycles 22: GB2021

Hirata M, Kishikawa S, Kondoh A, Yamanaka T, Kaihotsu I, Damdin B, Hongo A (2009) Effects of some environmental factors on the plant growth in the central Mongolian Plateau. J Arid Land Studies 19-2:403–411 (in Japanese)

Heijmans MMPD, Arp WJ, Chapin FS III (2004) Carbon dioxide and water vapour exchange from understory species in boreal forest. Agric For Meteorol 123:135–147

Iizuka S, Yamanaka T, Tanaka T (2004) Extraction of water from a plant for isotopic measurement: pot experiments. Bull Terr Environ Res Ctr Univ Tsukuba 5:81–86 (in Japanese)

Kurc SA, Small EE (2004) Dynamics of evapotranspiration in semiarid grassland and shrubland ecosystems during the summer monsoon season, central New Mexico. Water Resour Res 40:W09305

Larcher W (2003) Physiological plant ecology: ecophysiology and stress physiology of functional groups. Springer

Li S-G, Asanuma J, Kotani A, Davaa G, Oyunbaatar D (2007) Evapotranspiration from a Mongolian steppe under grazing and its environmental constraints. J Hydrol 333:133–143

Nakano T, Manabu M, Shinoda M (2007) Measurement of nighttime water vapor flux using a closed chamber technique. J Agric Meteorol 63:103–107 (in Japanese)

Pratt M, Rasmussen GA (2001): Determining your stocking rate. Utah State University Extension Electric Publishing, NR/RM/04, 7 p.

Romero-Saltos H, Sternberg L, Da SL, Moreira MZ, Nepstad DC (2005) Rainfall exclusion in an eastern Amazonian forest alters soil water movement and depth of water uptake. Am J Bot 92:443–455

Schwartz FW, Zhang H (2003) Fundamentals of ground water. Wiley, New York

Smith DM, Jarvis PG, Odongo JCW (1997) Source of water used by trees and millet in Sahelian windbreak systems. J Hydrol 198:140–153

Satoh T (2010) Study on vapor transfer processes into the atmosphere from vegetated surface in an arid region. Master's thesis, Graduate School of Life and Environmental Sciences, University of Tsukuba, Japan

Socki RA, Romanek CS, Gibson EK Jr (1999) On-line technique for measuring stable oxygen and hydrogen isotopes from microliter quantities of water. Anal Chem 71:2250–2253

Stannard DI, Weltz MA (2006) Partitioning evapotranspiration in sparsely vegetated rangeland using a portable chamber. Water Resour Res 42:W02413

Sugita M, Asanuma J, Tsujimura M, Mariko S, Lu M, Kimura F, Azzaya D, Adyasuren T (2007) An overview of the rangelands atmosphere-hydrosphere-biosphere interaction study experiment in northeastern Asia (RAISE). J Hydrol 333:3–20

Sugita M, Yoshizawa S, Byambakhuu I (2012) Is water a limiting factor for the nomadic pastoralism in Mongolian steppe? (in preparation)

Tsujimura M, Sasaki L, Yamanaka T, Sugimoto A, Li S-G, Matsushima D, Kotanib A, Saandar M (2007a) Vertical distribution of stable isotopic composition in atmospheric water vapor and subsurface water in grassland and forest sites, eastern Mongolia. J Hydrol 333:35–46

Tsujimura M, Abe Y, Tanaka T, Shimada J, Higuchi S, Yamanaka T, Davaa G, Oyunbaatar D (2007b) Stable isotopic and geochemical characteristics of groundwater in Kherlen River basin, a semi-arid region in eastern Mongolia. J Hydrol 333:47–57

Udo SO, Aro TO (1999) Global PAR related to global solar radiation for central Nigeria. Agric For Meteorol 97:21–31

Weltzin JF, McPherson GR (1997) Spatial and temporal soil moisture resource partitioning by trees and grasses in a temperate savanna, Arizona, USA. Oecologia (Berl) 112: 156–164

Yamanaka T, Shimada J (1997) Simple but reliable method to extract soil water for stable isotope analysis. J Jpn Soc Hydrol Water Resour 10(2):181–184 (in Japanese)

Yamanaka T, Kaihotsu I, Oyunbaatar D (2003) Temperature effect on soil water content measured by time domain reflectometry and its correction using in situ data. Bull. Terr. Environ. Res. Ctr., Univ. Tsukuba., 16:246–254 (in Japanese)

Yamanaka T, Kaihotsu I, Oyunbaatar D, Ganbold T (2007a) Summertime soil hydrological cycle and surface energy balance on the Mongolian steppe. J Arid Environ 69:65–79

Yamanaka T, Tsujimura M, Oyunbaatar D, Davaa G (2007b) Isotopic variation of precipitation over eastern Mongolia and its implication for the atmospheric water cycle. J Hydrol 333:21–34

Yepez EA, Huxman TE, Ignace DD, English NB, Weltzin JF, Castellanos AE, Williams DG (2005): Dynamics of transpiration and evaporation following a moisture pulse in semiarid grassland: a chamber-based isotope method for partitioning fluc components. Agric For Meteorol 132:359–376

Yoshizawa S (2010) Study on sustainable water use in arid and semiarid of Mongolia. Master thesis, Graduate School of Life and Environmental Sciences, University of Tsukuba, Japan

Zenchich SJ, Froend RH, Turner JV, Gailitis V (2002) Influence of groundwater depth on the seasonal sources of water accessed by *Banksia* tree species on a shallow, sandy coastal aquifer. Oecologia 131:8–19

Chapter 6
A Possible Future Picture of Mongolian Forest-Steppe Vegetation Under Climate Change and Increasing Livestock: Results from a New Vegetation Transition Model at the Topographic Scale

Reiichiro Ishii and Noboru Fujita

Keywords Climate change • Forest-steppe transition model • Livestock grazing pressure • Topographic scale • Vegetation–soil moisture positive feedback

6.1 Introduction

To know how Mongolian vegetation changes under global changes of climate and human activities, a new modeling scheme is needed. It must be able to incorporate both these two different types of drivers in a comparative way and also to be validated by observation in the real ecosystem. Generally, we do not have long-term observation data of vegetation change along with these drivers, and therefore it is difficult to quantify the correlations between the vegetation and the drivers. In such a case, it is reasonable to do it by comparing the distribution of spatial vegetation patterns with the distribution of environmental parameters altered by the drivers. Therefore, at first, to develop such a model we need to determine the spatial scale and the vegetation pattern on which we focus because vegetation exhibits various spatial patterns reflecting the distribution of the essential environmental conditions but it also varies depending on the spatial scale of observation. To date, vegetation dynamics under global environmental changes have been studied mainly in two different spatial scales and contexts. At larger scales, such as a global or continental scale, we observe vegetation distribution with extent of thousands of kilometers

R. Ishii (✉)
Research Institute for Global Change, Japan Agency for Marine-Earth Science and Technology, Yokosuka, Japan
e-mail: r.ishii@jamstec.go.jp

N. Fujita
Research Institute for Humanity and Nature, Kyoto, Japan
e-mail: fujita@rihn.go.jp

N. Yamamura et al. (eds.), *The Mongolian Ecosystem Network: Environmental Issues Under Climate and Social Changes*, Ecological Research Monographs,
DOI 10.1007/978-4-431-54052-6_6,

with resolutions of 1 km to 1 degree. At this scale, vegetation patterns are characterized on the basis of the dominant plant function types of vegetation covers at the resolution (i.e., biome). In Mongolia, as easily recognized with satellite images, vegetation cover changes gradually from evergreen conifer forest (i.e., taiga) to steppe, then dry steppe, and finally to desert as latitude decreases. This spatial vegetation pattern matches well with the distribution pattern of annual precipitation at the country scale, high in the north and low in the south (Sugita et al. 2007), and therefore the large-scale vegetation transition pattern in Mongolia can be accounted for by the distribution of precipitation, the major source of soil water, or the limiting factors for plant growth in central Asia. As a general fact, in global vegetation distribution, climatic conditions are the major factor that determines the distribution of vegetation types as biomes (Whittaker 1962). These facts tell us that climate change must have impacts on global vegetation change. From this aspect, the vegetation would be altered toward that of drier climate conditions if the global climate change alters the regional climate to be drier and/or hotter in the coming 100 years (IPCC-AR4 2007). Because the climate data are based on interpolation of sparse weather station (less than 300 points in the whole of Mongolia), the spatial resolution is not sufficiently high to apply to phenomena at finer spatial scales where variation of microscopic environmental factors must be taken into consideration.

On the other hand, vegetation exhibits different spatial patterns at smaller scales. Consider the case where we observe plant community structure in the field. The spatial resolution is high enough to know the density or location of plants, based on which we identify different type of spatial patterns within a biome. Different from observation at larger scale, in a plain small plot, however, the variation of physical (abiotic) conditions for each plant are often difficult to detect and can be regarded as negligible. For sufficiently small plots, therefore, we assume that they are homogeneous. For example, climatic or soil conditions (e.g., air temperature, precipitation, soil structure) can be assumed identical within a plot smaller than 100–1,000 m (hereafter we call such spatial scale as the *local scale*). There, even if the heterogeneity of abiotic environmental conditions cannot be detected or is negligible, we still observe various spatial vegetation patterns consisting of clear transition, dots, patches, stripes, etc. (e.g., Klausmeier 1999), which tells us that the dominant factors to determine vegetation patterns are not always abiotic conditions. Instead, at the local scale, the distributions of biological factors, that is, a plant's interactions with surrounding organisms (i.e., reproduction, dispersal, resource competition, herbivory, and also human activities) might be the major factors. Of these, grazing pressure by livestock is considered to have substantial impacts on the vegetation status in Mongolia. This opinion is supported by previous experimental manipulations of the livestock grazing pressure, that is, heavy grazing pressure lessens the grassland biomass (Fujita et al. 2009).

Because livestock distributes widely in the country, change in livestock density and distribution must also have substantial effects on the vegetation pattern at the country scale. However, the pattern of population distributions at resolutions of soum or aimag (prefecture) does not show clear correlation with the vegetation pattern at the country scale as does climate conditions (Saizen et al. 2010); this does not necessarily mean the inferiority of herbivory effect to that of climate on the vegetation, rather, it shows the difference in spatial resolutions at which the effects of these two drivers on the vegetation pattern become apparent (dominant). One of the reasons is that herders and livestock do not distribute monotonically (or homogeneously) within a soum or an aimag. Their density exhibits a more concentrated pattern reflecting the heterogeneity of topography, transportation network, land ownership, and so on at finer scales where in turn the gradient of climate becomes more difficult to detect. This discrepancy in spatial scales between these drivers' distributions has prevented us from comparing the effects of the two major drivers on the vegetation. To tackle this problem and model the vegetation dynamics mechanistically on the basis of plant growth and plant loss processes, we chose here to set the spatial resolution to about 100 m (i.e., set local scale to a square grid of 100 m) for the following reasons. (1) This is fine enough to incorporate the heterogeneous distribution of livestock density that is finer than soum scale and can be observed by field observations (Nachin et al., in this book). (2) This is fine enough to assume abiological conditions to be homogeneous within a cell grid, which can be estimated from large-scale climatic conditions using a hydrological modeling method reflecting topography. (3) Furthermore, in Mongolia, vegetation exhibits conspicuous spatial patterns that have been considered to relate to these two drivers.

6.1.1 An Interesting Pattern at the Topographic Scale Observed in Northern Mongolia

At the country scale, Mongolian vegetation has its transition zones between boreal forest and steppe in the northern part. When one observes this transient zone closer at the topographic scale, it is generally recognized that the vegetation exhibits discontinuous transition, depending on the slope aspects (Fig. 6.1); that is, the slopes with a north aspect are covered with larch forest (*Larix sibirica*) whereas those with a south aspect are mostly covered with grass species (Fujita et al. 2009). This transition implies that the vegetative structure of the north and south slopes could be regarded as the Siberian boreal forest (i.e., taiga) and the typical steppe vegetation covering the central Mongolia, respectively. These two types of vegetation cover change from one to another clearly on every ridge and valley line in the region. Hereafter, we call this pattern the *target vegetation pattern*, for which we identified two general questions to be solved. Q1: Why are such discontinuous transitions observed at this scale? Q2: Why are the vegetation states slope-aspect dependent?

As we have seen, to answer these questions and reconstruct the pattern at the topographic scale, we need a new mechanistic modeling framework for this specific scale because they are obviously impossible with the previously developed models for larger or smaller spatial scales. The following is the essential background knowledge concerning these questions.

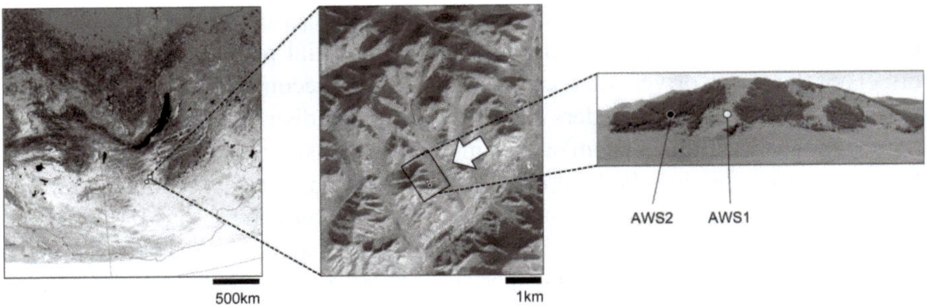

Fig. 6.1 Vegetation patterns in our study observed at different spatial scales. *Left panel*: Overall Mongolian vegetation at the continental scale. *Dark gray* represents forest cover, typically taiga forest; *light gray* represents steppe vegetation. Ulaanbaatar (*open circle*) is located in the transition zone of these two types of vegetation. *Open rectangle* in the northeast of Ulaanbaatar indicates our study area in Gachuurt. *Middle panel*: Close-up view of Gachuurt study site by high-resolution satellite remote sensing (Landsat ETM+). Larch forest (*dark gray*) and steppe cover (*light gray*) distribute continuously on the north-facing and south-facing slopes, respectively. *Rectangle* indicates the intensive research area where plant composition and biomass are investigated (see Fujita et al. 2009). Two automated weather systems (AWS) are set here: *AWS1* on the steppe (*white circle*) and *AWS2* in the forest (*black circle*). *White thick arrows* indicate the direction of the photograph shown in the right panel. *Right panel*: In situ view of the intensive research area in Gachuurt

6.1.2 Previous Explanations for the Target Pattern

Both the foregoing two questions about the target pattern cannot be answered either by precipitation distribution or by livestock density alone, and thus another mechanism and set of conditions relevant to this specific scale must be taken into consideration. These characteristics of the vegetation transition themselves have been widely known and documented in previous studies about the vegetation in the same regions (Ishikawa et al. 2005; Gunin et al. 1999), which only provide verbal explanations for the target patterns.

From hydrological aspects, they suggest the following explanations concerning the two questions. The discontinuity should be caused by the mutualistic relationships between forest cover and permafrost. That is, the existence of forest keeps the ground surface temperature lower, to help keep the surface of permafrost shallower in summer, while in turn, the shallower permafrost surface keeps the soil water from percolating into the deeper layer and helps the tree species to utilize the limited precipitation. This difference suggests that once this cooperating relationship breaks down, either the forest or permafrost could not be held and, thus, the terrain turns into the other stable state, i.e., grassland without permafrost. For the slope aspect dependence, they consider the difference in the received solar radiation. Because both the south ends of permafrost distribution and boreal forests lie in the northern part of Mongolia (Ishikawa et al. 2005), their hypothesis of the two stable vegetation states is consistent with the large-scale pattern and therefore sounds reasonable. Their view is also supported by our preliminary observation conducted in Gachuurt valley (N 48.0°, E 107.1°), in which volumetric soil water content (SWC) was significantly higher in the forest than in the adjacent grassland with the same precipitation, although we could not confirm that the permafrost to be the necessary condition for forest existence or vice versa at this scale. One of the clearest differences between the two vegetations we observed in situ were much thicker soil depth (resulting from richer litter input) and much slower wind speed in the forest zone than in steppe zone (see Fig. 6.2 for the preliminary results of the field observations). Thus, even though the specific effect of permafrost is still not clear, it is natural and reasonable to assume a positive feedback interaction between the dynamics of plant biomass and SWC. From previous theoretical studies, it is known that a positive feedback between the states of organisms and their environmental conditions might cause catastrophic changes of ecosystem status, and the discontinuous spatial patterns of vegetation in arid regions are considered to be one of the typical phenomena generated by positive feedback interaction between soil water and plant biomass (e.g., plants enhance water infiltration and reduce evaporation from the soil surface) (Klausmeier 1999).

On the other hand, it is also suggested by previous studies that the way of pastoralization plays substantial roles for the formation of the target pattern in Mongolia. For example, Hil argues that it is the herders' preference for clearing forests (by cutting and burning) on south slopes that results in the slope aspect dependence of the vegetation, that it is the grazing by livestock which prevents the grassland from being overcome by forest cover (Hilbig 2000). These arguments are also consistent with historical records and documents, although it is not clear whether both these processes are essential for the pattern formation (see discussion by Dulamsuren et al. 2005b). There exists a rich literature about grazed savanna (van de Koppel et al. 1997), where two major mechanisms for the bistability of forest and steppe covers have been suggested. The first is based on the difference of edibility of tree and grass species. Trees could be easily grazed when young and small but become more difficult for livestock to eat after become tall, stiff, and hard, so that they suffer less mortality from livestock herbivory. The other possible mechanism is density-dependent accessibility; that is, once a dense forest is established, even small seedlings could not be easily eaten because their surroundings constrain livestock from approaching (Walker 1989). Both these traits of grazing pressure on trees result in their positive density dependence, which might cause bistable vegetation states. These studies tell that we cannot ignore the effect of livestock on bistability.

These previous discussions support the significance of a vegetation transition model at the topographic scale for elucidation of the ecological interactions and environmental conditions and its relevance for generating the target pattern incorporating the effects of climate change and grazing pressure by livestock.

6.2 The Model

The model consists of two steps. First, we examine a set of simple mathematical models to analyze the behavior of the qualitative dynamics of interactions among plant growth under soil water limitation and the livestock grazing pressure at the local scale. For the second step, based on the mathematical model, we develop a numerical spatial simulation model parameterized to observed values in a Gachuurt plot (Fujita et al. 2009). In this chapter, we describe the essence of the model analysis and some examples of predictions under scenarios with different precipitation and livestock density, which are fully described somewhere else (Ishii, in preparation).

6.2.1 The Mathematical Model

6.2.1.1 Dynamics of Biomasses of the Two Plant Types

We formulate the dynamics of substantial interactions at the local scale as simply as possible. Consider a homogeneous area in an arid region in which grass species and tree species compete for the soil water. Hereafter, $P1$ and $P2$ denote the biomass of grass and tree species, respectively, the local dynamics of which we analyze. The growth rates of both species are Michaelis–Menten (MM) type functions of SWC (W), where the maximum growth rate of species i (g_i) and half-saturation coefficient of (k_i) is in a trade-off relationship, that is, the tree species has a larger maximum growth rate ($g_1 < g_2$) but the grass species has a larger growth rate for small W ($k_1 < k_2$). Both $P1$ and $P2$ are reduced by their mortality and increased by intraspecific or interspecific competition and predation by their herbivores. We also formulate the dynamics of SWC, W. W increases by infiltration of precipitation, W_{in}, and decreases by the drainage/runoff from the soil, W_{out}, and by uptake by the plants, W_{uptake}. All these elements are functions of precipitation φ and the plant biomass, $P1$ and $P2$. From these assumptions, the simultaneous equations of the dynamics of $P1$, $P2$, and W could be expressed as:

$$\frac{dP1}{dt} = g_1 \frac{W}{W+k_1} P1 - \left(m_1 + m_{11}P1 + m_{21}P2 + h_1\right)P1 \quad (6.1a)$$

$$\frac{dP2}{dt} = g_2 \frac{W}{W+k_2} P2 - \left(m_2 + m_{12}P1 + m_{22}P2 + h_2\right)P2 \quad (6.1b)$$

$$\frac{dW}{dt} = W_{in}\left(P1, P2, \varphi\right) - W_{uptake}\left(P1, P2, \varphi\right) - W_{runoff}\left(P1, P2W\right) \quad (6.1c)$$

where g_i, k_i, m_i, and h_i are the maximum growth rate, half-saturation constant of the MM function, natural loss rate, and the loss caused by herbivory for plant species i, respectively. m_{ij} denotes the coefficient for the competition effect of species j on species i (if i=j, it implies intraspecific competition). The soil water content W is defined as the volume of water in the active soil layer per unit surface area that could be accessed by plant roots. Therefore, we do not assume a fixed soil depth. For the analysis, here we employ the following formulations for the three terms of (6.1c): W_{in}, W_{out}, and W_{uptake}. W_{in} can be assumed as the product of the precipitation and the rate of precipitation that enters the soil. Although part of the precipitation enters from the soil surface directly, water could infiltrate more easily through the surfaces of stems and roots. Thus, we assume a MM-type function of the total biomass of the two plants, that is, an increasing function of plant biomass of either species. This assumption implies that a plant has a *facilitation effect* on the other plants by enhancing the soil water infiltration to improve their growth, which is in contrast to the *competition effect* over soil water [indicated by parameters m_{ij} in (6.1)]. W_{out} is proportional to W and the rate changes with soil properties (e.g., porosity) and increases with the total plant biomass. W_{uptake} is the sum of the uptake rates of the two plants, each of which is a MM-type function of W. From these, we substitute the following for the three terms in the right-hand side of (6.1c):

$$W_{in}\left(P1,P2,\varphi\right)=\varphi\frac{P1+P2+k_s w_0}{P1+P2+k_s} \tag{6.2a}$$

$$W_{runoff}\left(P1,P2W\right)=r_w\left(1-r_a\frac{P1+P2}{P1+P2+r_b}\right)W \tag{6.2b}$$

$$W_{uptake}\left(P1,P2,\varphi\right)=\frac{u_1 W}{W+u_1}P1+\frac{u_2 W}{W+u_2}P2 \tag{6.2c}$$

where w_0, k_s, r_w, r_a, r_b, and u_i denote the infiltration rate achieved without plants, a parameter to define the slope of MM function for infiltration of plant biomass, the runoff rate driven by the gravity, the half-saturation constant of the MM function for water-holding capacity of plant biomass, and the half-saturation constant for water uptake by species i of W.

6.2.2 Results of the Mathematical Model

We analyzed the submodel 1 to find the equilibria of the system and investigate its stability. By setting each of the equations (6.1a–c) equal to 0 and solving them, the equilibria of the simultaneous (6.1) could be obtained in the form of $\{P1^*, P2^*, W^*\}$.

We here feature the change of W^* with precipitation parameter φ to clarify the effects of vegetation states to SWC. Note that values less than zero are not feasible for $P1^*$ and $P2^*$. Therefore, if $P1^*$ and/or $P2^*$ have smaller values than 0 for a given φ, we set $P1$ and/or $P2$ in (6.1) to 0 and recalculate the equilibrium for W^*. The feasible equilibria and their stability could be analyzed by the numerical method. Figure 6.2 summarizes the typical relationships between W^* and φ, where thick lines and dots represent the stable and unstable feasible equilibria, respectively. Because the stability of W^* always matches the stability of the vegetation state, the stability of $P1^*$ and $P2^*$ are also depicted as the line type of W^* in Fig. 6.2 (we use "SWC-biomass" when we refer to conjugation). The qualitative traits of the dynamics of the system could be explained graphically and intuitively

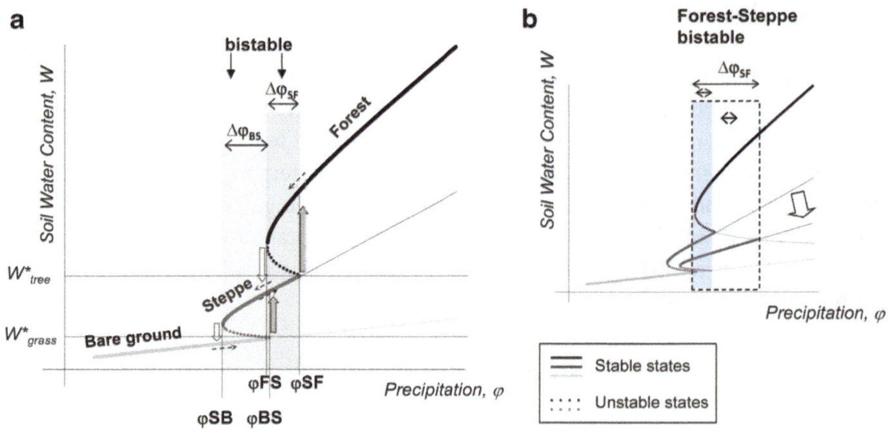

Fig. 6.2 Relationships between soil water content and precipitation. (**a**) Soil water content equilibria W^* of the mathematical model (6.1). *Thick solid/broken curves* indicate the stable/unstable W^* for stable vegetation states. The curves for forest and steppe are *black* and *gray*, respectively. The critical precipitation vegetation transitions are indicated as φ, where the slight increase or decrease of precipitation causes discontinuous vegetation transitions, indicated by *gray* or *white thick arrows*. The horizontal distance of the two *coupled arrows* is the precipitation range of vegetation bistability precipitation, $\Delta\varphi$. (See text for details.) (**b**) Effect of uneven diet preference by herbivores on vegetation bistability. Two *gray curves* represent steady states of SWC for steppe vegetation with different grazing pressures: the *higher curve* is for $h_1 = 0$, identical to (**a**); the *lower curve* is for $h_1 > 0$ (*thick arrow*). *FS/SF*, forest steppe; *SB/BS*, bare ground

with these isoclines and their intersections. First, we investigate how the vegetation and SWC states would be altered under changing precipitation alone, and then we investigate the dependence on the herbivory pressure later. We can identify three phases along the change in φ reflecting the status of the vegetation: (1) bare ground with no plants in the lowest precipitation, and (2) two stable and one unstable steady states for W^* with feasible $P1^*$, where we might observe two stable but discontinuous states for a given precipitation, one without vegetation and the other with grass cover (indicated as "Steppe" in Fig. 6.2a). (3) For larger φ, we would have two other feasible equilibria for W^* with tree species, $P2^*$, where the larger one is stable and the smaller is unstable. Again, W^* has the steady states corresponding to the states of $P2^*$ as illustrated in Fig. 6.2 (indicated as "Forest" in Fig. 6.2a). Hence, along the gradient of precipitation, φ, there always exists at least one stable steady state of SWC and vegetation, and in several intermediate precipitation ranges, there may exist two stable states with an unstable equilibrium in between. This solution implies that two completely different types of vegetation might be observed under the same precipitation condition, which is consistent with the target pattern in Mongolian forest-steppe transition zone. However, the stability analysis of the feasible steady states does not tell which one should be realized.

6.2.2.1 Bistable Steady States and the Transition Between Vegetation Types

Assuming that the fluctuations of other environmental conditions are negligible, the vegetation we observe could be considered as that around the steady state for a given precipitation value. As stated earlier, we first ignore the herbivory effect by setting $h_i = 0$. For the range of precipitation values where only one feasible and stable equilibrium exists, that would be the vegetation state we observe. Then, how about the precipitation range with two stable steady states? Assuming no disturbance by external factors, in case the precipitation range shifts from one steady-state range to the bistable steady-state range (see arrow 1 in Fig. 6.2a), the one continuous with the previous steady state would be realized. In contrast, when the precipitation range shifts from the bistable steady-state range to one steady-state range (see arrow 2 in Fig. 6.2a), two possible changes might occur: if the new single stable steady state is continuous with the previously realized steady state (one of the two stable steady states, or SSS), only a quantitative change would be observed (type i). On the other hand, if the new single SSS is discontinuous with the previously realized steady state, a discontinuous and sudden transition should occur (type ii). In other words, in case of transition type i, the change in vegetation under precipitation change looks quantitative and is actually reversible, whereas if it is type ii, the vegetation change would exhibit abrupt and catastrophic transition between the two steady states of qualitatively different vegetation types. The transitions of ecosystem states in type ii are often observed in various ecosystems; this is well known as the ecosystem regime shift (Scheffer et al. 2001).

6.2.2.2 Vegetation Regime Shifts and Hysteresis Loop Are the Keys for the Target Pattern

Let us see possible transitions of vegetations and SWC along the change in precipitation in Mongolia using Fig. 6.2a, from bare ground to steppe, steppe to forest, or vice versa. Consider the initial phase to be the stable state of the bare ground under low precipitation. The vegetation state would shift in the following sequence: with an increase in precipitation, the corresponding SWC increases gradually, which lasts even after another stable state (grassland) emerges (type i), until the SWC value equals to that of the unstable equilibrium for grassland. As precipitation exceeds this critical value, SWC becomes high enough for grass species to grow and sustain a positive value of biomass ($P1^* > 0$); therefore, the W^* value at this point can be regarded as the critical value for the invasion of the grass species, W^*_{grass}. For higher precipitation, the number of stable states reduces to one again, and this new stable state is not continuous with the previous one (type ii), implying the previous vegetation status becomes unstable and would shift drastically to the new one, the steppe state. We call the precipitation at which the drastic transition from bare ground to steppe occurs as φ_{BS}, and hereafter the same naming manner is applied.

A similar sequence of vegetation transition would occur for a higher range of precipitation, that is, the transition from grassland to forest. The grassland increases in biomass and hence the SWC during farther increase in precipitation, even after the emergence of another steady states and we will observe forest cover only after the precipitation exceeds the critical SWC for tree species invasion, W^*_{tree}. The forest state would be the unique stable vegetation cover beyond this point (type ii) and, therefore, vegetation drastically changes into forest around this critical precipitation, φ_{SF}, and it dominates for the rest of the precipitation range. Thus, we might observe qualitatively different vegetation cover for only a slight increase in precipitation.

Then, what will happen if precipitation decreases? Let us see the transition in the opposite direction by setting our start point at the high end of precipitation, where forest cover dominates the land. The biomass and SWC of the forest decrease gradually (upper solid curve, Fig. 6.2a) even after another stable state, steppe, appears. When precipitation falls below the value at which the stable and unstable forest states merge, the steady states decrease to one in type ii, where vegetation states would change abruptly from forest to steppe with decrease in precipitation (φ_{FS}). Here, note that the precipitation values of the abrupt vegetation transitions are different when precipitation increases and when it decreases (i.e., $\varphi_{\mathrm{FS}} < \varphi_{\mathrm{SF}}$), implying that it exhibits a hysteresis loop for precipitation change; this occurs because SWC is not determined only by precipitation but is also dependent on the vegetation cover. This also implies that either steppe and forest could have stable vegetation at an intermediate precipitation between the two critical precipitation values, φ_{FS} and φ_{SF}, suggesting that we might observe a pair of qualitatively different vegetation systems, forest and steppe, for the same precipitation. Of course, a similar sequence of vegetation transition would be observed from steppe to bare ground under further decrease in precipitation, where another hysteresis loop between φ_{BS} and φ_{SB} could be recognized.

Therefore, in such a system with facilitation effect of plant growth over SWC, and hence a positive feedback interaction between plant growth and SWC, it would be possible to expect a spatially discontinuous vegetation pattern under similar precipitation if the initial vegetative conditions are sufficiently heterogeneous. Here, we also notice that the larger the difference of the two critical precipitations, $\Delta\varphi_{\mathrm{S,F}}$ ($=\varphi_{\mathrm{SF}}-\varphi_{\mathrm{FS}}$) or $\Delta\varphi_{\mathrm{B,S}}$ ($=\varphi_{\mathrm{SB}}-\varphi_{\mathrm{BS}}$), becomes, the greater the chance to observe the target pattern becomes because precipitation fluctuates considerably within the time scale of vegetation change. If $\Delta\varphi$ is much smaller than interannual fluctuation of precipitation, the vegetation becomes more homogeneous at the topographic scale. Then, what makes the value of $\Delta\varphi$ larger? Other than the strength of positive feedback between plant growth and SWC, grazing pressure by livestock might have a substantial effect.

6.2.2.3 Uneven Diet Preference of Herbivores Might Alter the Precipitation Range of Vegetation Bistability

By altering the values of h_i in (6.1), we could investigate how $\Delta\varphi$ changes with herbivore preference on plant species. The simplest but a realistic assumption is that livestock eat grass species *P*1 ($h_1>0$) but never eat trees ($h_2=0$). Figure 6.2b summarizes how the uneven grazing pressure alter the diagram in Fig. 6.2a. Because livestock only increases the loss rate of grass biomass, $P1$, it causes the smaller grass biomass for the same precipitation, hence causes lower SWC, lowering the position of the isocline of $P1$ in Fig. 6.2a, leaving that of $P2$ as it was. As the consequence of decrease in W_{in}, higher precipitation is required for SWC to reach W_{tree}, making the value of φ_{SF} larger. Once the forest is established at higher precipitation, the steady state is identical to the case without herbivory because herbivory has no effect on forest. This state also implies that the critical precipitation φ_{SFS} would not be altered by this type of herbivory. From these we could predict the precipitation range of the steppe-forest bistability, $\Delta\varphi_{S,F}$, should be expanded to the higher precipitation range by increase in grazing rate on grass species by livestock, which should increase the chance to observe different vegetation states under the same climate conditions. The results of this simple mathematical model also tell us the possibility of catastrophic vegetation transition as a consequence the synergetic effect of climate change and the livestock grazing effects in Mongolia. In the next section, we introduce our preliminary results of a numerical model to explain that the target spatial pattern could be generated by the same mechanism shown above with hydrological heterogeneity at the topographic scale.

6.3 Application to the Real Ecosystem: A Preliminary Result of Our Trial

6.3.1 Spatial Heterogeneity in Soil Water Content Caused by the Topography Effect in the Forest-Steppe Zone

The bistability of the vegetation states is necessary to answer the question "Q1" Sect. 6.1.1 but it is not sufficient because it requires the heterogeneity of initial condition, which must be related to the topography. To estimate all the parameters concerning the effects of the topography on water condition at the local scale is quite difficult without precise field measurements. As the first step, therefore, we here simply estimate the potential evaporation from a unit area from the soil surface caused by direct radiation, which is clearly dependent on slope aspect. To quantify the spatial heterogeneity in SWC, we employ the equation of Hargreaves–Samani (1982, 1985), which can estimate the potential evapotranspiration rate from air temperature and radiation. The equation contains parameters to be optimized for specific ecosystem, and we use those for the Mongolian steppe estimated by Tuya et al. (2006): $ET_0=0.0023*0.408RA*(T_{av}+17.8)*TD0.5$, where RA, T_{av}, and TD are extraterrestrial radiation, average daily temperature, *and* the temperature range, respectively. The temperature value could be obtained from the weather stations distributed in the whole country (Institute of Meteorology and Hydrology (IMH), Mongolia). The direct solar radiation could be estimated from the relationships of solar angle and the slope and aspect of the area of concern according to Corripio (2003). Since the precipitation is concentrated in the summer season (between June and September) and plant growth is also limited to the same period, we calculate the potential evaporation rate by solar radiation for the season. The topography data are obtained from the database of Shuttle Radar Topography Mission (SRTM, NASA), a remote sensing-based digital elevation model, which covers the globe with the resolution of 3 sec, that is, almost 90 m at the latitude of Mongolia. Note that we calculate the average slope and aspect of each quasi-square with its four corner points with an exactly measured set of {latitude, longitude, altitude}: we omit the microtopography at a smaller scale than this unit. By comparing the vegetation status of each cell in the area, we investigate the relationships between the estimated evaporation and the vegetation.

6.3.2 Vegetation Distribution Estimated by Satellite Remote Sensing Data

To estimate the current spatial distribution of the two vegetation types at a sufficiently fine scale to compare the result of submodel 2, we use the image data obtained by LANDSAT-ETM (NASA) during the month of July in 2006–2009. From the data of two bands, 3 (red) and 4 (near infrared), we obtained the distribution of Normalized Difference Vegetation Index (NDVI) for each pixel. The NDVI value in summer can be used as a criterion to classify evergreen forest from steppe, so that we map larch forest as the cell the maximum NDVI value of which is larger than 0.6, according to previous studies conducted in the same region (Sugita et al. 2007; Davaasuren 2001) at least once a year during the 4 years.

6.3.3 Study Site in the Forest-Steppe Zone

To estimate the distribution of the potential evaporation and compare it with the vegetation cover, we set a sample study site of rectangle area of 8 km × 11 km in Gachuurt valley (N 47.9765°-N 48.0823°, E 10.7.126°-E 107.234°, approximately 30 km northeast of Ulaanbaatar (Fig. 6.1), which includes both the forest and steppe vegetation cover to exhibit the target pattern, and we conducted the validation of the vegetation classification by satellite remote sensing. This sample site has a range of altitude of 1,430–2,083 m. The slopes are 9° on average, which have no correlation with slope aspect.

6.3.4 Preliminary Results of Field Measurements at Our Study Site

In September 2009, we could obtain preliminary meteorological and hydrological data through a summer season measured by two automated weather systems

(AWS) on the study site. One AWS is set in the steppe on a slope facing southeast (indicated by "AWS1" in Fig. 6.1) and the other in the forest zone on an adjacent slope facing northeast ("AWS2" in Fig. 6.1). The AWSs are installed at almost the same altitude (AWS1, 1,618 m; AWS2, 1,612 m) with a horizontal distance of 280 m between them. Comparing the data of the two AWSs, there was no significant difference in air temperature and precipitation at the daily base (Fig. 6.3a,b). These data suggest that climatic conditions could be regarded as almost identical among the bare slopes before vegetation is established. On the other hand, significant differences were observed in soil temperature at 10-cm depth, T_{soil} (Fig. 6.3a), wind speed (Fig. 6.3c), and SWC (Fig. 6.3d), all of which are considered to reflect the difference in the vegetation cover. Presumably, barrier and shade effects of tree stands slow wind speed and decrease T_{soil} in the forest, respectively, which in turn help in maintaining litter and increasing soil depth. With the effect of slower wind speed, lower soil temperature also contributes to maintain higher SWC in the forest. All these factors support the vegetation–soil symbiosis hypothesis already stated.

6.3.5 Comparison Between Estimated Potential Evaporation and Vegetation Distribution at a Sample Site in the Forest-Steppe Zone

Figure 6.4 shows the distributions of estimated daily potential evaporation (PE) of July and vegetation cover classifications. By overlaying these two distributions, we can see the forest cover distributing in the area of relatively lower PE while steppe in the relatively higher PE category. In the middle range of PE, however, either vegetation type might exist. This result shows a similar pattern with the right half of Fig. 6.2a in which forest and steppe dominate in moist and dry conditions, respectively, while bistable vegetation states are expected for the intermediate condition. With Fig. 6.2a we have discussed possible vegetation transitions at a local scale along with the temporal change in precipitation, but here, to add the spatial transition into consideration at the topographic scale, we incorporate the spatial distribution of PE. Remember that there was no difference in precipitation or topographic traits between the slopes. Therefore, let us assume variation of PE to be the major factor to generate heterogeneity of the initial soil water condition without forest cover and, thus, it could be used as a surrogate of potential physical factors to determine spatial heterogeneity in SWC. By substituting φ with φ–PE in (6.1), we can redraw Fig. 6.2a (converting the horizontal axis into φ–PE) to describe the relationship SWC/biomass and spatial difference in hydrological conditions, that is, the drier condition as you go to the right (Fig. 6.5; note that some of the parameters have not been obtained from our field study and were manually tuned to fit the vegetation boundary). In Fig. 6.5, two sets of curves, each of which represents the equilibria of SWC–biomass against PE for different precipitation values for July (black for

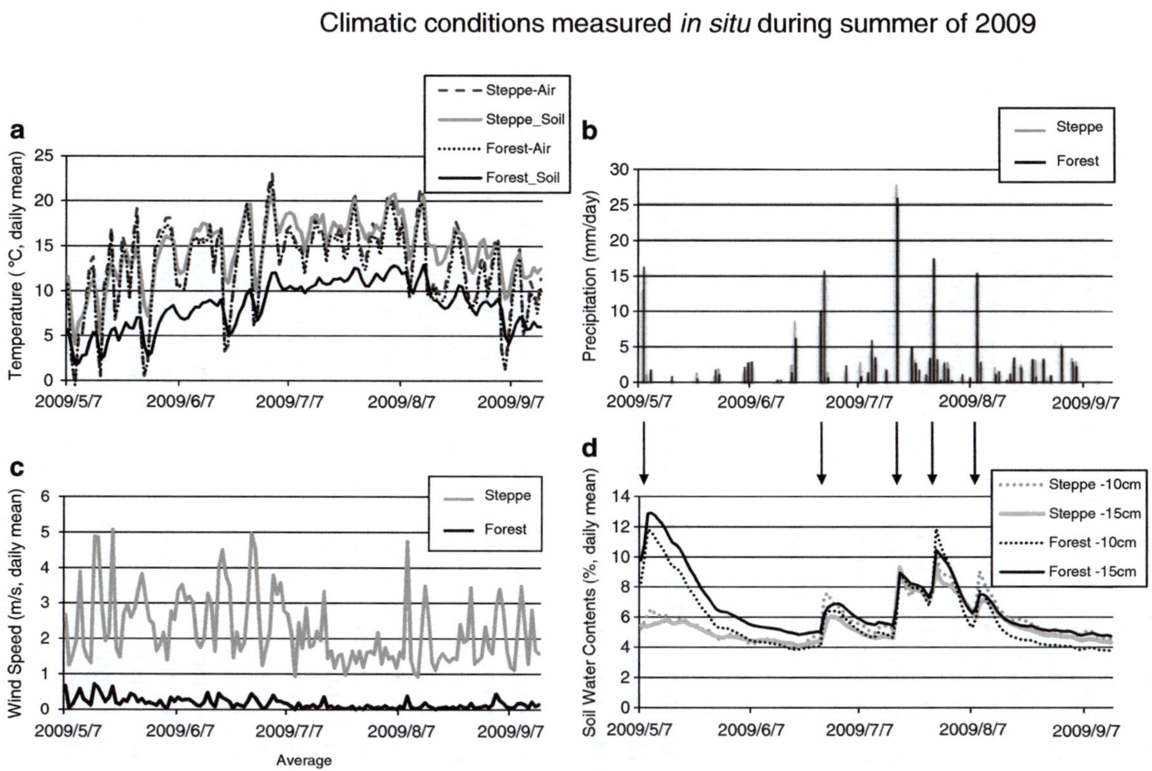

Fig. 6.3 Climatic and hydrological data obtained by two AWSs set in different vegetation covers in the Gachuurt site during summer of 2009. *Black and gray* indicate data of AWS set in the forest and steppe sites, respectively. (**a**) Daily mean temperature at the levels of 2.0 m above (air) and −10 cm beneath ground level (soil). (**b**) Daily total precipitation for each AWS. (**c**) Daily mean wind speed at 2.3 m. (**d**) SWC at −10 cm and −15 cm. *Downward arrows* indicate large rain events (>15 mm/day) shown in (**b**)

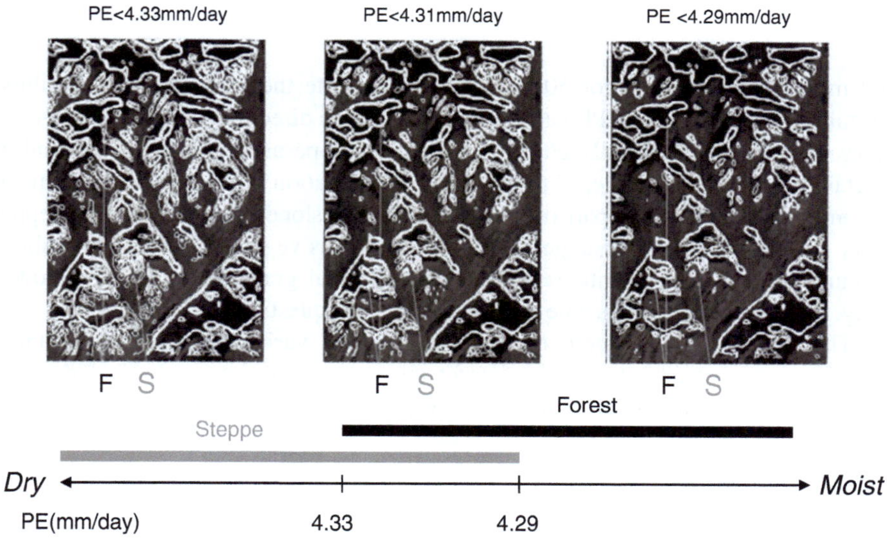

Fig. 6.4 Comparison between vegetation distribution and potential evaporation (PE) in the Gachuurt site. As the background, the Landsat image is set to show the vegetation distribution in the site, where *black and gray portions* indicate forest (F) and steppe (S) covers, respectively. Area surrounded by *white contours* indicates the area of PE smaller than the criteria shown above each panel. In the *left panel* only the steppe is outside the contours; in the *right panel*, only the forest can be seen inside the contours, which are distributed only on the north-facing slopes

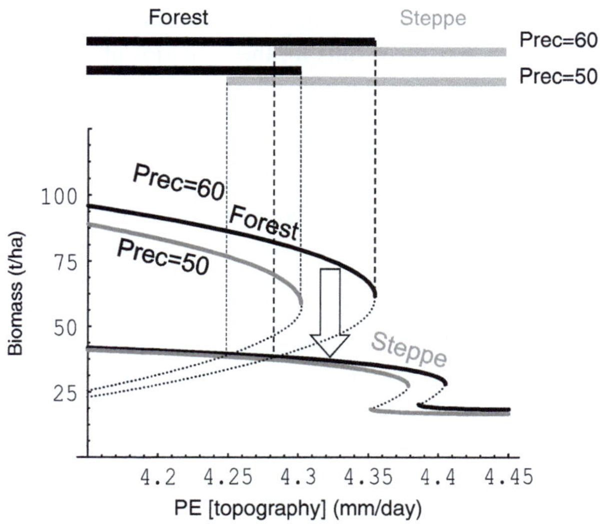

Fig. 6.5 Relationships between PE and plant biomass equilibria for given precipitation rate in summer. Estimated biomass for given precipitation (Prec, *black*, 60 mm/month; *gray*, 50 mm/month). PE is dependent on the topography and estimated as in Fig. 6.4. PE values on the horizontal axis correspond to the slope aspect in the region, i.e., north-facing slopes have small PE values and south-facing slopes have large PE. Biomass is estimated by setting each of the equations (6.1a–c) equal to 0 and solving them numerically in terms of $P1$ and $P2$ with all other parameters fixed. *Thick arrow* indicates a possible discontinuous vegetation transition in transient slopes after a significant change in precipitation. *Dotted line*, biomass estimation model

60 mm/month and gray for 50 mm/month), indicate the SWC/biomass exhibits bistability in intermediate PE values; hence, the slope direction, which with vegetation status might drastically change as PE (i.e., slope aspect) changes beyond a certain value. For example, if the monthly precipitation decreases from 60 m to 50 mm per month, some part of the forest-covered slopes would turn into steppe very shortly. This finding suggests that discontinuous vegetation transitions might occur with spatial gradients as well as the temporal gradient, as already shown (Fig. 6.2a), and gives a possible explanation to both questions, Q1and Q2.

The incorporation of both temporal and spatial variation of climate change condition thus enables us to reconstruct the vegetation patterns based on plant growth and loss, and therefore, to predict vegetation change with alteration of both drivers.

6.3.6 Preliminary Results from the Numerical Model for the Gachuurt Study Site

In what follows, we show preliminary results of the numerical simulation model at the topographic scale for the forest-steppe zone based on the methodology shown above (details are described in Ishii et al., in preparation). Vegetation dynamics are calculated for gridded cells using the mathematical model proposed in the present study where the heterogeneity in hydrological conditions is given by topography and climate data. In Fig. 6.6, the results for the scenarios of climate change, seven different precipitation rates (i.e., control, −5%, −10%, −15%, −20%, −25%, and −30% compared to the control) and two different livestock densities (control and +50%) are demonstrated. Figure 6a shows current vegetation in the study site in Gachuurt (location and dimension are as described above), whereas Fig. 6b,c shows the projected vegetation in 40 years in the future under only climate change and under both climate and livestock density changes, respectively. The upper panels show the vegetation map classified in three categories, forest, steppe and bare ground, for the most extreme climate scenarios. The lower panels show the relative coverage of these three land cover classes in the area. Although they are still qualitative, these preliminary results suggest some novel features of vegetation projections with following merits. Namely, with this method, we might project vegetation changes as transitions between qualitative vegetation classes (e.g., from forest to steppe), which is much clearer to detect and easier for validation than to observe quantitative change in biomass or vegetation indices at coarser resolution. Furthermore, this fine-scale mapping provides us with spatial information about the sensitivity of vegetation to the environmental changes. For example, in the upper panels of Fig. 6.6 b,c, it is suggested that deforestation and desertification would be more likely to occur in south-facing slopes and lower places under decreasing precipitation. The other merit of this method is that it would be able to provide the relative importance of the two different types of environmental changes.

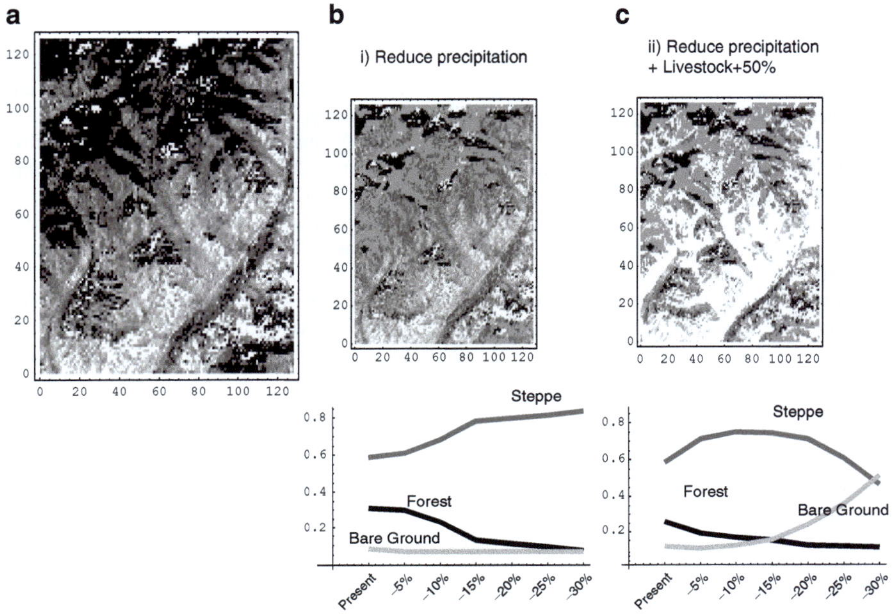

Fig. 6.6 Preliminary results of numerical model for 40 years future projection under various environmental scenarios. (**a**) Current vegetation distribution is classified into three states: forest (*black*), steppe (*gray*), and bare ground (*white*). The site is gridded into square cells of 90 m×90 m. Applying the biomass estimation method shown in Fig. 6.5 to each cell with its specific PE value and scenario-specific parameter values, we could estimate the steady states of spatial vegetation distribution under various scenarios. In the present parameter set, the steady states are reached within 40 years, which is reasonable in this region. As two different scenarios for climate change and human impacts, reduction in precipitation and increase in livestock density are incorporated. (**b**) Estimated vegetation status under precipitation reduction of –5% to –30% compared with the current conditions. *Lower panel*: Proportion of each vegetation type at the steady state. *Upper panel*: Predicted spatial vegetation distribution under –30% precipitation. (**c**) Estimated vegetation status under precipitation reduction of –5% to –30% and livestock increase of +50% compared with the current conditions. *Lower panel*: Proportion of each vegetation type at the steady state. *Upper panel*: Predicted spatial vegetation distribution under –30% precipitation and +50% livestock increase. Parameters used here are: g_1=0.05; g_2=0.08; m_1=0.01; m_2=0.015; m_{11}=0.01; m_{12}=0.0001; m_{21}=0.02; m_{22}=0.0015; h_1=0.01; h_2=0; k_s=10; r_w=0.2; u_1=1; u_2=3; w_0=0.04

Provided the parameterization of the model is sufficiently reliable, one might estimate appropriate upper limit of livestock density for different degrees of precipitation decrease to keep a certain percentage of forest cover in the area.

To make it quantitatively reliable, the model needs to be improved by updating and validating the parameters about plant dynamics, herbivory effects, and the climate. However, because most of these data could be collected without field observations, this methodology could be applied for a wider area with discontinuous vegetation patterns at the topographic scale, even for different types of vegetation observed in other parts of Mongolia.

References

Corripio JG (2003) Vectorial algebra algorithms for calculating terrain parameters from DEMs and solar radiation modelling in mountainous terrain. Int J Geogr Inf Sci 17:1–23.

Fujita, N, Amartuvshin, N, Yamada, Y., Matsui, K., Sakai, S., and Yamamura, N. (2009) Positive and negative effects of livestock grazing on plant diversity of Mongolian nomadic pasturelands along a slope with soil moisture gradient. Grassland Sci 55:126–134.

Davaasuren N (2001) Information System Design for Land Degradation Assessment, in Particular for Pasture Areas: Case Study of West Mongolia. Master Thesis to International Institute, Aerospace Survey and Earth Sciences.

Dulamsuren, Ch., Hauck M, Mühlenberg M (2005a) Vegetation at the taiga forest-steppe borderline in the western Khentej Mountains, northern Mongolia. Ann Bot Fenn 42:411–426.

Dulamsuren, Ch., Hauck M, Mühlenberg M (2005b) Ground vegetation in the Mongolian taiga forest-steppe ecotone does not offer evidence for the human origin of grasslands. Applied Vegetation Science 8:149–154.

Gunin, P. D., Vostokova, E. A., Dorofeyuk, N. I., Tarasov, P. E., Black, C. C. (1999) Vegetation dynamics of Mongolia. Geobotany, vol 26. Kluwer, Dordrecht.

Hargreaves, G.H. and Z.A. Samani. 1982. Estimating potential evapotranspiration. J Irrig Drain Eng ASCE 108(IR3):223–230.

Hargreaves, G.H., and Z.A. Samani. 1985. Reference Crop Evapotranspiration From Temperature. Appl Eng Agric 1:96–99.

Hilbig W (2000) Forest distribution and retreat in the forest steppe ecotone of Mongolia. Marburger Geographische Schriften 135: 171–187.

Ishikawa, M., Sharkhuu, N., Zhang, Y., Kadota, T., Ohata, T. 2005. Ground thermal and moisture conditions at the southern boundary of discontinuous permafrost, Mongolia. Permafrost and Periglacial Processes 16: 2089–216.

IPCC AR4 (2007).

Klausmeier C. A. (1999) Regular and Irregular Patterns in Semiarid Vegetation. Science 284: 1826–1828.

Saizen, I., Maekawa, A. and Yamamura, N. (2010) Spatial analysis of time-series changes in livestock distribution by detection of local spatial associations in Mongolia. Applied Geography 30:639–649.

Sugita, M., J. Asanuma, M. Tsujimura, S. Mariko, M. Lu, F. Kimura, D. Azzaya and T. Adyasuren. 2007. An Overview of the Rangelands Atmosphere-Hydrosphere-Biosphere Interaction Study Experiment in Northeastern Asia (RAISE). J Hydrol 333:3–20.

Scheffer, M., S. Carpenter, J. A. Foley, C. Folke & B. Walker (2001) Catastrophic shifts in ecosystems. Nature (Lond) 413 591–596.

Tuya, S., J.Batbayar, K. Kajiwara & Y. Honda (2006) A comparison of five potential evapotranspiration methods and relationship to NDVI for regional use in the Mongolian grassland. International Archives of the Photogrammetry, Remote Sensing and Spatial Information Science, Volume XXXVI, Part 6, Tokyo, Japan.

van de Koppel, J., M. Rietkerk, and F. J. Weissing. (1997) Catastrophic vegetation shifts and soil degradation in terrestrial grazing systems. Trends Ecol Evol 12:352–356.

Walker, B. H. (1989) In Conservation Biology for the Twenty-First Century (eds. Weston, D. & Pearl, M.). Oxford University Press, Oxford, pp 121–130.

Whittaker RH (1962) Classification of natural communities. Bot Rev 28(1):1–239.

Chapter 7
Deforestation and Reforestation of Degraded Forestland in Mongolia

Jamsran Tsogtbaatar

Keywords Deforestation • Mongolia • Reforestation • Tree planting

7.1 Introduction

Mongolia is located in the transition zone between the deserts of Central Asia and the Boreal taiga of southern Siberia. These zones reflect a general trend from drier and warmer at lower elevations in the south and to moister and colder in the north of the country at higher elevations. Situated far from the moderating influences of the world's oceans, Mongolia experiences a markedly continental climate with cold winters and warm summers. The high altitude and the relatively high latitude exacerbate the semiarid continental climate, resulting in long winters. The mean temperature is below freezing point during about 7 months of the year.

In general, the amount of precipitation in Mongolia is low. Annual precipitation ranges from an average of 400 mm in forest zones and 100 mm in desert steppe; about 85% of the total annual precipitation falls from April to September. Based on topographic features, the country can be divided into (1) the mountains located in the north and the west; (2) the intermountain basins where the major farming regions are located; (3) the plateau and desert belt located in eastern and southern Mongolia; and (4) rivers and lakes. In addition, four geographic regions are commonly mentioned: Altai Mountain on the western border, Hangai Mountain in the northwest and Hentey Mountain in the East Mongolian region, coinciding with the steppe region, and the Gobi region in the south and southeastern part of the country.

J. Tsogtbaatar (✉)
Institute of Geoecology, Mongolian Academy of Sciences, Ulaanbaatar, Mongolia
e-mail: geoeco@magicnet.mn

N. Yamamura et al. (eds.), *The Mongolian Ecosystem Network: Environmental Issues Under Climate and Social Changes*, Ecological Research Monographs,
DOI 10.1007/978-4-431-54052-6_7,

The highest and the longest spine of mountains is situated in the most western part of the country, the Altai, which extends about 1,500 km. Most of the southern part of the country is a vast rolling oasis-dotted plain, forming the northern fringe of the Gobi, which is predominantly stony with sands covering only 3% of the terrain. The Gobi occupies about one-third of Mongolia's territory. The great desert of central Asia extends from the south.

The borders of Mongolia contain the transition zone between the great *taiga* forests to the north, the Gobi desert to the south, the Altai Mountains to the west, and the central Asian steppe to the east. For this reason, Mongolia is important from ecosystem and species conservation perspectives. Its harsh climate, characterized by extreme temperature fluctuations, low precipitation, and severe winters produces fragile ecosystems that are easily degraded by human activity and are slow to recover once damaged.

7.2 Forest Resources of Mongolia

The territory of Mongolia is divided into six natural zones: desert, desert steppe, forest steppe, boreal forest, taiga, and high mountains (Fig. 7.1).

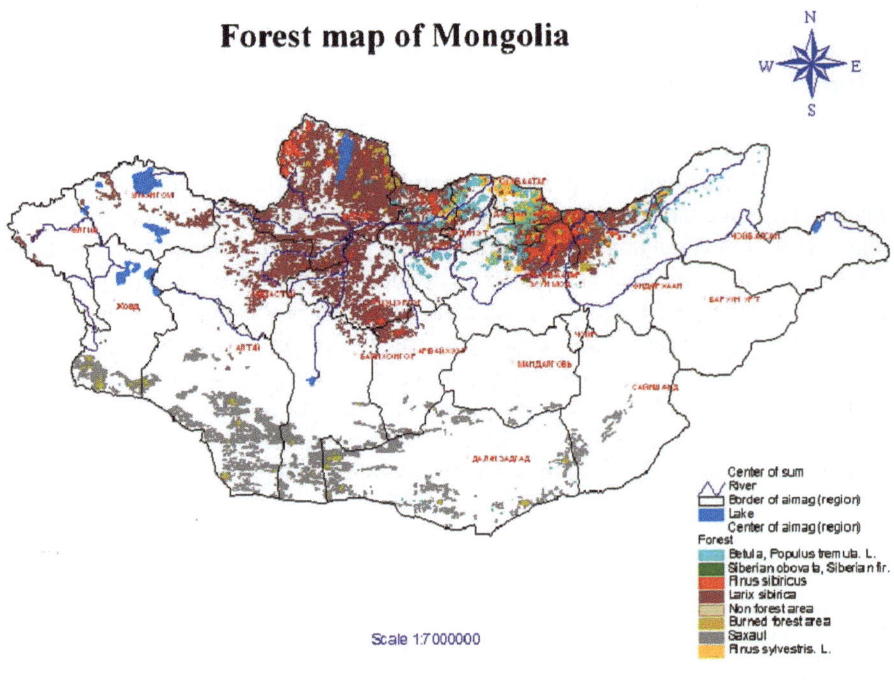

Fig. 7.1 Map of natural zones of Mongolia

Table 7.1 Distribution of natural grassland by vegetation zones

Vegetation zone	Percentage,%
High mountain zone	5.3
Mountain forest-steppe zone	27.1
Steppe zone	22.8
Desert steppe zone	19.5
Desert zone	19.4
Intervegetation zones	5.9
Total	100.0

Source: Ministry of Nature and Environment, Mongolia, 2002

Climatic and soil conditions are reflected in the vegetation patterns. Latitudinal and altitudinal belts of vegetation, often changing rapidly over a few miles, are probably the most obvious feature of the local Mongolian landscape. There are four basic vegetation zones from north to south of the country, namely, the forest-steppe zone of the mountains; the steppe zone; and the semidesert and desert zones in the south and extreme south, respectively. The *steppes* predominate, covering about three-quarters of the national territory.

The main forest region located in the northern part of the country along the Russian border forms a transition zone between the Siberian taiga forest and the central Asian steppe zones. Even though the extent of forests in Mongolia is relatively limited, it is distinctive and somewhat heterogeneous, because of the country's vast territory and geography, and it is of enormous importance in supporting the environmental stability of the country.

In broad terms of vegetation classes, Mongolia can be divided into grassland and shrubs (about 53%), forests (about 12%), and desert vegetation (about 34%); about 1% is used for human settlement and crop cultivation. Grazing takes place in all these land classes. Vegetation types of natural grassland, which cover about 83% of Mongolian territory, are distributed as shown in Table 7.1.

Based on altitudinal location of forests, the mountain forests in Mongolia can be differentiated into four belts named subalpine, taiga, subtaiga, and pseudo-taiga. The lowest boundary lies at an elevation of about 650 m and the upper timberline is found at about 2,600 m.

7.2.1 Subalpine Belt

The subalpine belt often covers the upper timberline of forest distribution in Mongolia. The regeneration process is going on slowly with cold-resistant plants and those that grow on shallow soils as well as moss and lichen. It has significant importance in keeping water sources and in protecting soil in upstream areas.

7.2.2 *Taiga Belt*

The taiga forests with underlying permafrost soils are found in Hentey, eastern Hovsgol, and the southeastern Hangai region of the country. Natural regeneration of forest stands in this region is adequate. However, coniferous tree species are often replaced by birch trees after forest disturbances such as forest fires and logging. Although birch and larch trees are dominant, cold-resistant tree species such as Siberian pine and fir trees are common in this belt.

7.2.3 *Subtaiga Belt*

The subtaiga belt comprises the most widely distributed forest type in Mongolia and has commercial importance for logging. In addition, it is intensely impacted by human activities and thus prone to forest fire because of its location at the lower timberline of forest distribution and bordering with the steppe vegetation zone. This type of forest consists of larch, pine, and birch stands with a rich vegetation composition compared to other forest types in Mongolia. Although natural regeneration of forest stands is sufficient, most coniferous tree species could be replaced by deciduous trees such as birch and poplar after the human negative impact of forest fire or logging. In some cases, the forest area can be replaced with the vegetation of dry steppe ecosystem. Rehabilitation and reforestation activities after human-induced disturbances have to be proposed on this belt.

7.2.4 *Pseudo-Taiga Belt*

This belt of mountain forests with specific trend of succession process and continuous permafrost soils occurs only in Mongolia, especially in the central part of the Hangai and Altai mountain ranges. The main dominant tree species in this belt is larch. The natural regeneration of forest in this belt could be maintained only through selective cutting or conditions with absence of negative human disturbances, but it is directly replaced by the dry steppe vegetation after clear-cutting. Forest patches in the Mongol Altai mountain were intensively cut down and then turned directly into steppe vegetation, missing the natural regeneration process because of the harsh climatic conditions in this region.

The basic vegetation form in the *taiga* belt is larch and cedar pine (Siberian pine). Forests grow thickest on the northern shady slopes, the most widespread tree being the larch, followed by the cedar, with a varying admixture of spruce, pine, and fir. Deciduous trees include species such as birch, aspen, and poplar.

The forest cover of Mongolia extends over an area of 17.5 million hectares (ha) or about 11.2% of the country's territory. Of this, the area actually under closed

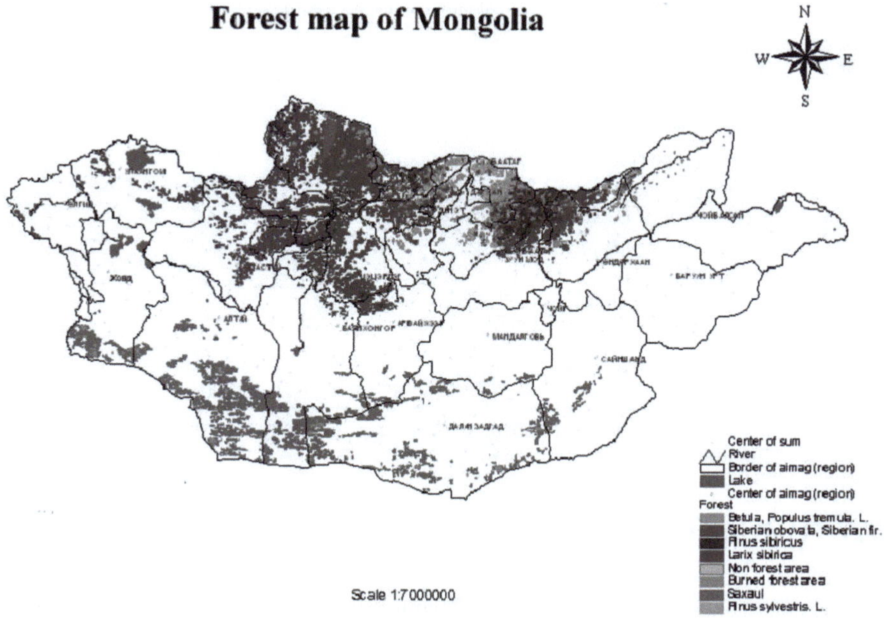

Fig. 7.2 Map of forest distribution in Mongolia

Table 7.2 The Mongolian forestland distribution by main tree species

No.	Tree species	Closed forest area (ha)	Percentage (%)
1.	Larch (*Larix siberica*)	7,526,899	61
2.	Scotch pine (*Pinus sylvestris*)	662,113	5
3.	Siberian pine (*Pinus siberica*)	984,658	8
4.	Spruce (*Picea obovata*)	27,872	<1
5.	Siberian fir (*Abies siberica*)	2,337	<1
6.	Birch and other (*Betula, Populus*)	1,198,720	10
7.	Saxaul (*Haloxylon*)	2,028,823	16
Total		12,431,422	100

forest cover is about 12.4 million ha, representing about 8.1% of the country's total area (Fig. 7.2).

The location and quality of forests vary greatly throughout the country. The distribution of tree species in the country is shown in Table 7.2. According to the composition of the forest resources, larch stands are the dominant tree species in the country: they are resistant to the coldest conditions in Mongolia.

It has been estimated that the average growing stock of wood in the coniferous and deciduous forests is about 103 m^3/ha. The standing stock of main tree species such as *Larix siberica*, *Pinus silvestris*, *Pinus sibica*, *Picea obovata*, *Abies siberica*, *Betula platyphylla*, *Populus tremula*, *Populus diversifolia*, and *Salix* is shown in Table 7.3.

Table 7.3 Distribution of forest stand stock by tree species in Mongolia

Name of tree species	Standing stock (per m^3)	Percentage,%
Siberian larch (*Larix siberica*)	1,033,016.6	74.90
Scotch pine (*Pinus silvestris*)	97,085.4	7.03
Siberian pine (*Pinus siberica*)	150,524.8	10.91
Spruce (*Picea obovata*)	3,434.2	0.24
Siberian fir (*Abies siberica*)	375.2	0.02
Birch (*Betula platyphylla*)	89,404.5	6.48
Poplar (*Populus*)	1,814.3	0.15
Aspen (*Populus tremula*)	1,509.3	0.13
Willow (*Salix*)	613.3	0.04
Saxaul (*Hallohylon*)	1,404.1	0.10
Total	1,379,181.7	100

Source: State of Environment of Mongolia, Ministry of Nature and Environment, 2007

7.3 Driving Factors of Deforestation

Deforestation and forest degradation in the country raise national and regional concern because of to their economic and ecological importance. The main causes of deforestation and forest degradation in Mongolia are forest and steppe fire followed by damage by pests and diseases, improper commercial and illegal wood logging for construction and for use as fuel, and uncontrolled grazing in forest areas and land degradation.

7.3.1 Climate Change Impact

Mongolia's continental climate is characterized by significant fluctuations in climatic parameters and weather phenomenon. Droughts can occur for several years in the southern Gobi desert areas of Mongolia, and concurrently, heavy rainfall and floods are the norm in the mountains of the north. In a considerable part of the territory of Mongolia, the annual precipitation is no more than 200 mm. At the same time, 90% of this precipitation is lost to evapotranspiration.

Since the 1940s, Mongolia has already experienced a significant change in its climate because of global warming. To some extent, the level of changes of climate parameters in Mongolia is higher than the global average because of its geographic location and elevation (Dagvadorj et al. 2009). The intensity of the warming trend differs in Mongolia's ecological and climatic zones and also varies by season. The main findings of climate change studies in Mongolia are given as follows:

- Annual mean temperatures rose by 2.1°C between 1940 and 2007. The warming has been most pronounced in winter, with mean temperature increase of 3.6°C, whereas spring, autumn, and summer mean temperatures have risen by 1.8°C, 1.3°C, and 0.5°C, respectively (Fig. 7.3).

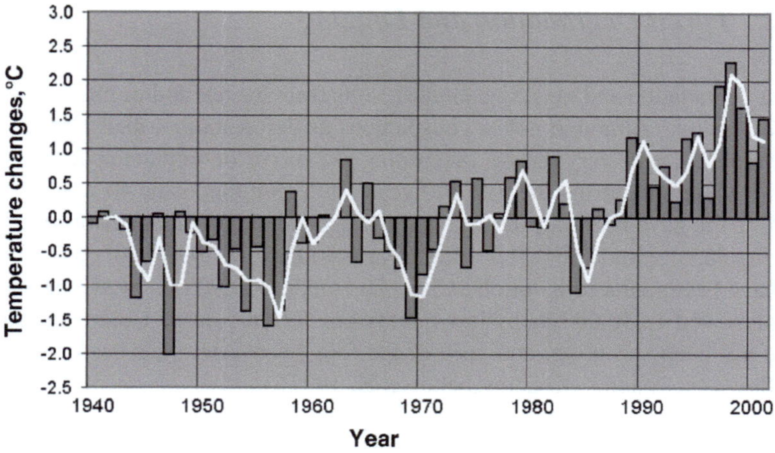

Fig. 7.3 Temperature changes compared from 1940 to 2000

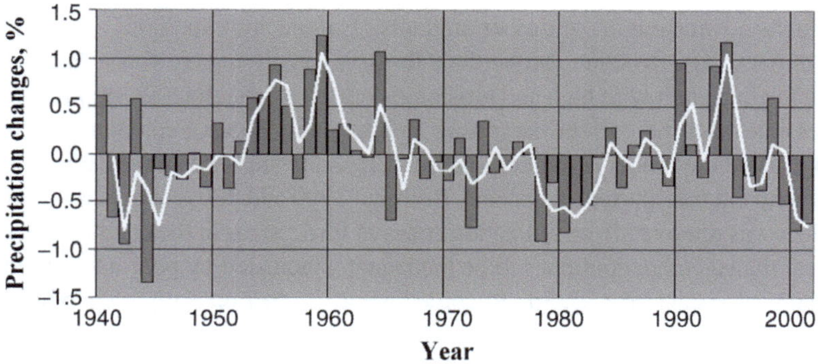

Fig. 7.4 Changes in precipitation amounts compared from 1940 to 2000

- Annual precipitation changes are quite variable, decreasing at one site and increasing at another site nearby. Seasonally, autumn and winter precipitation has increased by 4–9% while spring and summer precipitation has decreased by 7.5–10%.
- Spatially, annual precipitation decreased by 30–90 mm in the central region and increased by 2–60 mm in the most western part and by 30–70 mm in the most southeastern part of the country. The magnitude of changes in precipitation, regardless of increasing or decreasing, is 5–25% (Fig. 7.4).
- There were no statistically significant changes in the maximum number of consecutive dry days on average. Maximum number of consecutive dry days slightly increased in central Mongolia, where annual mean precipitation was decreased and decreased in southeastern Mongolia where annual mean precipitation was increased.
- Potential evapotranspiration has increased by 7–12%.

7.3.2 Forest Degradation and Logging

Forest degradation and logging damage results from misuse and/or mismanagement of the resource. Although not as conspicuous as deforestation, their impact on the environment can be very serious. According to a survey of human impact on ecosystems in Mongolia during the past 100 years, it is seen that some 40% of all forests in Mongolia have been impacted to some degree; 684,000 ha have not regenerated after fire damage and 250,000 ha after clear-cutting; 1,737,000 ha of coniferous forests have been replaced by birch and poplar stands, 159,000 ha by steppe and sand or stones, and 1,230,000 ha by low-quality coniferous forests. Cold-resistant taiga forest has been shrinking; and 16% of the forest ecosystem has been replaced by non-forest ecosystems (Krasnoschekov et al. 1992).

In the 1950s the Government of Mongolia, following the Soviet model, had established 12 forestry state enterprises in several of the northern-central *aimags* including settlements for workers ('forestry villages'). To intensify forest management, mechanical harvesting operations and electronic sawmilling were introduced. These sawmills need large volumes of timber, and by the mid-1980s cutting volume had reached 2 million m^3 of timber annually (Erdenechuluun 2006).

In terms of age structure, Mongolia's forest is mostly mature forest that is distributed in the taiga belt with high soil erosion risk or on steep mountain slopes, which are inaccessible with the current harvesting technologies. Forest exploitation in the past has exhausted the timber resources in more accessible areas given the capacity of their equipment and transportation systems. Between 1990 and 2009, about 13.4 million m^3 of timber was removed from utilization zones of forest areas in the country. The quality of forest management continues to be inadequate, indicated by poor natural regeneration and ecological retrogression. Recorded wood production is reported to have fallen (Fig. 7.5) as a result of the reduction in the area of designated utilization forests.

In general, clear-cutting practice in the mountain forests of Mongolia is not applicable for the natural condition of these forests because it leads to the succession change of coniferous forests by deciduous forests and to the replacement of forest ecosystems by steppe or dry meadow-steppe ecosystems, enhances surface and slope runoff, and develops the erosion process and reduction of the forest areas. Hence, silvicultural application and forest exploitation activities in Mongolia require special methods and systems of timber cutting that will not have a destructive effect on the protective function of the forest ecosystem and at the same time enables natural regeneration of the forests.

The selective and shelter wood-cutting system with moderate intensity of stand stock removal and reduction of stand density down to 40–50% in the forests enables preservation of basic characteristics of the forest environment and the protection function of the forest, not causing sharp changes in the plant communities, and forming favorable conditions for growth of trees and forest regeneration, which is important for establishment of the next generation of forests.

However, because harvesting is confined to limited and easily accessible areas and monitoring and control of the harvesting operations are barely exercised, then wasteful overharvesting takes place. Illegal and unaccounted removals of trees are also common in the country.

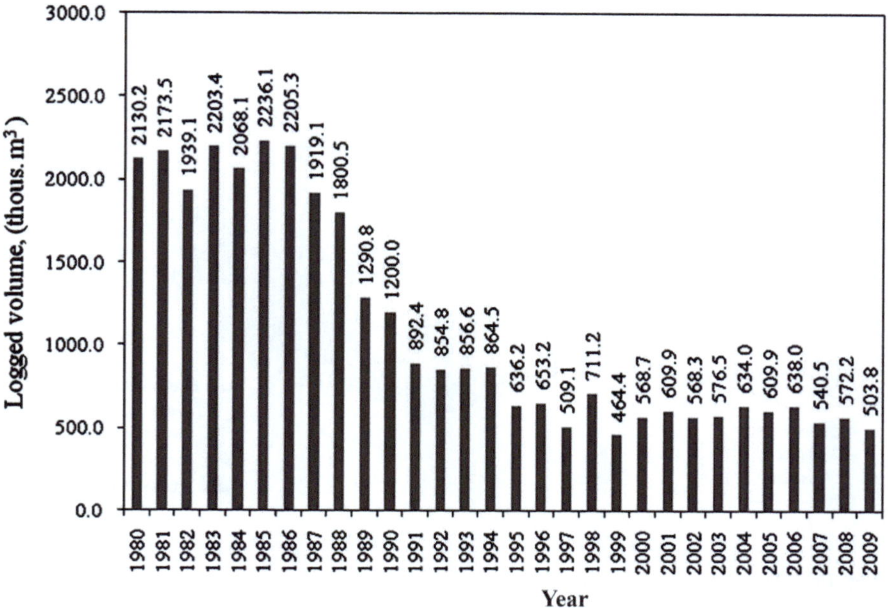

Fig. 7.5 Annual wood harvesting volume of Mongolia

Some of the factors and impacts linked to deforestation and forest degradation such as forest fires, grazing, loss of biodiversity, and desertification are relevant in this connection (Tsogtbaatar 2004).

7.3.3 Desertification and Land Degradation

Loss of forests, coupled with overgrazing and inappropriate land use practices, leads to desertification and formation of barren, eroding wastelands. During the past 40 years, the desert areas of Mongolia have increased by some 38,000 ha and deserts now make up some 41% of the total land area. It has been estimated that the size of usable pastureland has decreased by 6.9 million ha during the past 30 years as a consequence of desertification; and that 30% of the total pasture area in Mongolia has been degraded by misuse (Fig. 7.6).

7.3.4 Uncontrolled Grazing and Land Destruction

The main basis of Mongolia's economy, pasturing livestock husbandry, still plays an important role in the economy, employment, and export revenues of Mongolia. For the statistical report of 2009, about 21.1% of gross domestic product (GDP)

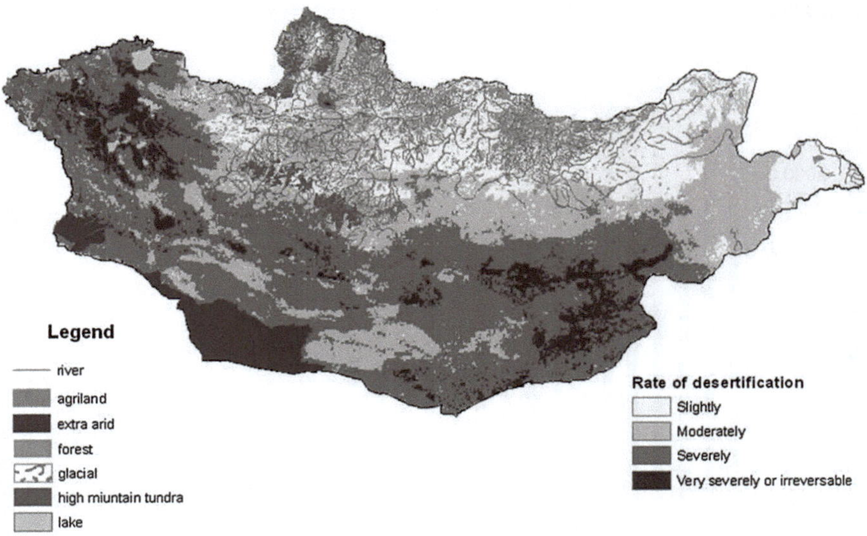

Fig. 7.6 Map of desertification of Mongolia

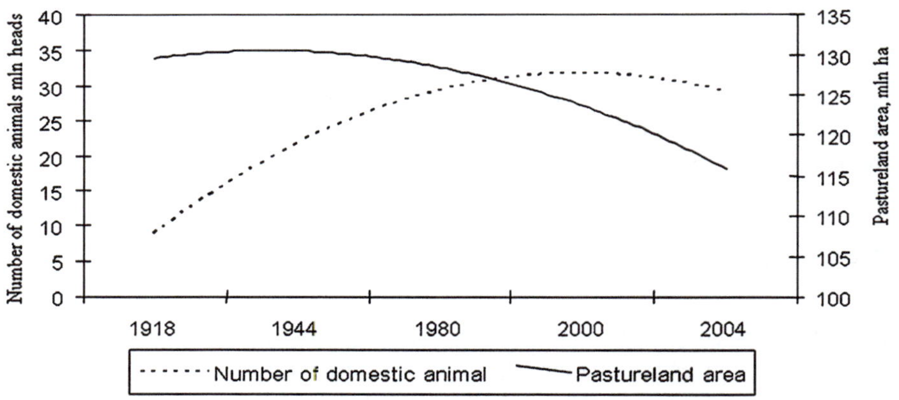

Fig. 7.7 Changes in livestock numbers and rangeland in Mongolia

was produced by the agriculture sector, of which 82.5% accounts for livestock husbandry. According to the livestock census for 2009, in total 44.0 million head of livestock were counted, which was higher by 735.4 thousand head or 1.7% compared with the previous year.

The general health condition of pasturelands in Mongolia is poor with unregulated grazing of some 30 million head of livestock, such that the grass does not get a chance to regrow (Fig. 7.7). It is shown that Mongolia had about 9.6 million head of livestock at 130.0 million ha of grazing land area by 1918; this figure was then

Fig.7.8 Free grazing in forest area

Fig. 7.9 Gathering of animals in degraded area

drastically changed, to having 28.0 million head of livestock at 111.2 million ha area of grazing land area by 2004.

Some 35–40% of the livestock population (about 12 million) is allowed to graze in and near forest areas, trampling young trees and saplings, inflicting heavy environmental toll on the land (Figs. 7.8 and 7.9). Grazing occurs also in protected areas having open access and free grazing in intact areas.

According to some statistical report indicates that total pasture land area of Mongolia was 130.0 million ha for 9.6 million livestock in 1918, but it was changed

Table 7.4 Grazing impact on tree survival of different plantation sites

Site no.	Distance from forest (km)	Altitude (m)	Grazing intensity	Survival rate (%)
1	0.1	1,560	Light	83.3
2	5.0	1,680	Heavy	5.0
4	0.1	1,750	Medium	50.5

as total pasture land area changed to 127.3 million ha for 30.2 million livestock in 2000 (Avaadorj et al. 2006).

To save forests from grazing damage, it is clear that action needs to go beyond the forest boundary, by promoting good pastureland management: improvement of pastures, controlled and rotational grazing, fodder development, and other science-based measures.

These results in Table 7.4 show that grazing intensity and distance from forest edge strongly affected the survival rate of planted seedlings. It is indicated that prevention of grazing is the most important factor for the success of the plantation. Plantation sites would be established near the natural stand if possible, because natural stands located near a plantation provide a barrier from wind and a water supply in water-limited conditions (Dae 2005).

7.3.5 Forest Fire

Forest fires, by far, have the most serious impact on the forests of Mongolia. Forest fires are mostly of incendiary nature, caused by herders and local people. Low-intensity forest fires that cover the lower layers of the forest (ground fires) lightly burn the tree stems and cause less damage, whereas fires that cover the upper layer of the forest (crown fire) burn the branches and leaves of the trees, which essentially stops growth. More serious fires will completely burn and destroy the forest vegetation. In some cases fire-damaged areas recuperate through revival of the affected plants or through natural regeneration.

Even light forest fires can attract pests and diseases, and plants weakened by fire become easily susceptible. Forest fires often result in replacement to coniferous forest with birch or to steppe vegetation. Fire, coupled with grazing, leads to increased soil erosion.

The forest area damaged by fire in 1980 was 107,200 ha (compared with an area of 115,000 ha damaged by insects). In 1990, the corresponding figures were 649,800 and 33,100 ha; in 1994 these were 120,000 and 135,000 ha, respectively (Fig. 7.10). Since then, the area of fire damage has increased because of dry climatic conditions and the carelessness of local people. Historically, a number of massive forest fires took place, for instance, in the periods of 1990–1992 and 1996–1999.

In 1996, the total area affected by fire was 10.20 million ha, of which the forest area involved was 2.30 million ha, the rest being pasture and steppe; 25 people were

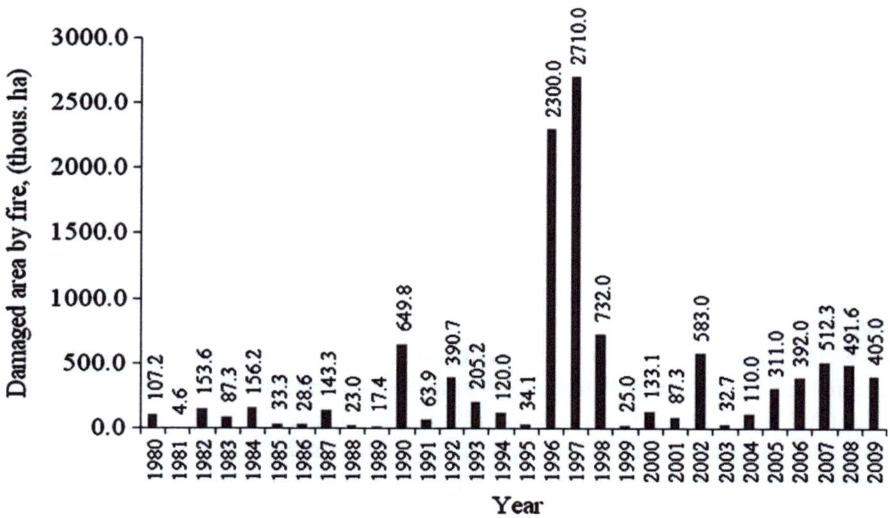

Fig. 7.10 Forest fire-affected areas in Mongolia

Table 7.5 Changes in plant community after forest fire or logging

Vegetation community	Natural stand	Burned stand	Logged stand
	Number of species (%)	Number of species (%)	Number of species (%)
Taiga	11 (30.6)	5 (11.4)	0 (0)
Forest meadow	21 (58.3)	29 (65.9)	25 (67.6)
Steppe	4 (11.1)	10 (22.7)	12 (32.4)
Total	36 (100.0)	44 (100.0)	37 (100.0)

killed and 700 rendered homeless in the 1996 fire. This fire burned 307 million m^3 of forest growing stock of which some 22 million m^3 was totally lost. In 1997, fires burned 12.4 million ha; the extent of forest area affected was 2.7 million ha, and 600,000 head of livestock and unknown numbers of wildlife were destroyed.

Forest fire, logging, and grazing activities were the major factors changing plant community and succession in Mongolia (Tsogtbaatar 2004). The patterns of stand development after fire or logging at *L. siberica* stands mainly progressed in four stages: (1) secondary *L. siberica* stand, (2) birch stand, (3) bush stand, and (4) grassland (steppe). It depended strongly on the intensity of degradation and the potential of natural regeneration (Table 7.5).

However, bush stand and steppe were influenced by the progress of heavy degradation intensity, and natural regeneration developed very slowly or was impossible. Thus, artificial rehabilitation may be urgently needed instead of natural restoration progress.

7.4 Forest Rehabilitation and Tree Planting

A major issue related to forest management in Mongolia is reforestation because of the difficult site conditions. Considering the specific habitat factors existing in Mongolia, sufficient natural regeneration of desired species such as pine and larch cannot often be achieved after logging operations or fire incidences. The share of nonconiferous species such as birch and aspen increases, and some areas may turn into grassland. Under such a situation, artificial planting or sowing with desired species is necessary to supplement the establishment of natural regeneration.

Reforestation activities in Mongolia commenced in 1971, some 48 years after the first sawmill was established and commercial timber extraction started (Fig. 7.11). The total area planted in 1971 was only 67 ha, and the rate of planting has since increased considerably; the area planted in 2008 was 9,512 ha. The main species planted are pine (*Pinus silvestris*), larch (*Larix siberica*), poplars (*Populus* spp.), and elm (*Ulmus pumila*). The total plantation area recorded by the end of 2009 is about 141,490 ha.

The area planted over a period of some 40 years is very small compared to the deforestation rate that has taken place (1.2 million ha during the past 20–30 years) and the degradation level of the remaining accessible forests. Even if only the area harvested during the period is considered, the reforestation rate is inadequate. It is clear that annual planting must increase in the future.

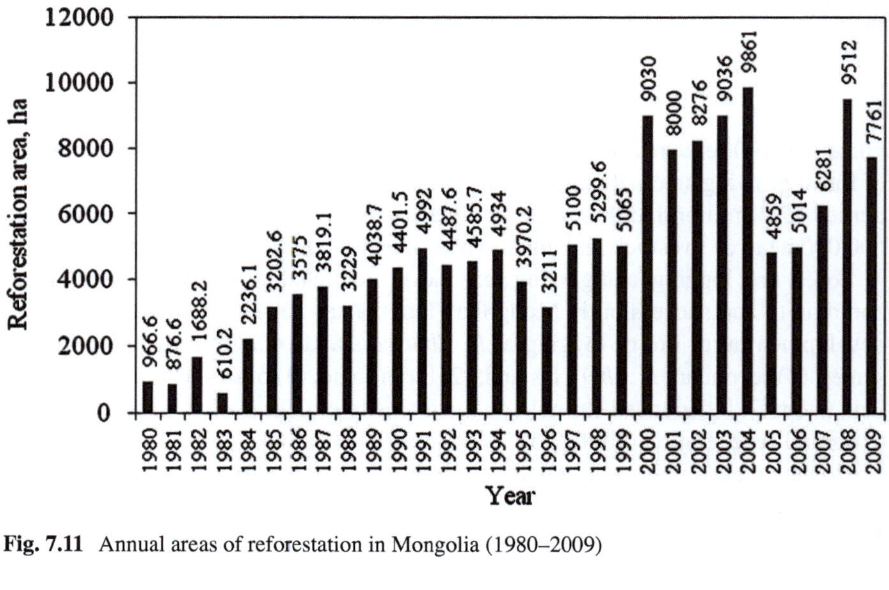

Fig. 7.11 Annual areas of reforestation in Mongolia (1980–2009)

7.5 Issues on Forest Rehabilitation and Tree Planting

Current problems and limitations in reforestation work or tree planting in Mongolia are more related to difficulty in collecting qualified seeds and producing normal seedlings; lack of supportive inputs and incentives for plantation activities and application of time-consuming methods; usage of labor intensive technique/ insufficient skilled labor and no effective application of participatory approaches; high conflict between forestry and livestock sectors; limited financial allotment; and poor maintenance and weak system of land management.

Steps for successful reforestation have not been considered, such as proper selection of tree species, species–site matching, timely planting and maintenance in the context of the changing climate in Mongolia, and protection and maintenance of reforestation sites as well. To increase the survival rate of reforestation, it is necessary to take into attention the criteria for selection of tree species including good adapting ability, adaptation to drought condition, and adaptability to poor soil conditions.

7.6 Recommendations

An analysis of the forest situation in Mongolia indicates that the existing policy has not succeeded in achieving sustainable forest management. It is necessary to review the policy and to identify the underlying causes for policy failures. The main key indicators for sustainable forest management in Mongolia have been identified as listed below:

- Increase in the extent of forest and tree cover
- Conservation and maintenance of soil and water resources
- Maintenance and enhancement of ecosystem function and vitality of forests
- Adequacy of policy and legal and institutional framework

These points are suggested in this connection that forest policy is to be approached in a comprehensive/holistic manner, covering, inter alia, the following:

- Forest land use and management (ownership and functional classification of forest resource base; forest resource expansion; improvement of productivity; management planning; wide-based participation of local people)
- Forest protection and land rehabilitation (protection functions of forests; tree planting for protection and land rehabilitation; protection of forests)
- Environmental conservation (protected area system; improving the standard of environmental conservation; environmental conservation and income generation)
- Forest products utilization (forest harvesting; forest-based processing industries; non-wood forest products and services; trade and marketing; efficiency in forest products consumption; demand, supply, and scarcity)

- Socioeconomic contributions of forestry (basic human needs; generation of employment and income and poverty alleviation; entrepreneurship and people's participation)
- Institutional arrangements (institutional restructuring; changes in laws, rules and regulations; planning, monitoring and evaluation; investment and financial matters)

7.7 Conclusion

The loss of forests coupled with overgrazing and inappropriate land use practices leads to desertification and land degradation in Mongolia. Regarding this point of view, forest rehabilitation and forest conservation are important aspects in context of climate change and human well-being. According to the review of current status of deforestation and land degradation, the reforestation and plantations should be established on a manageable scale with the participation of the local community, which could play a crucial role for successful implementation of plantation work in Mongolia.

To implement reforestation and tree planting in the country, it is necessary to establish a forest community for long-term community-based forest rehabilitation, introducing forestry professional services and technical support. In all cases, following aspects such as objective of the planting activity and factors influenced on success and failure of forest rehabilitation in selected sites must be considered in accordance with relevant silviculture techniques.

References

Avaadorj D, Badrakh S, Baasandorj Y (2006) Changes of physical property of pastureland soil in Mongolia. Ulaanbaatar: Khokh Sudar Printing (in Mongolian).

Dagvadorj, D., L.Natsagdorj, J.Dorjpurev, and B.Namkhainyam. Assessment Report on Climate Change: Mongolia. 2009. Ministry of Nature, Environment and Tourism of Mongolia.

Erdenechuluun, T. 2006. The Legal and Illegal Economies: Wood Supply in Mongolia. Discussion papers. World Bank.

Krasnoschekov Yu.N, Korotkov A.E, and Dugarjav Ch. 1992. Principles of Ecological Assessment of Forest Ecosystem of Mongolia. Scientific Proceedings on Ecology and Nature Management. Puscshino. pp 70–81 (in Russian).

Park Yoeng Dae. 2005. Ecological Studies on Rehabilitation of Forests Degraded by Fire or Logging in Southern Khentii, Mongolia. A Dissertation for Degree of Doctor of Philosophy. Seoul. Korea.

Tsogtbaatar J (2004) Deforestation and reforestation needs in Mongolia. For Ecol Manag 201:57–63.

State of Environment. 2002. Ministry of Nature and Environment. Mongolia [in Mongolian].

State of Environment. 2007. Ministry of Nature and Environment. Mongolia [in Mongolian].

Chapter 8
Soil Diversity in Mongolia

Kenji Tamura, Maki Asano, and Undarmaa Jamsran

Keywords Kastanozems • Soil degradation • Soil formation • Steppe • Vegetation–soil sequence

8.1 Introduction

Mongolia is located in a transition zone between the Siberian taiga and Central Asian desert region. Thus, it contains diverse natural zones and landscapes. In addition, Mongolia has an immensely varied combination of geological conditions and topographies. One of the natural wonders of Mongolia is that it includes the southernmost distribution of permafrost, as well as the northernmost area of desert. These variations in the environmental factors in different parts of the country specify the soil distribution of Mongolia.

Soil distribution in Mongolia from north to south changes in line with the following longitudinal zonal schema: (1) mountain taiga zone with cryomorphic taiga soils; (2) mountain forest-steppe zone with Chernozems, dark Kastanozems, forest dark-colored soils, and derno taiga soils; (3) dry-steppe zone with Kastanozems; (4) semidesert zone with brown semidesert soils; and (5) desert zone with gray-brown desert soil and extra-arid desert "borzon" soils (Dorzgotov 2003). Dorzgotov (2003) also showed that the horizontal vegetation–soil sequence is represented in particular in the central plain of Mongolia, with widely distributed Kastanozems in the steppe area, and brown semidesert soil in the semidesert to desert area. The study of the vegetation–soil sequence in the plain areas is taken up later in this chapter.

K. Tamura (✉) • M. Asano
Graduate School of Life and Environmental Sciences, University of Tsukuba, Tsukuba, Japan
e-mail: tamura@agbi.tsukuba.ac.jp

U. Jamsran
Center for Ecosystem Study, Mongolian State University of Agriculture, Ulaanbaatar, Mongolia

N. Yamamura et al. (eds.), *The Mongolian Ecosystem Network: Environmental Issues Under Climate and Social Changes*, Ecological Research Monographs,
DOI 10.1007/978-4-431-54052-6_8, © Springer Japan 2013

Table 8.1 Classifications of Mongolian soils (Dorzgotov 2003)

Mountain soils	
1. Mountain tundra soils	1.1 Mountain tundra ochro soil
	1.2 Mountain tundra peaty-gleyed soil
	1.3 Mountain tundra cryoturbated soil
2. Mountain meadow, meadow-steppe soil	2.1 Mountain meadow brunizem soil
	2.2 Mountain meadow dark colored raw-humic soil
	2.3 Mountain meadow peaty raw-humic soil
	2.4 Mountain meadow-steppe mull-humic soil
3. Mountain forest soils	3.1 Mountain taiga cryomorphic ochro soil
	3.2 Mountain taiga cryomorphic peat-muck humic soil
	3.3 Mountain taiga podozolic soil
	3.4 Mountain derno-taiga soil
	3.5 Mountain forest, dark colored derno soil
	3.6 Forest slightly podozolic sandy soil
4. Mountain steppe soil	4.1 High mountain steppe raw humic soil
	4.2 Mountain chernozem
	4.3 Mountain kastanozems
5. Mountain desert-steppe and desert soils	5.1 Mountain semi-desert brown soil
	5.2 Mountain desert gray-brown soil
Plain and valley soils	
6. Steppe soils	6.1 Chernozem
	6.2 Kastanozem
7. Semidesert and desert soils	7.1 Semidesert brown soil
	7.2 Desert gray-brown soil
	7.3 Extra-arid desert "Borzon" soil
	7.4 Takyr
8. Hydromorphic soils	8.1 Meadow dark humic gleyed soil
	8.2 Meadow-boggy raw humic gleyed soil
	8.3 Boggy peaty gleyed soil
9. Halomorphic soil	9.1 Solonchak automorphic
	9.2 Solonchak hydromorphic
	9.3 Solonetz automorphic
	9.4 Solonetz hydromorphic
10. Floodplain soils	10.1 Alluvial boggy gleyed soil
	10.2 Alluvial meadow gleyed soil
	10.3 Alluvial derno soil

The soil classification system of Mongolia represents local soil diversity. Under this classification system, first, all zonal soils are divided into two categories: (1) soils of mountains, and (2) soils of plains areas and intermountain depressions (valleys). These classifications are then divided into ten groups, according to topography, vegetation, and moisture. Furthermore, the lower-level categories define the soil character in detail. The lower-level categories have type, subtype, and genus, which are based on soil diagnostic horizons, features, colors, and materials. Table 8.1 shows a systematic list of Mongolian soils by soil type.

8.2 Characteristics of Soil Formation in Mongolia

Mongolian soils are characterized by a long winter season and very little soil moistening. In addition, the majority of Mongolia is at an altitude of 1,500 m or more above sea level. Figure 8.1 shows the monthly soil temperature at a soil depth of 60 cm under forest steppe, steppe, and desert steppe. The figure shows significant differences between summer (May–October) and winter (November–April) soil temperatures in each vegetation region, and it is peculiar in that the freezing period is long and the depth is deep. Only in the short summer, the rainy season, does biological activity become high and soil formation progress. The characteristics of soil formation are summarized as (1) slow process of chemical weathering and clay formation, (2) specific process of geochemical migration elements, (3) slow biological cycling of elements and comparatively easy change of elements from the biological to the geological cycle of migration, (4) short period of biological activity, caused not only by a lack of moisture, but also by long periods of below-freezing soil temperatures, (5) sharp differences and the peculiarities of the seasonal process of soil forming, even during the warm period of the year, caused by the contrast in moisture, the very clear monsoon characteristic of the precipitation, (6) basic soil-forming process taking place at minus temperatures or low plus temperatures, and (7) short soil profile and humus layer (Dorzgotov 2003).

In addition, when describing the soil profile morphology in Mongolia, we should take into account not only the current soil-forming process but also the paleo-environment. Horiuchi et al. (2000) reported that the last deglacial ended around 18 KY BP, and that the taiga was developed after 12 KY BP (1,000 years before present). After this period, there was an increase in vegetation biomass along with an extensive development of soil layers. Maximum postglacial warmth in the Mongolian

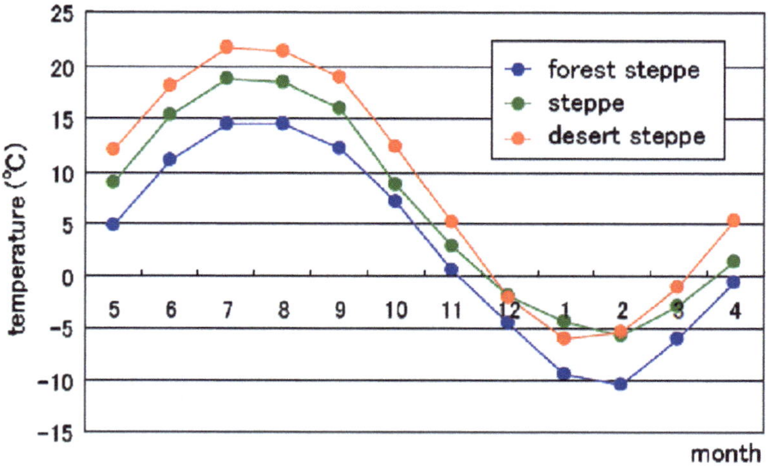

Fig. 8.1 Soil temperature at the soil depth of 60 cm under vegetation sequence from 1993 to 1997. (Data obtained by Institute of Meteorology and Hydrology of Mongolia)

plateau was reported to have been between 9 and 4 KY BP (Feng et al. 2005). Therefore, some relict features were observed in the soil profiles. Such properties included (1) difference of soil-forming sediment in the plain area, by geological age and stone and soil texture, (2) presence of a strongly carbonated layer and the absence of basic sediments of soluble salts and gypsum, (3) paleocryogenic wedges or cracks in the plain soils, that formed hidden cryolithogenic mottles in the soil profile, (4) residual paleocryomorphic and paleohydromorphic features in the steppe soils, and (5) residual solonetic properties (Dorzgotov 2003).

8.3 Soils of the Mongolian Grassland

In the eastern plain, as mentioned in the first section, the horizontal vegetation–soil sequence is well represented, where there are widely distributed Kastanozems in the steppe area. The rangelands of Mongolia are primarily a region of semiarid steppes, so we can say Kastanozems are the most important soil in Mongolia. Kastanozems have a brownish humus-rich surface horizon (less deep and less black than that of the Chernozems) and they show a prominent accumulation of secondary carbonates in the sub(surface) soil. The chestnut brown color of the surface soil gave these soils

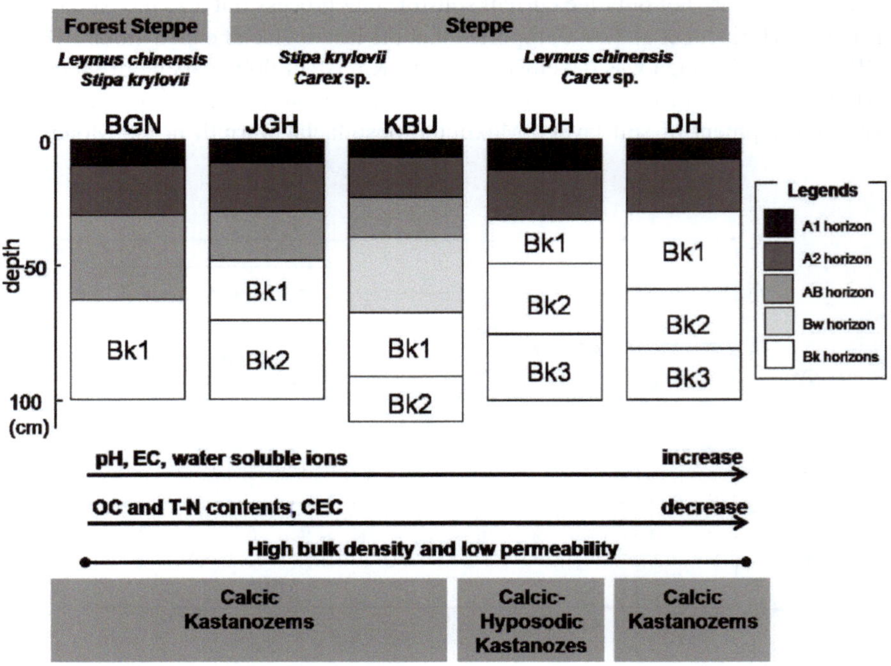

Fig. 8.2 Schematic diagram of the morphology and physicochemical characteristics of the soils (Asano et al. 2007)

their name "Kastanozem"; common international synonyms are (Dark) Chestnut Soils (Russia), (Dark) Brown Soils (Canada), and Ustolls and Borolls in the Order of the Mollisols (USDA Soil Taxonomy). Kastanozems are potentially rich soils; and the periodic lack of soil moisture is the main obstacle to high yields. Irrigation is nearly always necessary for high yields; and care must be taken to avoid secondary salinization of the surface soil. Small grains and (irrigated) food and vegetable crops are the principal crops grown. Wind erosion is a problem on Kastanozems, especially on fallow land (Driessen et al. 2001).

Figure 8.2 summarizes the morphology and physicochemical characteristics of Mongolian steppe soils (Asano et al. 2007). Although the variability of physicochemical properties of A horizons were relatively smaller among the study sites, there was a significant difference in the inorganic carbon content, pH, EC, exchangeable bases, and water-soluble ions in the Bk horizons. The amount of exchangeable bases and water-soluble ions tended to increase, especially Na^+, SO_4^{2-}, and Cl^- in the Bk horizon at study sites with a higher aridity index. The soils at forest-steppe sites and steppe sites were classified as Calcic Kastanozems and Calcic Hyposodic Kastanozems, respectively, and the soil at desert steppes was classified as Haplic Calcisols. It was suggested that these differences in the morphology and physicochemical characteristics reflect the amount of precipitation and potential evapotranspiration as well as the biological activities of root and microorganisms at each study site. At drier sites, soil degradation caused by grazing is more advanced as a result of low levels of precipitation, thin vegetation cover, and soil compaction (Opp et al. 2000). These differences reflect the differences in the thickness of the A horizon and the depth of the $CaCO_3$ accumulation horizon, which is characterized by alkalinity and high salt content (Asano et al. 2007).

References

Asano M, Tamura K, Kawada K, Higashi T (2007) Morphological and physico-chemical characteristics of soils in a steppe region of the Kherlen River basin, Mongolia. J Hydrol 333: 100–108.

Dorzgotov D. (2003) Soils of Mongolia. Admon, Ulaanbaatar (in Mongolia with English abstract)

Driessen P., Deckers J., Spaargaren O. (2001) Lecture notes of the major soils of the world. World Soil Resources Reports 94, FAO, Rome, pp233–236.

Feng, Z.-D., Wang, W.G., Guo, L.L., Khosbayar, P., Narantsetseg, T., Jull, A.J.T., An, C.B., Li, X.Q., Zhang, H.C., MaZ, Y.Z., 2005. Lacustrine and eolian records of Holocene climate changes in the Mongolian Plateau: preliminary results. Quat. Int. 136, 25–32.

Horiuchi, K., Minoura, K., Hoshino, K., Oda, T., Nakamura, T., Kawai, T. 2000. Paleoenvironmental history of Lake Baikal during the last 23,000 years: evidence from sedimentary and mineralogical analyses. Palaeogeogr., Palaeoclimatol. Palaeoecol. 157, 95–108.

Opp, C., Haase, D., Khakimov, V. 2000. Soils and soil processes in the Tyvinian part of the Uvs-noor basin. Polish. J. Soil Sci. 8, 71–80.

Chapter 9
Effects of Climate and Grazing on Surface Soil in Grassland

Muneto Hirobe and Junji Kondo

Keywords Climate • Grassland soil • Grazing • Net N transformations

9.1 Introduction

Because of the long history of nomadic activity, Mongolian grasslands have been maintained under grazing condition by livestock animals, such as sheep, goats, and horses. However, after the change of the political system in 1991, the numbers of livestock have continued to increase rapidly, except for the serious damage caused by cold winters in 2000 and 2001. In addition, nomadic activity has been concentrating more in regions near urban areas or main roads (Batjargal 1997). Therefore, it is important to understand the effects of livestock grazing on Mongolian grasslands under the present nomadic conditions for considering the sustainable management of grassland ecosystems.

Primary productivity and nutrient cycling are crucial properties to express the condition of an ecosystem, and they could be accelerated or decelerated by herbivory by large mammals (Ritchie et al. 1998; Augustine and McNaughton 2006). Many studies have suggested that herbivory decreases primary productivity and decelerates nutrient cycling (Augustine and McNaughton 2006; Pastor et al. 1993), because selective herbivory for the higher palatability species often leads to the dominance of lower palatability species with low relative growth rates and higher concentrations of anti-herbivore defense chemicals (Coley et al. 1985; Ritchie et al. 1998; Sasaki et al. 2007), and the decline in the quality and quantity of litter (Pastor et al. 1993; Ritchie et al. 1998). For example, Ritchie et al. (1998) reported

M. Hirobe (✉) • J. Kondo
Laboratory of Forest Ecology, Graduate School of Environmental Science, Okayama University, Okayama, Japan
e-mail: mhirobe@cc.okayama-u.ac.jp

N. Yamamura et al. (eds.), *The Mongolian Ecosystem Network: Environmental Issues Under Climate and Social Changes*, Ecological Research Monographs,
DOI 10.1007/978-4-431-54052-6_9, © Springer Japan 2013

that grazing decreased leguminous species coverage and aboveground biomass, resulting in the decline of soil N mineralization rate in an oak savanna. Similarly, Pastor et al. (1993) reported similar results, in that grazing declined soil N mineralization rate accompanied with decreasing total amount and N content in supplied plant litter in boreal forests. Such a mechanism was also confirmed by a model analysis (Pastor and Cohen 1997).

On the other hand, the increase of primary production and acceleration of nutrient cycling by herbivory have been reported in some ecosystems (Augustine and McNaughton 2006; Frank and Groffman 1998; Holland and Detling 1990; Tracy and Frank 1998). For example, Frank and Groffman (1998) reported that large herbivores increased N cycling in the grasslands of Yellowstone National Park (Frank and Groffman 1998). Similarly, Augustine and McNaughton (2006) found that the N cycling rate was accelerated if there was enough moisture in the soil with high nutrient levels in glades in central Kenya (Augustine and McNaughton 2006). In these cases, the proportion of primary production through herbivory increases as a consequence of the rapid regrowth of plants during periods without grazing (Augustine and McNaughton 2006), and the waste products of herbivores contribute to higher soil nutrient availability (Frank and Groffman 1998; Shand and Coutts 2006; Yamauchi and Yamamura 2004).

A wide range of factors has been suggested as the possible determinants of herbivore effects on ecosystem primary productivity and nutrient cycling (Ritchie et al. 1998; Bardgett and Wardle 2003). Soil fertility is suggested to be a major determinant, because plants require critical levels of nutrient supply to tolerate herbivory and to maintain their primary productivity (Bardgett and Wardle 2003; Augustine and McNaughton 2006). Climate is also suggested to be a major determinant by affecting the plant tolerance to drought that is connected to the tolerance to herbivory (Coughenour 1985; Milchunas et al. 1988). In arid and semiarid regions, herbivore effects may sometimes be unclear because precipitation varies greatly both spatially and temporally, and herbivores in natural ecosystems cannot follow productivity fluctuation (Ellis et al. 1993). In addition, the interaction between soil fertility and climate could be important (Augustine and McNaughton 2006). In a semiarid savanna, Augustine and McNaughton (2006) reported that herbivory reduced primary productivity irrespective of soil fertility in a low-rainfall year; however, in a high-rainfall year, herbivory increased primary productivity and soil N availability in sites with high soil fertility sites and decreased them in low soil fertility sites.

In Mongolian grasslands, there is a latitudinal gradient in the degree of aridity. The degree of aridity increases from north to south, and the type of grassland changes from forest steppe to steppe, and further to desert steppe. Therefore, the effects of livestock grazing may change under different climatic conditions. Our primary objective of this study is to clarify the effects of livestock grazing and climate and their interactions on Mongolian grasslands. To examine this, we compared the chemical properties of the surface soils (0–10 cm) at the inside and outside of the fence in the three types of grasslands.

9.2 Materials and Methods

9.2.1 Study Sites

We established three study sites along an aridity gradient from north to south in Mongolian grasslands, which were in forest-steppe, steppe, and desert-steppe zones. Details of this study were presented elsewhere (Kondo et al. 2011). The forest-steppe site was in Gachuurt (48°05′ N, 107°14′ E, about 1,745 m in elevation), the steppe site was in Mandalgobi (45°43′ N, 106°16′ E, about 1,370 m in elevation), and the desert-steppe site was in Dalanzadgad (43°34′ N, 104°25′ E, about 1,455 m in elevation), respectively.

The forest-steppe site is generally dominated by grasses, with forbs on south-facing slopes, and is dominated by the trees *Larix sibirica* and *Betula platyphylla* on ridges and on north-facing slopes. The main livestock around the site are cattle, sheep, and goats. The climate is characterized by dry and extremely cold winters and relatively humid and cool summers. Mean annual precipitation was 298.9 mm from 1998 to 2007, and about 80% of precipitation falls as rain from July to September (National Climate Data Center 2008). Mean annual air temperature was −0.6°C, and mean monthly air temperature ranged from −22.3°C in January to 16.9°C in July (National Climate Data Center 2008). The soils were Mollisols (Soil Survey Staff 2006) and the texture was clay-loam.

The steppe site is flat and generally dominated by grasses, forbs, and shrubs. The main livestock around the site are sheep, goats, and cattle (Sasaki et al. 2008). The climate is characterized by being very dry, with strong winds with an annual mean wind velocity of 5.1 $m\,s^{-1}$ and mean maximum wind velocity of 14.3 $m\,s^{-1}$ (National Climate Data Center 2008). Mean annual precipitation was 156.2 mm from 1998 to 2007, and about 80% of precipitation falls as rain during July and August (National Climate Data Center 2008). Mean annual air temperature was 3.0°C, and mean monthly air temperature ranged from −17.3°C in January to 21.1°C in July (National Climate Data Center 2008). The soils were Aridisols (Soil Survey Staff 2006), and the texture was sandy loam with abundant gravels.

The desert-steppe site is flat and generally dominated by grasses and forbs. The main livestock around the site are sheep, goats, and camels. The climate is characterized by being extremely dry, with strong winds of annual mean wind velocity of 4.0 $m\,s^{-1}$ and mean maximum wind velocity of 14.6 $m\,s^{-1}$ (National Climate Data Center 2008). Mean annual precipitation was 142.3 mm during 1998 to 2007, and about 80% of the precipitation falls as rain from July to September. Mean annual air temperature was 6.0°C, and mean monthly air temperature ranged from −19°C in January to 28.0°C in July (National Climate Data Center 2008). The soils were Aridisols (Soil Survey Staff 2006) and the texture was sandy loam with abundant gravels.

In the grasslands in the three study sites, a total of six plots were established inside and outside the enclosure fence. The fence was established in 1994 in the forest-steppe site, 1981 in the steppe site, and in the 1960s in the desert-steppe site. Some previous studies indicated that the period of enclosure is one of the important factors determining the effects of grazing on soil properties (Ritchie et al. 1998; Van Wijnen et al. 1998; Pei et al. 2008). In our study, the enclosure had been in place at least 13 years, and hence we treated the area inside the fence as "no grazing."

The vegetation in the forest-steppe site was dominated by *Carex duriuscula*, *Poa pratensis*, or *Potentilla mongolica* inside the fence and by *Calamagrostis mecilentha*, *Astragalus mongholicus*, or *Artemisia laciniata* outside the fence. The vegetation in the steppe site was dominated by *Allium pollyrhizum*, *Stipa glareosa*, or *Caragana leucophloea* inside the fence and by *A. pollyrhizum* and *S. glareosa* outside the fence. The vegetation in the desert-steppe site was dominated by *A. pollyrhizum* and *S. glareosa* inside the fence and by *A. pollyrhizum*, *S. glareosa*, or *Ajania fructicosa* outside the fence.

9.2.2 Soil Sampling and Analyses

At each plot, we located ten sampling points that were at least 2 m apart from each other. At each sampling point, a soil sample was taken from the surface mineral soil layer (0–10 cm) using a hand trowel, on 18 August in the forest-steppe site, 13 August in the steppe site, and 14 August in the desert-steppe site in 2007. Soil samples from the forest-steppe site were sieved through 2-mm mesh, and coarse roots were removed by hand. Inorganic N pools were measured using undried soil. Remaining soil samples were dried at 40°C for more than 48 h. Because soil samples from steppe and desert-steppe sites were in nearly air-dried condition at the time of sampling, these soil samples were dried at 40°C for more than 48 h, then sieved through 2-mm mesh and coarse roots removed by hand. For steppe and desert-steppe sites, inorganic N pools were measured using dried soil samples. Dried soil samples were used for the analyses of soil electrical conductivity (EC), pH (H_2O), and concentrations of inorganic N (except for forest-steppe), total N, available P ($NaHCO_3$ extraction; Olsen and Sommers 1982), and exchangeable Na, K, Ca, and Mg. As an index of N cycling, we also analyzed laboratory rates of net N transformation rates of air-dried soil (28 days in the dark at 25°C with moisture held constant at 20%, 40%, and 60% of the field capacity).

9.2.3 Statistical Analysis

Effects of grassland type and grazing, and their interaction on soil properties, were clarified by two-way analysis of variance at the significance level of $P<0.05$ (Statistica 2003).

9.3 Results and Discussion

Two-way analysis of variance (ANOVA) showed that there could be five groups in the examined soil properties (Tables 9.1 and 9.2; all data are from Kondo et al. 2011), based on the effects of aridity and the presence or absence of grazing. The first group was the concentrations of total N, available P, and exchangeable K and Mg. These properties significantly decreased with the greater degree of aridity ($P<0.05$) but were not affected by the presence or absence of grazing ($P>0.05$). The second group was the concentration of exchangeable Na and Ca. These properties were significantly different among the three grasslands ($P<0.05$) irrespective of aridity and were not affected by the presence or absence of grazing ($P>0.05$). The third group was inorganic N concentration, which was not affected by either aridity or the presence or absence of grazing ($P>0.05$). The fourth group was EC and pH (H_2O). These properties showed significant interactions, and the presence or absence of grazing had different influences under different aridity conditions ($P<0.05$). The fifth group was net soil N transformation rates, which decreased with the greater degree of aridity and decreased in the presence of grazing ($P<0.05$).

Regarding the aridity effects on grassland soil, previous studies have shown that with the greater degree of aridity, primary productivity (Milchunas et al. 1989; Sala et al. 1988) and litter input to soil (Lane et al. 2000; Scholes et al. 2002; McCulley et al. 2005) decrease. Consequently, the concentrations of soil organic matter and organic N decrease, and the mineralization rates of C and N are retarded (Vinton and Burke 1997; Aranibar et al. 2004; McCulley et al. 2005). Moreover, under the greater degree of aridity, soil base cations are not leached substantially, and soil EC and pH are often higher (Chytrý et al. 2007; Dregne 1976).

The soil properties of the first group in this study are essential macronutrients, and the decrease with the greater degree of aridity was probably the result of the decreases of primary productivity and litter input to soil (Vinton and Burke 1997; Aranibar et al. 2004; McCulley et al. 2005). Those of the second group are nonessential elements (Na) and nonlimiting essential macronutrients (Ca), and it has been suggested that the dynamics of these elements is affected more by physical processes such as water movement in soil, than by biological processes, in arid and semiarid regions (Schlesinger et al. 1996; Jobbágy and Jackson 2001). However, there are no data available directly concerned with water movement in soil, and the cause of the difference is still unclear.

The inorganic N concentration was the third group, and this was not different at a particular moment when the soil was sampled. However, it is highly dynamic both spatially and temporally (Burke 1989; Hirobe et al. 2003), because it is the sum of inorganic N input, production, and output including plant uptake (Binkley and Hart 1989). Therefore, the inorganic N concentration may not always be one of the best indices to express soil chemical environments. The fourth group was characterized

Table 9.1 Soil chemical properties (mean ± 1SE) in the three types of Mongolian grasslands

	Desert steppe		Steppe		Forest steppe	
Soil property	Not grazed	Grazed	Not grazed	Grazed	Not grazed	Grazed
EC (mS m^{-1})	39.5±2.92	31.3±1.22	33.4±5.16	52.4±11.3	48.7±2.82	43.8±1.58
pH (H_2O)	8.57±0.04	8.83±0.01	7.72±0.08	8.06±0.09	6.21±0.05	6.24±0.04
Inorganic N pool (mg-N kg-$soil^{-1}$)	7.20±1.71	5.15±0.34	13.1±1.02	26.6±15.0	18.8±1.18	15.1±1.16
Total N pool (g-N kg-$soil^{-1}$)	0.95±0.03	0.82±0.02	1.09±0.05	0.98±0.05	8.43±0.29	8.34±0.26
Available P pool (mg-P kg-$soil^{-1}$)	5.38±0.52	4.84±0.25	9.11±0.47	9.88±0.61	14.5±1.21	12.3±0.72
Exchangeable Na pool (mg-Na kg-$soil^{-1}$)	21.4±1.42	19.8±1.00	32.6±5.13	57.4±13.3	19.4±2.06	19.9±2.79
Exchangeable K pool (mg-K kg-$soil^{-1}$)	233.8±15.3	231.7±14.4	223.2±12.07	222.0±38.9	473.2±28.2	486.7±24.9
Exchangeable Ca pool (g-Ca kg-$soil^{-1}$)	3.47±0.08	3.22±0.07	1.09±0.09	1.51±0.28	3.90±0.13	3.60±0.30
Exchangeable Mg pool (mg-Mg kg-$soil^{-1}$)	199.4±9.95	150.5±5.76	377.2±36.8	368.3±57.7	624.0±17.6	564.2±25.1
Net N mineralization rate (20%) (mg-N kg-$soil^{-1}$ day^{-1})	0.63±0.19	−0.02±0.22	0.16±0.12	−0.26±0.26	7.21±0.51	5.02±0.49
Net N mineralization rate (40%) (mg-N kg-$soil^{-1}$ day^{-1})	0.98±0.29	0.14±0.03	0.33±0.13	−0.17±0.12	7.18±0.41	6.50±0.41
Net N mineralization rate (60%) (mg-N kg-$soil^{-1}$ day^{-1})	1.44±0.30	0.30±0.06	0.77±0.14	0.18±0.13	6.40±0.54	5.93±0.56
Net nitrification rate (20%) (mg-N kg-$soil^{-1}$ day^{-1})	0.64±0.19	0.09±0.01	0.40±0.13	0.40±0.16	4.60±0.38	3.83±0.31
Net nitrification rate (40%) (mg-N kg-$soil^{-1}$ day^{-1})	0.96±0.30	0.14±0.01	0.52±0.15	0.34±0.23	6.62±0.51	6.30±0.43
Net nitrification rate (60%) (mg-N kg-$soil^{-1}$ day^{-1})	1.47±0.31	0.54±0.05	1.01±0.15	0.94±0.40	6.06±0.68	5.99±0.56

EC electrical conductivity
Source: All data are from Kondo et al. (2011)

Table 9.2 Results of two-way analysis of variance (ANOVA) on the effects of aridity and grazing, and their interaction on soil chemical properties in the three types of Mongolian grasslands

	F value		
Soil property	Aridity	Grazing	Aridity × grazing
EC (mS m^{-1})	2.13^{NS}	0.20^{NS}	3.78*
pH (H_2O)	994***	20.1***	4.10*
Inorganic N pool (mg-N kg-$soil^{-1}$)	2.68^{NS}	0.26^{NS}	1.18^{NS}
Total N pool (g-N kg-$soil^{-1}$)	1415***	0.50^{NS}	0.03^{NS}
Available P pool (mg-P kg-$soil^{-1}$)	68.2***	0.83^{NS}	1.69^{NS}
Exchangeable Na pool (mg-Na kg-$soil^{-1}$)	11.0***	2.31^{NS}	3.09^{NS}
Exchangeable K pool (mg-K kg-$soil^{-1}$)	74.9***	0.00^{NS}	0.00^{NS}
Exchangeable Ca pool (g-Ca kg-$soil^{-1}$)	103***	0.10^{NS}	2.47^{NS}
Exchangeable Mg pool (mg-Mg kg-$soil^{-1}$)	91.5***	2.30^{NS}	0.35^{NS}
Net N mineralization rate (20%) (mg-N kg-$soil^{-1}$ day^{-1})	196***	12.5***	2.67^{NS}
Net N mineralization rate (40%) (mg-N kg-$soil^{-1}$ day^{-1})	339***	6.51*	0.28^{NS}
Net N mineralization rate (60%) (mg-N kg-$soil^{-1}$ day^{-1})	199***	6.96*	0.37^{NS}
Net nitrification rate (20%) (mg-N kg-$soil^{-1}$ day^{-1})	183***	4.70*	1.19^{NS}
Net nitrification rate (40%) (mg-N kg-$soil^{-1}$ day^{-1})	229***	2.35^{NS}	0.68^{NS}
Net nitrification rate (60%) (mg-N kg-$soil^{-1}$ day^{-1})	47.5***	1.19^{NS}	0.32^{NS}

Source: All data are from Kondo et al. (2011)
NS not significant
*$P<0.05$; **$P<0.01$; ***$P<0.001$

by the significant interaction of aridity and the presence or absence of grazing, but the pattern was different between EC and pH. The EC was expected to increase with the greater degree of aridity (Dregne 1976), but we could not detect any significant effects of aridity and grazing. The significant interaction found in this study may be the result of temporal variation induced by the behavior of livestock animals under the presence of grazing. The increase of pH with the greater degree of aridity was caused by less leaching of base cations (Dregne 1976; Chytrý et al. 2007), and the presence of grazing increased pH, probably because the loss of aboveground plant biomass could lead to increased evaporation (Laio et al. 2001; Maestre et al. 2002), decrease of water demand by plants (Jobbágy and Jackson 2001), and the lesser water permeability of soil (Rietkerk et al. 2000; Maestre et al. 2002).

As already discussed, the soil chemical environment deteriorated with the greater degree of aridity, whereas the effects of grazing were not prominent, except for pH. However, for net soil N transformation rates, an index of N cycling, both aridity and the presence or absence of grazing were significant, especially under drier incubation conditions. Net soil N transformation rates were retarded with the greater degree of aridity, because of the decreased organic N resource, as suggested by the decrease of total N and the suppressed microbial activity by the increase of pH (Killham 1994; Hirobe et al. 1998). The presence of grazing might retard net soil N transformation rates mainly by raising pH and suppressing microbial activity (Killham 1994). These results suggested that N cycling was retarded by the presence of grazing irrespective of the degree of aridity in Mongolian grasslands, even if the effects of grazing on soil environment were not prominent. It was also suggested that greater grazing intensity than occurs in present conditions would retard N cycling further, and might induce the serious degradation of the grasslands, especially under the greater degree of aridity.

Acknowledgments We thank Dr. N. Fujita, Ms. K. Uugantsetseg, Mr. N. Amartuvshin, and Mr. A. Enkhbaatar for their help in the field and laboratory. We are grateful to Drs. K. Yoshikawa and K. Sakamoto for their valuable suggestions. This research was partly supported by the Research Institute for Humanity and Nature (project number D-04).

References

Aranibar JN, Otter L, Macko SA, Feral CJW, Epstein HE, Dowty PR, Eckardt F, Shugart HH, Swap RJ (2004) Nitrogen cycling in the soil-plant system along a precipitation gradient in the Kalahari sands. Global Change Biol 10:359–373.

Augustine, DJ, McNaughton, SJ (2006) Interactive effects of ungulate herbivores, soil fertility, and variable rainfall on ecosystem processes in a semi-arid savanna. Ecosystems 9: 1242–1256.

Bardgett, RD, Wardle, DA (2003) Herbivore-mediated linkages between aboveground and belowground communities. Ecology 84: 2258–2268.

Batjargal, Z (1997) Desertification in Mongolia. RALA report 200.

Binkley, D, Hart, SC (1989) The components of nitrogen availability assessments in forest soils. Adv Soil Sci 10: 57–112.

Burke, IC (1989) Control of nitrogen mineralization in a sagebrush steppe landscape. Ecology 70: 1115–1126.

Chytrý, M, Danihelka, J, Ermakov, N, Hájek, M, Hájková, P, Kočí, M, Kubeová, S, Lustyk, P, Otýpková, Z, Popov, D, Roleček, J, Řezníčková, M, Šmarda, P, Valachovič M (2007) Plant species richness in continental southern Siberia: effects of pH and climate in the context of the species pool hypothesis. Global Ecol Biogeogr 16: 668–678.

Coley, PD, Bryant, JP, Chapin, FS (1985) Resource availability and plant antiherbivore defense. Science 230: 895–899.

Coughenour, MB (1985) Graminoid response to grazing by large herbivores :adaptations, exaptations, and interacting processes. Ann MO Bot Gard 72: 852–863.

Dregne HE (1976) Soils of arid regions. Elsevier, New York.

Ellis, JE, Coughenour, MB, Swift, DM (1993) Climate variability, ecosystem stability, and the implications for range and livestock development. In: Behnke, R, Scoones, I, Kerven, C (eds.) Range ecology at disequilibrium. Overseas Development Institute, London, pp. 31–41.

Frank, DA, Groffman, PM (1998) Ungulate vs. landscape control of soil C and N processes in grasslands of Yellowstone National Park. Ecology 79: 2229–2241.

Hirobe , M, Tokuchi, N, Iwatsubo, G (1998) Spatial variability of soil nitrogen transformation patterns along a slope in a *Cryptomeria japonica* D. Don plantation. Eur J Soil Biol 34: 123–131.

Hirobe, M, Tokuchi, N, Wachrinrat, C, Takeda, H (2003) Fire history influences on the spatial heterogeneity of soil nitrogen transformations in three adjacent stands in a dry tropical forest in Thailand. Plant Soil 249: 309–318.

Holland, E.A. and Detling, J.K. 1990. Plant response to herbivory and belowground nitrogen cycling. Ecology 71: 1040–1049.

Jobbágy, EG, Jackson., RB (2001) The distribution of soil nutrients with depth: Global patterns and imprint of plants. Biogeochemistry 53: 51–77.

Killham, K (1994) Soil Ecology. Cambridge University Press, Cambridge.

Kondo, J, Hirobe, M, Uugantsetseg, K, Amartuvshin, N, Fujita, N, Sakamoto, K, Yoshikawa, K (2011) Effects of livestock grazing on the surface soil properties in Mongolian Steppe ecosystems along an aridity gradient. J Jpn Soc Reveg Technol 36: 406-415 (in Japanese with English summary).

Laio, F, Porporato, A, Ridolfi, L, Rodriguez-Iturbe, I (2001) Plants in water-controlled ecosystems: active role in hydrologic processes and response to water stress: II. Probabilistic soil moisture dynamics. Adv Water Resour 24 707–723.

Lane, DR, Coffin, DP, Lauenroth, WK (2000) Changes in grassland canopy structure across a precipitation gradient. J Veg Sci 11: 359–368.

Maestre, FT, Huesca, M, Zaady, E, Bautistaa, S, Cortina, J (2002) Infiltration, penetration resistance and microphytic crust composition in contrasted microsites within a Mediterranean semi-arid steppe. Soil Biol Biochem 34: 895–898.

McCulley, RL, Burke, IC, Nelson, JA, Lauenroth, WK, Knapp, AK, Kelly, EF (2005) Regional patterns in carbon cycling across the Great Plains of North America. Ecosystems 8: 106–121.

Milchunas, DG, Lauenroth, WK, Chapman, PL, Kazempour, MK (1989) Effects of grazing, topography, and precipitation on the structure of a semiarid grassland. Vegetatio 80: 11–23.

Milchunas, DG, Sala, O, Lauenroth, W (1988) A generalized model of the effects of grazing by large herbivores on grassland community structure. Am Nat 132: 87–106.

National Climate Data Center (2008) Global Summary of the Day, http://www7.ncdc.noaa.gov.

Olsen, SR, Sommers, LE (1982) Phosphorus. In: Page, AL, Miller, RH, Keeney, DR (eds) Methods of Soil Analysis, Part 2. ASA/SSSA, Madison, WI, pp 403–430.

Pastor, J, Dewey, B, Naiman, RJ, MacInnes, PS, Cohen, Y (1993) Moose browsing and soil fertility in the boreal forests of Isle Royale National Park. Ecology 74: 467–480

Pastor, J, Cohen, Y (1997) Herbivores, the functional diversity of plants species, and the cycling of nutrients in ecosystems. Theor Popul Biol 51: 165–179.

Pei, S, Fu, H, Wan, C (2008) Changes in soil properties and vegetation following exclosure and grazing in degraded Alxa desert steppe of Inner Mongolia, China. Agric Ecosyst Environ 124: 33–39.

Rietkerk, M, Ketner, P, Burger, J, Hoorens, B, Olff H (2000) Multiscale soil and vegetation patchiness along a gradient of herbivore impact in a semi-arid grazing system in West Africa. Plant Ecol 148: 207–224.

Ritchie, ME, Tilman, D, Johannes, MHK (1998) Herbivore effects on plant and nitrogen dynamics in oak savanna. Ecology 79: 165–177.

Sala, OE, Parton, WJ, Lauenroth, WK, Joyce, LA (1988) Primary production of the central grassland region of the United States. Ecology 69: 40–45.

Sasaki, T, Okayasu, T, Shirato, Y, Jamsran, U, Takeuchi, K (2007) Quantifying the resilience of plant communities under different grazing intensities in a degraded shrubland: A case study in Mandalgobi, Mongolia. Grassland Sci 53: 192–195.

Sasaki, T, Okayasu, T, Jamsran, U, Takeuti, K (2008) Threshold changes in vegetation along a grazing gradient in Mongolian rangelands. J Ecol 96: 145–154.

Schlesinger, WH, Raikes, JA, Hartley, AE, Cross, AF (1996) On the spatial pattern of soil nutrients in desert ecosystems. Ecology 77: 364–374.

Scholes, RJ, Dowty, PR, Caylor, K (2002) Trends in savanna structure and composition along an aridity gradient in the Kalahari. J Veg Sci 13: 419–428.

Shand, CA, Coutts, AG (2006) The effects of sheep faeces on soil solution composition. Plant Soil 285: 135–148.

Soil Survey Staff (2006) Keys to soil taxonomy, 3rd edn. USDA, Washington, DC.

Statistica (2003) STATISTICA 06J, Statsoft, Inc.

Tracy, BF, Frank, DA (1998) Herbivore influence on soil microbial biomass and nitrogen mineralization in a northern grassland ecosystem: Yellowstone National Park. Oecologia (Berl) 114: 556–562.

Van Wijnen, HJ, Van Der Wal, R, Bakker, JP (1998) The impact of herbivores on nitrogen mineralization rate: consequences for salt-marsh succession. Oecologia (Berl) 188: 225–231.

Vinton, MA, Burke, IC (1997) Contingent effects of plant species on soils along a regional moisture gradient in the Great Plains. Oecologia (Berl) 110: 393–402.

Yamauchi A, Yamamura N (2004) Herbivory promotes plant production and reproduction in nutrient-poor conditions: effects of plant adaptive phenology. Am Nat 163:138–153.

Chapter 10
Degradation of Mongolian Grassland Vegetation Under Overgrazing by Livestock and Its Recovery by Protection from Livestock Grazing

Indree Tuvshintogtokh and Dorjgotov Ariungerel

Keywords Degradation • Dominant species • Protection of grazing • Steppe vegetation • Succession

10.1 Introduction

In Mongolia, the largest biome is steppe, covering 83.4% of the total land area of the country, which corresponds to roughly 2.6% of the global grassland vegetation (World Resources Institute 2003). Mongolia has 126.0 million hectares of pasture area, including 6.2 million hectares of bluffs and gullies that are not valuable for pasture area and 3.0 million hectares of sand-covered areas, and saline soil areas of extra-arid desert occupy 20.5 million hectares. According to this estimation, only 76.5%, or 96.8 million hectares (ha), of total pasture area is currently used for grazing purposes. Approximately 70% of pastureland area has been degraded environmentally: around 40% is degraded ecologically, 20% is moderately degraded, 5% is severely degraded, and 2–3% is classified as very severely degraded (Awaadorj and Badrakh 2007).

The main reasons of this degradation are overgrazing and climate changes. In the period between 1940 and 2004, the mean annual temperature of Mongolia increased by 1.9°C, which is comparatively higher than that cited in earlier research

I. Tuvshintogtokh (✉)
Institute of Botany, Mongolian Academy of Sciences, Ulaanbaatar, Mongolia
e-mail: tuvshintogtokh@yahoo.com

D. Ariungerel
MercyCorps, Gobi Forage Project, Ulaanbaatar, Mongolia

N. Yamamura et al. (eds.), *The Mongolian Ecosystem Network: Environmental Issues Under Climate and Social Changes*, Ecological Research Monographs,
DOI 10.1007/978-4-431-54052-6_10,

papers (1.56°C). The total annual precipitation has decreased by 2% during those 65 years, which is considered insignificant in light of the current global changes. The sum of effective precipitation during the vegetative growth period is decreased by 13.8 mm, or 7.2%. In Eastern Mongolia, the sum of effective precipitation during the vegetative growth period is increased by 2.7–9.5%. In the central and Gobi parts of the country, it is decreased by more than 10% (Natsagdorj and Gomboluudev 2005).

Anthropogenic factors such as the nomadic herding system have the most important impact for changing the steppe vegetation cover. The most recent changes in the socioeconomic status have forced new changes in pastoral management because of the relaxation of central government control and implementation of a more "free-enterprise" system. Livestock numbers increased to 43 million from 25.8 million between 1990 and 2009. The carrying capacity of Mongolian pastureland can support only 60 million sheep units during the winter–spring seasons, but right now our livestock population is 65 million sheep units. The grazing pressure is increasing year by year in Mongolia because of overgrazing and degradation of the rangeland ecosystem. Introduction of heavy grazing often initiates a regressive succession, including a decrease in biomass and in structural complexity. Grazing intensity has altered the species composition of the grasslands. The extent of grassland degradation can be assessed by investigating the floristic composition of the plant community, and the floristic composition has been demonstrated to be an accurate indicator of grassland condition (Clements 1928; Dyksterhuis 1949; Nakamura et al. 2000). The floristic composition of the steppe is decreased such that mesophytes and unpalatable species have increased to 50–70%, and xerophytes have increased through increased grazing, as shown by results of the investigations by Mongol and neighbors Ovor Baigal and Inner Mongolia (Chognii 2001; Tong et al. 2004).

On the other hand, protection of grazing often leads to a progressive succession with changes in floristic composition (Clements 1916; Chognii 2001). The effect of protection of grazing varies with, for example, floristic composition and physiognomy, soil conditions, and precipitation, and grazing history may lead to changes in the phenology of species and to increases in canopy cover and standing biomass (Browns and Bagley 1986; Smith and Rushton 1994). Depending on whether the dominant species are palatable, the direct influence of grazing may lead to either an increase or a decrease in species diversity. It is generally assumed that the highest species richness is found at intermediate levels of disturbance (Collins and Barber 1985; Maarel 1993) and under conditions of moderate grazing (Helle and Aspi 1983).

Objectives of our survey are these:

1. Define succession, as affected by a long-term change on the vegetation communities of the steppe, dry steppe, desert steppe, and desert of Mongolia resulting from grazing pressure and climate changes.
2. Define pasture restoration after giving up grazing. Our study is focused on the following characteristics: (a) floristic composition and species richness, (b) vegetation coverage, and (c) dominant and subdominant species of community.

10.2 Materials and Methods

The geobotanical survey was conducted in the east, southeast, and south parts of Mongolia (Dornod, Sukhbaatar, Dornogobi, Dundgobi, and Umnugobi aimags) between 2001 and 2009. The typical steppe, dry steppe, desert steppe, and desert cover the territories of these aimags. We determined the floristic composition, height, phenological stage, and species coverage in quadrats of 10 m^2. All covered points are shown in (Fig. 10.1). In all, 291 geobotanical releves were analyzed, then compared with 1:1,500,000 scale vegetation maps from 1975 compiled for previous research purposes. We also defined succession stages in the two dominant vegetation communities of Eastern Mongolia through grazing.

We were studied nine fenced plots (Fig. 10.2). When designing our surveys, we made use of the right-of-way zone through the Ulaanbaatar to the Zamiin Uud. The width of the right of way, which is fenced off, is normally about 25–30 m, and in some places up to 1–1.5 km. The period of stringent isolation from grazing exceeded more than 50 years. We also used field research plots of the Institute of Botany, MAS (fenced for 6–10 years) (Table 10.1). We compared the vegetation community in pastures that were moderately or heavily grazed and those where grazing has been abandoned. Seven of our nine sampling sites are located in desert steppe, one is in typical steppe, and one is in the mountain steppe belt.

At each sampling site, 20 quadrats (10 are inside and 10 are outside the fenced plots) were sampled along the transect, using a frame of 1×1 m^2 in the grassland and 10×10 m^2 in the shrublands, for collecting data of floristic composition,

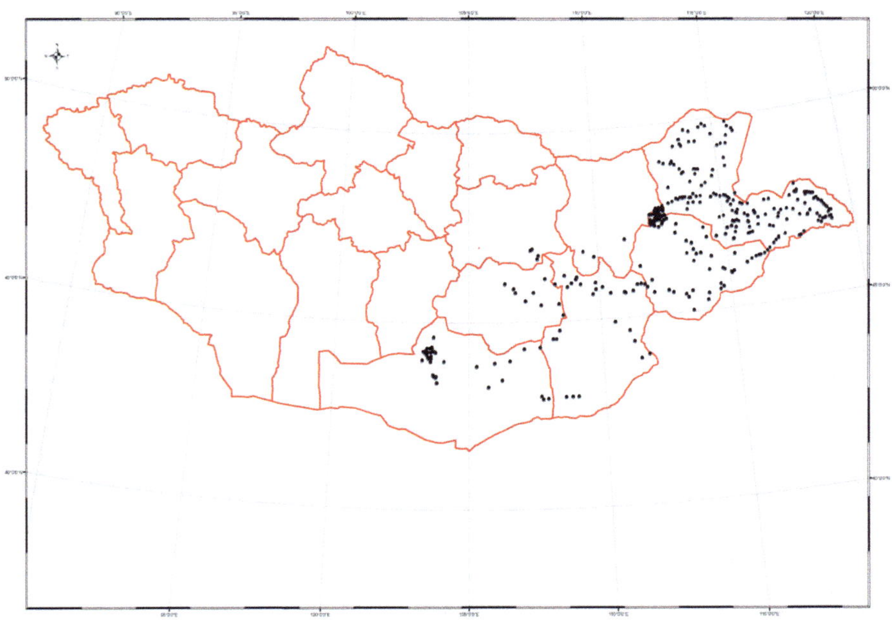

Fig. 10.1 Points of geobotanical releves

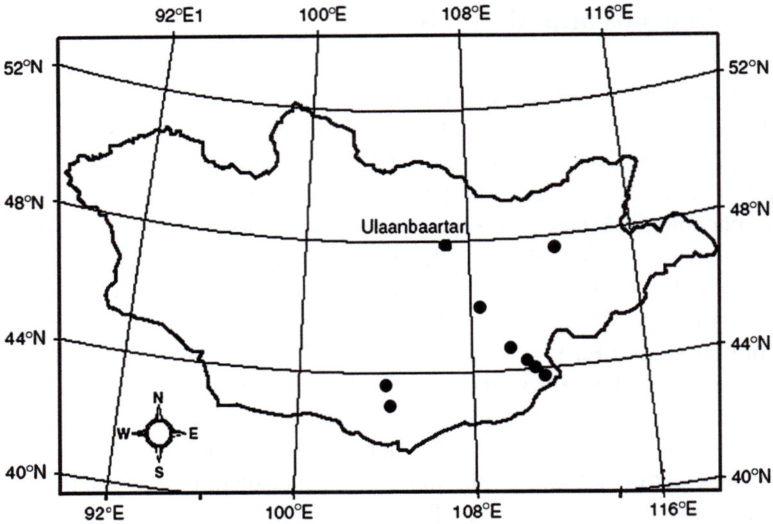

Fig. 10.2 Location fenced from livestock grazing

Table 10.1 Characteristics of sites fenced from livestock grazing

No.	Site name	Longitude	Latitude	Elevation (m)	Vegetation	Grazing level	Years fenced	Fenced site
04-1	Tumentsogt	47,79	112,51	985	Typical steppe	Moderate	6	Research site of Institute of Botany, MAS
01-1	Dalanjargalan	46,04	109,05	1,107	Desert steppe	Moderate	>50	Along the railway
01-2	Urgun-West	44,89	110,51	942	Desert steppe	Moderate	>50	Along the railway
01-3	Urgun-East	44,63	111,03	1,032	Desert steppe	Moderate	>50	Along the railway
03-1	Gurvan Saikhan	43,58	103,65	1,936	Mountain steppe	Moderate	7	Research site of SPA "Gurvan Saikhan"
02-1	Bulgan-North	44,18	103,73	1,312	Desert steppe	Heavy	20	Research site of Agricultural Institute
02-2	Bulgan-South	44,11	103,41	1,302	Desert steppe	Heavy	10	Research site of Institute of Botany, MAS
02-3	Erdene-North	44,29	111,48	1,012	Desert steppe	Heavy	>50	Along the railway
02-4	Erdene-South	44,03	111,66	991	Desert steppe	Moderate	>50	Along the railway

coverage, abundance, height, and phenological stage of species. All dominant species were classified into C3 and C4 plants through photosynthetic pathways (Wang and Yin 1997; Wang 2002, 2003; Pyankov et al. 2000).

Similarity between plant communities is defined by Jaccarda's formula (ISj) (Mueller-Dombois and Ellenberg 1974) as $ISj = c/(A + B - c) \times 100\%$, where A is the number of species of the fenced site, B is the number of species outside the fence, and c is the number of similar species of both sites.

10.3 Results

Dominant species display more abundance and greater coverage than other species and play a main role in composing the vegetation community and changes of pasture quality; these species act to confer environmental conditions. Therefore, changes of dominant species can express changes of the vegetation community and environmental conditions. We can identify changes of the dominant species for the years 1975–2009 (Table 10.2).

The vegetation of typical steppe, dry steppe, desert steppe, and desert in the vegetation map of Mongolia with scale 1:1,500,000 (1975) was compared with our research results between 2001 and 2009, as follows.

Our research points involved 32 contours of the vegetation map, and the dominant species of all contours changed. In eight contours, all dominant species were changed.

Table 10.2 Comparison of dominant species and vegetation subtype, 1975–2009

Number of map contours	Dominant species		Vegetation subtype	
	Remaining (1975–2009)	Current (2009)	Previous (1975)	Current (2009)
35a	–	*Cleistogenes squarrosa, Allium polyrrhizum, Stipa krylovii, S.gobica*	Typical steppe	Dry steppe Desert steppe
35a	–	*Stipa gobica, Allium polyrrhizum, Cleistogenes squarrosa, C. songorica*	Dry steppe	Desert steppe
38a	*Stipa grandis, S. baicalensis*	*Filifolium sibiricum, Leymus chinensis, Cleistogenes squarrosa*	Typical steppe	Typical steppe
38a	*Caragana leucophloea*	*Cleistogenes squarrosa, Allium polyrrhizum, Artemisia frigida, Reaumuria soongorica*	Dry steppe	Dry steppe Desert steppe
39a	*Stipa klemenzii*	*Allium polyrrhizum*	Dry steppe	Dry steppe
40a	*Stipa baicalensis*	*Leymus chinensis, Cleistogenes squarrosa, Artemisia frigida, Poa attenuata*	Typical steppe	Typical steppe Dry steppe
40a	–	*Stipa grandis, Cleistogenes squarrosa, Leymus chinensis*	Typical steppe	Typical steppe

(continued)

Table 10.2 (continued)

Number of map contours	Dominant species		Vegetation subtype	
	Remaining (1975–2009)	Current (2009)	Previous (1975)	Current (2009)
41	*Stipa krylovii, Cleistogenes squarrosa, Leymus chinensis*	*Allium polyrrhizum, Reaumuria soongorica*	Dry steppe	Dry steppe Desert steppe
42a	*Caragana microphylla*	*Stipa krylovii, Cleistogenes squarrosa, Leymus chinensis, Carex duriuscula, Allium polyrrhizum, Artemisia adamsii*	Typical steppe	Dry steppe
42a	*Caragana microphylla*	–	Typical steppe	Dry steppe
42a	*Caragana microphylla*	*Allium polyrrhizum, Stipa gobica, Cleistogenes squarrosa, Artemisia frigida, Reaumuria soongorica, Salsola passerina*	Dry steppe	Dry steppe Desert steppe
46	–	*Reaumuria soongorica*	Dry steppe	Desert steppe
47	*Stipa krylovii*	*Allium polyrrhizum, Cleistognes squarrosa, Artemisia pectinata*	Dry steppe	Dry steppe
48a	*Stipa klemenzii, Artemisia frigida*	*Stipa gobica, Allium polyrrhizum, Cleistogenes songorica*	Dry steppe	Desert steppe
49a	*Stipa gobica*	*Achnatherum splendens Ajania trifida*	Desert steppe	Desert steppe
49a	*Stipa gobica, Artemisia xerophytica*	*Allium polyrrhizum, Eurotia ceratoides, Caragana korshinskii, Anabasis brevifolia, Achnatherum splendens*	Desert steppe	Desert steppe
51	*Stipa gobica, Cleistogenes songorica*	*Caragna korshinskii, Allium polyrrhizum*	Desert steppe	Desert steppe
52a	*Stipa gobica*	*Allium polyrrhizum, Cleistogenes squarrosa, Caragana korshinskii, C. pygmaea*	Desert steppe	Desert steppe
54	*Allium polyrrhizum, Stipa gobica*	*Cleistogenes songorica, Allium mongolicum*	Desert steppe	Desert steppe
55	*Stipa glareosa, Anabasis brevifolia, Allium polyrrhizum*	*Artemisia frigida, Cleistogenes songorica*	Desert steppe	Desert steppe

(continued)

Table 10.2 (continued)

Number of map contours	Dominant species		Vegetation subtype	
	Remaining (1975–2009)	Current (2009)	Previous (1975)	Current (2009)
58a	*Stipa gobica, S.glareosa, Anabasis brevifolia, Caragana leucophloea*	*Allium polyrrhizum, Cleistogenes songorica, Haloxylon ammodendron*	Desert steppe	Desert steppe Desert
59a	*Stipa gobica, Caragana pygmaea*	*Allium polyrrhizum, Artemisia frigida, Cleistogenes squarrosa, C. songorica, Ajania achilleoides, Convolvulus ammanii, Reaumuria soongorica*	Desert steppe	Desert steppe
60a	–	*Caragana korshinskii, Stipa gobica, Cleistogenes songorica*	Sandy vegetation	Desert steppe
62	*Stipa glareosa*	*Allium polyrrhizum, Cleistogenes squarrosa, Caragana leucophloea, Convolvulus ammanii*	Desert steppe	Desert steppe
63	*Anabasis brevifolia*	*Salsola passerina*	Desert	Desert
64	–	*Anabasis brevifolia*	Desert	Desert
71	*Haloxylon ammodendron*	*Ajania achilleoides, Stipa gobica, Artemisia xerophytica, Convolvulus ammanii*	Desert	Desert Desert steppe
79	*Allium polyrrhizum*	*Cleistogenes songorica, C. squarrosa, Stipa gobica, Artemisia frigida*	Desert steppe	Desert steppe
80	–	*Kalidium gracile*	Desert steppe	Solonchak
82	*Reaumuria soongorica*	*Brachanthemum gobicum, Anabasis brevifolia*	Desert	Desert
83	–	*Artemisia pectinata, Chenopodium acuminatum, Artemisia frigida*	Solonchak	Dry steppe
84	*Haloxylon ammodendron*	–	Desert	Desert

The remaining 37 dominant species consist of 46% *Stipa* species (*Stipa gobica, S. glareosa, S. klemenzii, S. krylovii, S. baicalensis, S. grandis*), 16% *Caragana* species (*Caragana microphylla, C. leucophloea, C. pygmaea*), and 8.1% of *Allium polyrrhizum* and *Salsola passerina*.

Of the new 94 (overlapped) dominant plant species, there were 21 *Cleistogenes squarrosa, C. songorica*, 16 *Allium polyrrhizum*, 10 *Stipa*, 7 *Artemisia frigida*,

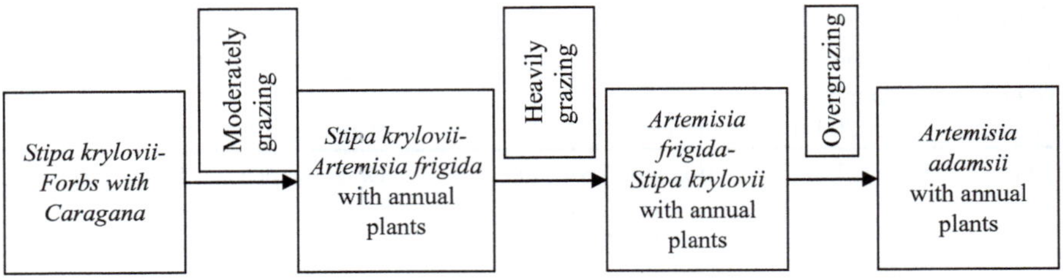

Fig. 10.3 Degradation pathways of the *Stipa krylovii* community

5 *Reaumuria soongorica*, 6 *Caragana* (*Caragana korshinskii* 4, *C. pygmaea* 1, *C. leucophloea* 1), 4 *Leymus chinensis*, 3 *Convolvulus ammanii*, and 3 *Anabasis brevifolia* registered.

Our study has shown that in the first stage, forbs and mesophyte plants disappeared because of the steppe degradation. The typical steppe is changed to dry and desert steppe, and dry steppe is changed to desert steppe; desert steppe and desert plant species are comparatively stable.

Nowadays, *Cleistogenes squarrosa*, *Stipa krylovii*, *Artemisia frigida*, and *Leymus chinensis* are the main dominants in the dry steppe, and *Allium polyrrhizum*, *Stipa gobica*, *Reaumuria soongorica*, *Cleistogenes songorica*, *Caragana korshinskii*, *Anabasis brevifolia*, and *Convolvulus ammanii* are dominant in the desert steppe.

We considered how floristic composition, vegetation coverage, and dominant species changed in the slightly, moderately, or heavily degraded community with *Stipa krylovii* widely scattered over the whole dry steppe zone of Mongolia. As seen from our results, the *Stipa krylovii-forbs-Caragana* community is changed to *Stipa krylovii–Artemisia frigida* with annual plants through moderate grazing and to *Artemisia frigida–Stipa krylovii* with annual plants community through heavy grazing, and to the *Artemisia adamsii* with annual palants community in the overgrazed condition (Fig. 10.3).

The total vegetation coverage of the community was not changed, but species diversity decreased from 44 species to 7 species, and *Stipa krylovii* coverage decreased to 2% from 15%; also, coverage of annuals increased to 87% from 14% (Table 10.3).

Our study was carried out in the *Stipa grandis* steppe, which can represent the typical steppe of Eastern Mongolia. In particular, the aforementioned principle is frequent in this community. Consequently, succession is going through changes in dominant species in response to the increased grazing effect (Fig. 10.4).

In this case, general coverage was not changed, but species diversity was reduced to seven species from 35 plant species whereas the coverage of *Stipa grandis* is decreased to 15% from 35%. In the overgrazed area with colony of Brandt's, *Stipa grandis* disappeared, and annuals such as *Ephedra sinica* are increased to 65% from 9% (Table 10.4).

The climax steppe is changed by the degraded community with unpalatable species and has five- to sixfold reduced species diversity.

In our results, when grazing ceases and the pasture is fenced, community structure and floristic composition are changed (Table 10.5). The coverage of community was increased by 1.5- to 2.5-fold, and the height of the community was also increased, by 1.4- to 2.4-fold.

Table 10.3 Change in vegetation community of *Stipa krylovii* through grazing level

Vegetation community	Total vegetation coverage (%)	Species numbers	Dominant plants: Species	Dominant plants: Coverage (%)	Annual plants	Annual percent for the plant community (%)
Stipa krylovii-Forbs-Caragana	50	44	*Stipa krylovii*	15	*Chenopodium viride*	14
			Caragana microphylla	2	*Salsola collina*	
			Caragana stenophylla	5		
			Serratula centauroides	4		
			Filifolium sibiricum	4		
			Potentilla acaulis	2		
Stipa krylovii-Artemisia frigida with annual plants	52	23	*Stipa krylovii*	7	*Chenopodium viride*	67
			Artemisia frigida	5	*Salsola collina*	
					Axyris amaranthoides	
Artemisia frigida-Stipa krylovii-with annual plants	54	22	*Artemisia frigida*	10	*Chenopodium viride*	74
			Stipa krylovii	3	*Salsola collina*	
Artemisia adamsii with annual plants	47	7	*Artemisia adamsii*	10	*Chenopodium viride*	87
			Stipa krylovii	2	*Salsola collina*	
					Chenopodium aristatum	

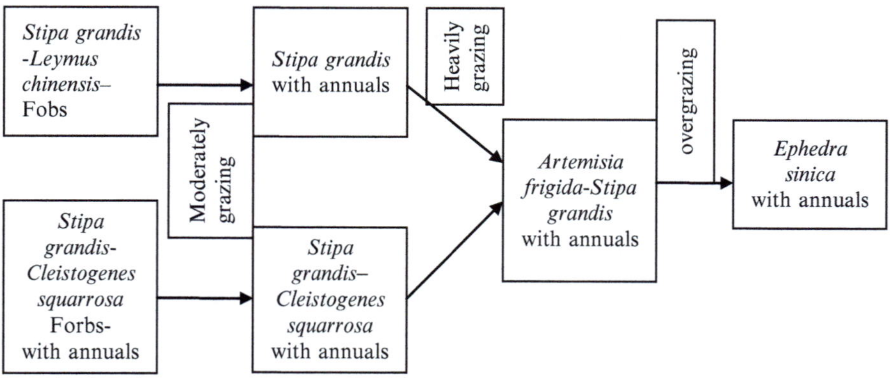

Fig. 10.4 Degradation pathways of the *Stipa grandis* community

Also, species richness is increased in the seven fenced sites, and floristic composition is very different between grazed and ungrazed pastures. Although the number of species has not changed in some sites, such as "Urgun-East" and "Gurvan Saikhan," the floristic composition is different. ISj has higher value in the site 1 "Tumentsogt" when protected from grazing for a few years and lower value in sites "Bulgan-North" and "Bulgan-South" with heavy grazing.

The coverage of dominant and subdominant species is recovering through the long-term abandonment of grazing (Tables 10.5 and 10.6). Although the dominant

Table 10.4 Changes in vegetation community of *Stipa grandis* through grazing level

Vegetation community	Total vegetation coverage (%)	Species number	Dominant plants: Species	Dominant plants: Coverage (%)	Annual plants	Annual percent for the plant community (%)
Stipa grandis-Leymus chinensis-Pulsatilla turczaninovii	66	35	*Stipa grandis*	30	*Chenopodium aristatum*	9
			Leymus chinensis	8		
			Pulsatilla turczaninovii	5		
Stipa grandis with annual plants	45	25	*Stipa grandis*	25	*Chenopodium viride*	22
			Leymus chinensis	2		
			Caragana microphylla	2		
			Kochia prostrata	2		
Stipa grandis-Cleistogenes squarrosa-Serratula centauroides with annual plants	60	35	*Stipa grandis*	20	*Chenopodium viride*	33
			Cleistogenes squarrosa	5		
			Serratula centauroides	3		
Stipa grandis-Cleistogenes squarrosa with annual plants	64	29	*Stipa grandis*	16	*Chenopodium viride*	54
			Cleistogenes squarrosa	5	*Salsola collina*	
Artemisia frigida- Stipa grandis with annual plants	60	12	*Artemisia frigida*	20	*Chenopodium viride*	21
			Stipa grandis	15	*Salsola collina*	
Ephedra sinica with annual plants	70	7	*Ephedra sinica*	15	*Chenopodium viride*	65
			Potentilla bifurca	5	*Chenopodium acuminatum*	
			Carex duriuscula	2	*Salsola collina*	

Table 10.5 Vegetation treatment of research sites comparing ungrazed and grazed pastures

Site	Vegetation coverage, %		Average height, cm		Species richness		IS_j (%)	Dominant species	
	In	Out	In	Out	In	Out		Ungrazed	Grazed
Tumentsogt	46.1	30.3	18.6	12.3	30	28	75.8	*Stipa grandis –Polygonum divaricatum*	*Stipa grandis*
Dalanjargalan	62	34	30	15	31	21	62.5	*Stipa gobica +Allium polyrrhizum*	*Stipa gobica +Allium polyrrhizum*
Urgun-West	41	22	28.5	16	25	19	57.1	*Artemisia anethifolia +Artemisia xerophytica +Eurotia ceratoides*	*Artemisia anethifolia +Artemisia xerophytica +Eurotia ceratoides*
Urgun-East	20	15	15.9	9.7	20	20	53.8	*Stipa glareosa +Cleistogenes songorica*	*Stipa glareosa +Cleistogenes songorica*
Gurvan Saikhan	64	30.5	15.8	8.6	11	11	57.1	*Agropyron cristatum + Artemisia frigida +Allium polyrrhizum*	*Allium polyrrhizum +Artemisia frigida +Convolvulus ammanii*
Bulgan-North	10.5	6.8	4.5	2	16	10	50,0	*Stipa gobica +Convolvulus ammanii +Artemisia frigida*	*Stipa gobica +Artemisia frigida +Cleistogenes songorica*
Bulgan-South	18	7.2	7.5	5.4	24	13	48.0	*Salsola collina +Stipa gobica +Eurotia ceratoides*	*Stipa gobica +Cleistogenes songorica +Allium polyrrhizum*
Erdene-North	17.3	11.8	14.1	5.8	19	15	57.1	*Cleistogenes songorica +Allium polyrrhizum +Bassia dasyphylla +Salsola collina +Stipa glareosa*	*Allium polyrrhizum +Stipa glareosa*
Erdene-South	45.4	25.1	15.1	10.8	20	19	62.5	*Salsola pestifera +Stipa glareosa +Cleistogenes songorica*	*Stipa glareosa +Cleistogenes songorica*

Table 10.6 Comparison of the coverage of dominant species in fenced sites

Site	Dominant species	Coverage (%)		C3/C4
		Ungrazed	Grazed	
Tumentsogt	*Stipa grandis* P. Smirn.	16	13	C3
	Polygonum divaricatum L.	5	1	C3
	Serratula centauroides L.	2.5	1.3	C3
Dalanjargalan	*Stipa gobica* Roshev.	35	13	C3
	Allium polyrrhizum Turcz.	20	10	C3
	Artemisia pectinata Pall. Polsak.	1	5	C3
Urgun-West	*Artemisia anethifolia* Weber	15	15	C3
	Artemisia xerophytica Krasch.	12	4	C3
	Eurotia ceratoides (L.) C.A. Mey	10	2	C3
Urgun-East	*Stipa glareosa* P. Smirn.	8	5	C3
	Cleistogenens songorica (Roshev.) Ohwi.	4	4	C4
Gurvan Saikhan	*Agropyron cristatum* (L.) Gaertn.	22	4	C3
	Allium polyrrhizum Turcz.	15	15	C3
	Artemisia frigida Willd.	17	8	C3
	Stipa krylovii Roshev.	4	1.5	C3
	Convolvulus ammanii Desr.	1.5	5.5	C3
Bulgan-North	*Stipa gobica* Roshev.	5.6	3.1	C3
	Convolvulus ammanii Desr.	1.7	0.1	C3
	Artemisia frigida Willd.	1.3	1.6	C3
	Cleistogenes songorica (Roshev.) Ohwi.	0.4	1	C4
Bulgan-South	*Stipa gobica* Roshev.	3.5	4	C3
	Eurotia ceratoides (L.) C.A. Mey	3	0.2	C3
	Allium polyrrhizum Turcz.	1.5	1.6	C3
	Cleistogenes songorica (Roshev.) Ohwi.	0.6	2.2	C4
	Salsola collina Pall.	5	0.1	C4
Erdene-North	*Salsola pestifera* A. Nelson	16	3.5	C4
	Stipa glareosa P. Smirn.	13	7.5	C3
	Cleistogenes songorica (Roshev.) Ohwi.	3.5	3.5	C4
	Allium mongolicum Regel.	3	1.3	C3
Erdene-South	*Allium polyrrhizum* Tuzcz.	3.5	5.6	C3
	Cleistogenes songorica (Roshev.) Ohwi.	4	0.3	C4
	Bassia dasyphylla (Fisch.) O. Kuntze	3.5	–	C4
	Stipa glareosa P. Smirn.	1.6	2.5	C3
	Salsola collina Pall.	2	0.1	C4

species were not changed in "Dalanjargalan" and "Urgun-West" sites, their coverage is different. According to our study, the coverage of bunchgrass such as *Stipa grandis*, *Stipa krylovii*, and *Agropyron cristatum* and taproot forbs such as *Polygonum divaricatum* and *Serratula centauroides* in the typical steppe is being dramatically increased through being protected from grazing for the long term.

As for desert steppe, the coverage of bunchgrass such as *Stipa gobica* and *Stipa glareosa*, and of semi-shrubs, namely *Artemisia xerophytica* and *Eurotia ceratoides*, is increased. There is high coverage of some annuals in both the ungrazed and grazed areas in the desert steppe where are found *Salsola collina*, *Salsola pestifera*, and *Bassia dasyphylla* (C4 plants). However, for annuals including *Artemisia pectinata* (C3) and *A. anethifolia* (C3), no results could be determined.

Grazing does not strongly influence some plants such as *Allium polyrrhizum*, *Cleistogenes songorica*, and *Convolvulus ammanii.*

Sand coverage of "Bulgan-South" sites was increased to 72% from 28% through giving up grazing, and senescence of year-round stands accumulated as well. A 5-cm sand layer covered the bare ground, and a 10-cm sand layer appeared around the bunchgrass and shrubs. Grass and forbs such as *Stipa gobica*, *Cleistogenes songorica*, and *Allium polyrrhizum* do not like to grow in the sand, but shrubs increased in the sand area (Table 10.6).

10.4 Discussion

If annual mean temperature is increased by 1°C, natural zones can be moved as much as 200–300 km backward and forward, and zonal elevation changes by 100–200 m upward and downward as that principle is expressed (IPCC 1996). This progression can be influenced by increased overgrazing with dry climate changes. Our study has determined that a typical steppe changed to desert steppe and dry steppe to desert, as well as plants of the desert steppe and desert, will move to the north, notably *Reaumuria soongorica*, *Caragana korshinskii*, *C. pygmaea*, *C. leucophloea*, *Convolvulus ammanii*, *Anabasis brevifolia*, and *Salsola passerina*.

Those two factors strongly influence the vegetation community, particularly reduction of species diversity and change of the dominant species (Ichiroku et al. 2008; Kawada et al. 2008). At the beginning stage of degradation, forbs and mesophytes disappear, and in the next stage, only grass such as a drought-tolerant and grazing-resistant species can readily spread, for example, *Cleistogenes squarrosa*, *Allium polyrrhizum*, and *Ephedra sinica*. *Cleistogenes* occurs in 21 and *Allium polyrrhizum* in 16 of the 32 investigation contours on the vegetation map of Mongolia.

Scientists hypothesized that *Cleistogenes squarrosa* would increase through warming (Sanjid 2002). Studies have shown that *Cleistogenes squarrosa* is increased through grazing at light and moderate levels (Chognii 2001; Gunin et al. 2002), as well as after a steppe fire (Tuvshintogtokh and Urgamal 2007). Those results are ratified by our study. These results are dependent on the biomorphological characteristics of the *Cleistogenes squarrosa*, a C4 species, which has many buds and can regrow one or two times per year.

Allium polyrrhizum is widely distributed in the dry and desert steppe and covers 15 geobotanical regions of Mongolia. *Allium polyrrhizum* is scattered in Central Asia but is a rare species in the north of Mongolia because this is registered in the

"Red Book" of Russia and Kazakstan. Expansion of *Allium polyrrhizum* has occurred in past years (Tuvshintogtokh and Ariunbold 2007).

Allium polyrrhizum is fibrous rooted, a perennial of height 10–25 cm. This species has protein, but the part above ground would be dry in the autumn because it is not fodder for winter. It can be dormant for a long time, and after rain, it can quickly grow. It is tolerant of trampling and can grow from seeds and vegetative shoots.

Finally, *Ephedra* is extending its growing area. *Ephedra sinica* covers the center part of Mongolia, as noted by several studies (in Center and Dundgobi aimags) (Gunin 2001; Gunin et al. 2002). *Ephedra* is scattered in the pastures with *Stipa grandis* as noted in a study conducted in the Dornod aimag. The *Ephedra* communities can be scattered as spot- and point forming, and its growing area is in the Center region rather than Dornod. The lesser growing area of *Ephedra* is dependent on light and carrying capacity and climate factors such as lower temperature and high percipitation.

Total coverage of the plant community does not decrease, although species diversity can be changed, through overgrazing in Eastern Mongolia, but a palatable species can occur in place of annuals that is more related to current climate conditions. The summer pasture was good between 2008 and 2009, the year the study was conducted. Consquently, in rainy years pioneer speceis can cover all bare ground while palatable species are reduced. In drought years, soil erosion and degradation will increase because of reduced yield, coverage of the community, and fewer annuals. It seems that *Cleistogenes squarrosa*, *Allium ployrrhizum*, *Epdedra sinica*, and annuals can dramatically adapt to the dry and overgrazed steppe areas in Mongolia.

If pasture is not be used for a long time, it can support increasing species diversity and lead to change in dominant species for secondary succession. Cession of grazing in arid and semiarid steppes in Mongolia leads to changes in floristic composition and dominants and increases species diversity, vegetation cover, and community height. Canopy cover increase through the abandonment of grazing is reported by Browns and Bagley (1986), Smith and Rushton (1994), Zhang (1998), and Chognii (2001).

Differences in the vegetation in grazed and ungrazed areas is very clear, and protection from grazing directly influences changes in species composition, as reported by Zhang (1998) and Chognii (2001). Canopy cover and height of the plant community will increase through long-term cessation of grazing, leading to increased community biomass.

Consequently, species composition and dominant species were changed, which is expressed by progressive and regressive succession. Both types of succession consist of an old climax community (after passing through a variety of stages) and another climax community (after grazing is stopped).

After long-term protection from grazing in the overgrazed meadow and typical steppe (13 years), pasture recovered in three successional stages: (1) domination by perennial rhizome grasses (2–3 years); (2) domination by low bunchgrass and forbs (4–5 years); and (3) domination by high bunchgrass and forbs (6–10 years). Overgrazed pasture recovered through resting for more than 10 years (Chognii 2001).

After ending grazing in the desert steppe, changes occurred in community standing biomass and the environment of the "Bulgan-South" site (e.g., sand cover). If sand cover continues to increase, grass-dominated vegetation is replaced by a shrub-dominated community. The same result was reported by Zhang (1998), in that *Artemisia halodendron* is increased through protection of grazing for 5 years.

Annual C4 plants such as *Salsola collina*, *Salsola pestifera*, and *Bassia dasyphylla* were increased through the abandonment of grazing, which may be related to the C4 species, for example, *Chenopodaceae* and *Amaranthaceae*. However, annual grass and pioneer species can grow in disturbed and cultivated soil at early succession stages. C4 species have greater tolerance to environmental stress (Wang and Yin 1997), especially to dry and poor soil (Wang 2002).

Grazing was not a strong influence on some plants such as *Allium polyrrhizum*, *Cleistogenes songorica* (C4), and *Ephedra sinica*. It may be hypothesized that these species replaced the dominant species *Stipa gobica* and *S. glareosa* through heavy and moderate grazing in the desert steppe.

10.5 Conclusions

1. Within the past three or four decades, the steppe vegetation of Mongolia has become drier. In particular, a typical steppe is changed to dry steppe, and both steppe can change to desert steppe. However, the vegetation of desert steppe and desert is comparatively stable.
2. Degradation of the steppe and climate changes strongly influence the vegetation community by such factors as reduction of species diversity and change of the dominant species. Some species such as *Cleistogenes squarrosa*, *Allium polyrrhizum*, and *Ephedra sinica* have readily spread, and their area is expanding.
3. There is progressive succession as a result of the increased species diversity and coverage; as well, dominant species have recovered through the abandonment of grazing. Therefore, we consider that overgrazing can have a greater influence than climate change on degradation. We are considering, as a common and cheap approach, the rest-and-rotation system for pastures of Mongolia. The traditional style of nomadism is a better way to restore a degraded pasture.

References

Awaadorj D, Badrakh S (2007) Vegetation community changes in pastureland. Geoecological issues in Mongolia, pp 27–31

Browns JE, Bagley CF (1986) Vegetation responses to long-term sheep grazing on mountain ranges. J Range Manag 39: 431–434

Chognii O (2001) The feature of regeneration grassland with using nomadic grazing in Mongolia. Mongol sudlal, Ulaanbaatar (in Mongolian)

Clements FE (1928) Plant succession and indicators. Hafner, New York

Clements FE (1916) Plant succession: an analysis of the development of vegetation. Carnegie Inst. Washington Publ.

Collins SL, Barber SC (1985) Effects of disturbance on diversity in mixed-grass prairie. Vegetatio 64: 87–94

Dyksterhuis E.J. (1949) Condition and management of range land based on quantitative ecology. J Range Manag 2: 104–115

Gunin D (2001) Terrestrial biodiversity: Examples from plant communities' diversity in Central Asia. In: Biodiversity of Asia's dryland regions: Experiences and prospects. Jinst Hargana, Ulaanbaatar

Gunin D et al (2002) Changes in ecosystem integrity succession dynamics of rangeland ecosystems of Mongolia. In: Chuluun T and Dennis Ojima Fundamental Issues Affecting Sustainability of the Mongolian Steppe. IISNC, Ulaanbaatar

Helle T, Aspi J (1983) Effects of winter grazing by reindeer on vegetation. Oikos 40(3): 337–343

Ichiroku H., Kawada K., Kurosu M., Batjargal A., Tsendeekhuu Ts., Nakamura T. (2008) Grazing effects on floristic composition and above ground plant biomass of the grasslands in the Northeastern Mongolian steppes. J. Ecol. Field Biol. 31(2): 115–123

IPCC (1996) Climate Change 1995: The Science of Climate Change. Contribution of working group I to second assessment Report of the IPCC. New York, Cambridge University Press.

Kawada K., Kurosu M., Cheng Y., Tsendeekhuu Ts., Wuyunna, Nakamura T., Ichiroku H. (2008) Floristic composition, grazing effects and above-ground plant biomass in the Hulunbeier Grasslands of Inner Mongolia, China. J. Ecol. Field Biol. 31(4): 297–307

Maarel E (1993) Some remarks on disturbance and its relations to diversity and stability. J.Veg Sci 4 (6):733–736

Mueller-Dombois D., Ellenberg H. (1974) Aims and methods of vegetation ecology. Wiley, New York.

Nakamura T., Go T., Wuyunna., Hayashi I. (2000) Effect of grazing on the floristic composition of grasslands in Baiyinxile, Xillingole, Inner Mongolia. Grassl Sci 45: 342–350

Natsagdorj L, Gomboluudev P (2005) Evaluation of natural forcing leading to desertification in Mongolia-Mongolian geoscientist, Ulaanbaatar

Pyankov VI, Gunin PD, Tsoog S, Black CC (2000) C_4 plants in the vegetation of Mongolia: their natural occurrence and geographical distribution in relation to climate. Oecologia 123: 15–31

Sanjid J (2002) Impact on vegetation cover and plants through climate changes and adaptation study. In: The potential impact of Climate change and V&A assessment for the livestock sector in Mongolia. Admon, Ulaanbaatar

Smith RS, Rushton SP (1994) The effect of grazing management on the vegetation of mesotrophic (meadow) grassland in Northern England. J Appl Ecol 31: 13–24

Tong C, Wu j, Yong S et al (2004) A landscape-scale assessment of steppe degradation in the Xilin River basin, Inner Mongolia, China. J. of Arid Environments. 59(1): 133–149

Tuvshintogtokh I, Urgamal M (2007) Fire effect on Productivity and Community Dynamics of Mongolian grassland, International Forest Fire News (IFFN). 36: 67–75

Tuvshintogtokh I, Ariunbold E (2007) Changes dominant species in steppe zone of Mongolia. Proceedings of the Mongolian Academy of Sciences. 183(1): 41–53 (in mongolian)

Wang P, Yin L (1997) Ecological distribution and physiological adaptation to saline-alkali environment of C_3 and C_4 plants in Northeastern China prairie area. Chin J Appl Ecol 8(4): 407–411

Wang RZ (2002) Photosynthetic pathways, life forms and reproductive types for forage species along the desertification gradient on Hunshandake desert, North China. Photosynthetica 40 (3): 321–329

Wang RZ (2003) Photosynthetic pathway and morphological functional types in the steppe vegetation from Inner Mongolia, North China. Photosynthetica 41(1): 143–151

World Resources Institute (WRI) (2003) A Guide to World Resources 2002–2004: Decisions for the Earth Balance, Voice, and Power. World Resources Institute, Washinton, DC

Zhang W (1998) Changes in species diversity and canopy cover in steppe vegetation in Inner Mongolia under protection from grazing. Biodiversity and Conservation 7: 1365–1381

Chapter 11
Annual Production and Species Diversity of Mongolian Pasture Plants in Relation to Grazing Pressure by Livestock

Noboru Fujita, Narantsetsegiin Amartuvshin, and Erdenegerel Ariunbold

Keywords Cattle • Compensatory growth • Mowing height and interval • Overgrazing • Sheep and goats

11.1 Introduction

Among plant–animal interactions, those between plants and herbivores are of particular importance, and the effects of herbivores vary according to their body size and digestive capability (Olff and Ritchie 1998). Large herbivores have strong direct effects on plants, as well as indirect effects through grazing, browsing, trampling, defecation, and urination (Mysterud 2006). For example, deer affect the plant species composition and productivity in forests (Côtè et al. 2004). Livestock are usually the main large herbivores in pastures, and the diet preferences (Bryant et al. 1979; Squires 1982; Lu 1988) as well as body sizes of livestock differ. Sheep and goats are small livestock that graze pasture plants using their teeth, whereas cattle are large livestock that graze by entwining plants with their tongues. Such differences in mouth size and grazing habits result in varying plant heights after normal grazing by sheep, goats, and cattle. This difference may affect regrowth and competition among pasture plants after grazing, and thereby the production and survival of plants in the pasture.

N. Fujita (✉)
Research Institute for Humanity and Nature, Kyoto, Japan
e-mail: fujita@chikyu.ac.jp

N. Amartuvshin
Institute of Botany, Mongolian Academy of Sciences, Ulaanbaatar, Mongolia

E. Ariunbold
Institute of Geoecology, Mongolian Academy of Sciences, Ulaanbaatar, Mongolia

N. Yamamura et al. (eds.), *The Mongolian Ecosystem Network: Environmental Issues Under Climate and Social Changes*, Ecological Research Monographs,
DOI 10.1007/978-4-431-54052-6_11,

Grazing can be considered a negative occurrence for a plant in the sense of predation (Fleischner 1994), but this is not necessarily the case when grazing leads to overcompensation by the plant (Paige and Whitham 1987). It is well known that plant production can improve under intermediate grazing pressure (McNaughton 1979; Hilbert et al. 1981; Dyer et al. 1986), a phenomenon referred to as the grazing optimization hypothesis. However, some studies do not support this hypothesis (Belsky 1986; Belsky et al. 1993; Milchunas and Lauenroth 1993; Pucheta et al. 1998). Compensatory growth by plants in response to grazing has been observed following soil fertilization (Bryant et al. 1983; Georgiadis et al. 1989; Holland and Detling 1990; de Mazancourt et al. 1998, 1999; Leriche et al. 2001). However, it is not clear whether the optimization of pasture plant production is a direct result of grazing or is limited to fertilized conditions only.

Waste from grazing livestock can fertilize pasture plants. On the other hand, human mowing of pasture plants in exclosures that prevent livestock from entering does not fertilize the pasture when the trimmings are removed from the pasture. Thus, it is possible to examine the effects of grazing on the annual production of pasture plants lacking fertilization by conducting a mowing experiment.

It is well known that plant species diversity differs in pastures with differing livestock grazing pressures. Fujita et al. (2002) showed that intermediate grazing pressure by livestock maximized the species richness of plants in a Mongolian pasture. This is the same pattern as the intermediate disturbance hypothesis proposed for tropical rainforests and intertidal zones (Levin and Paine 1974; Connell 1978; Pacala and Crawley 1992). Both positive and negative effects of livestock grazing on the plant species diversity of pastures with different topographic positions, that is, with different soil moisture conditions, have been reported (Fujita et al. 2009). Livestock grazing of the same intensity can positively influence the species diversity of pasture plants in moist habitats but have a negative influence in dry habitats. However, it is not clear whether, within the same habitat, different pressures of livestock grazing may produce different effects on the species diversity of pasture plants.

Here, we examined the effects of livestock grazing on the annual production and species diversity of pasture plants using a mowing experiment in an exclosure to determine the effects of grazing under conditions of no fertilization. To alter the simulated grazing pressure, we mowed the pasture plants at different heights and intervals. The study sites included a flat valley bottom with high production and an upper slope of the valley with low production.

11.2 Materials and Methods

11.2.1 Study Sites

For the study sites, we used the same pasture as studied by Fujita et al. (2009) in a catchment area in the forest-steppe zone at Gachuurt near Ulan Bator, Mongolia. In the pasture, a stream flows only in spring from the north to the south in the flat

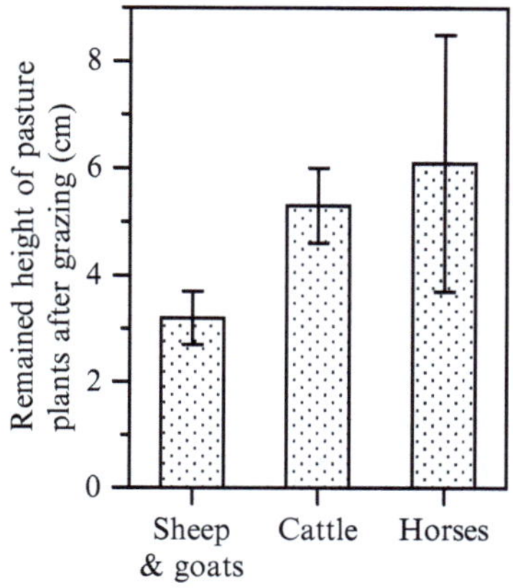

Fig. 11.1 Remaining height of pasture plants after normal grazing by sheep and goats, cattle, and horses. Each *bar* shows the average height with standard deviation

valley bottom, and the slope faces the east in the lower reaches. To examine the effects of grazing pressure on species diversity and the annual production of plants with different habitats, five 4×4 m^2 exclosures were set at two study plots near the upper slope of the hill (1,615 m alt.) and in the flat valley bottom (1,548 m alt.) in autumn of the year before each experiment. The exclosures were constructed using pillars and rectangular boards to avoid shading of light. In 2003, a land user constructed a large fence encompassing more than 200 ha that included our study site to prevent livestock from entering. This fence was usually effective at excluding livestock, although sometimes livestock were able to overcome it by human help.

As a preliminary experiment, we measured the typical grazing height by each livestock species at the corner of the large fence in 2004. In late June, more than 100 sheep and goats entered and grazed pasture plants for 30 min in the corner of the large fence outside the exclosure. Two cattle and two horses entered the large fence and grazed pasture plants from morning to 1600 (4 p.m.) in the corner of the large fence outside the exclosure in middle July and early August, respectively. After moving each livestock animal, we measured the heights of the grazed pasture plants that had been more than 10 cm tall before grazing at 30 randomly selected points in the corner (Fig. 11.1). The average grazing heights for sheep and goats was 3.2 mm, and that for cattle was 5.3 mm. Therefore, we chose to mow to heights of 3 and 5 cm to simulate grazing by sheep or goats and cattle, respectively.

11.2.2 Mowing Experiments

We randomly set a 1×1 m quadrat for each mowing condition in each exclosure. A total of five 1×1 m quadrats were used such that each mowing condition was duplicated in the five enclosures in each experiment. We conducted three different experiments: an interval experiment at the foot of the hill, a height experiment in two different habitats, and a height and interval experiment at the valley bottom.

11.2.3 Interval Experiment

We mowed pasture plants to 3 cm high at different intervals throughout the growing season in the flat valley bottom in 2005. The intervals were as follows: 5, 10, 20, 30, and 45 days, with a single mowing conducted at the end of the growing season as a control. We started mowing at 5- and 10-day intervals on 23 May, and the 20-, 30-, and 45-day intervals began on 2 June. The control quadrat was mowed on 21 August. Dry weight measurement after mowing was conducted in the same manner as for the height experiment. Specifically, after each mowing, we transported the trimmings from pasture plants to the laboratory, where they were dried at 80°C for 2 days to determine their dry weight. We calculated the total mass of the trimmings after mowing under each mowing condition.

11.2.4 Height Experiments in Two Different Habitats

We mowed pasture plants to 3 and 5 cm high at 1-month intervals from early June to early September 2004 on the upper slope and the flat valley bottom. For the control, pastures were not mowed during the growing season, and were mowed once to a 3-cm height at the end of the growing season in early August in each habitat. The 3- and 5-cm mowing heights represented the grazing heights for sheep or goats and cattle, respectively (Fig. 11.1). Grazed plant height only by horses was uneven, with a large standard deviation. Dry weight measurements after mowing were conducted in the same manner as in the interval experiments.

11.2.5 Height and Interval Experiments

We mowed to 3- and 0-cm heights at 1-month and half-month intervals from early June to early August in the flat valley bottom in 2006. For the control, we mowed only once to a 3-cm height in mid-August. The 0-cm mowing height and half-month mowing intervals were intended to simulate overgrazing conditions. Dry weight measurements of trimmings were conducted in the same manner as the height experiment. Just before the last mowing in mid-August, we recorded the plant species to determine the species number of pasture plants in each quadrat. This experiment was continued for 4 years from 2006 to 2009 in the same quadrat to examine the long-term effects of different grazing pressures. We dug up the underground parts of five quadrats for the 3-cm mowing height and 1-month mowing intervals, the 0-cm mowing height and half-month mowing intervals, and the control in early October 2009 to determine the difference in underground mass among treatments after 4 years of experiments. After washing rhizomes with buds, and roots, we measured their total dry weight of each quadrat in the same manner as the height experiment.

The Tukey HSD test was used to compare the means of the five samples. These statistical analyses were performed using JMP for Macintosh, ver. 5.1.2 (SAS Institute, USA).

11.3 Results

11.3.1 Effects of Different Mowing Intervals on Annual Production

Figure 11.2 shows the annual production of pasture plants at different mowing intervals in the flat valley bottom. A 1-month interval of mowing to a 3-cm plant height resulted in the maximum annual production. As the mowing intervals increased or decreased from 1 month, annual production was reduced. When compared with the control plot, which was not mowed during the growing season, all mowing intervals resulted in higher annual production, except for the 5-day interval.

11.3.2 Effects of Different Mowing Heights on Annual Production in Two Different Habitats

The annual production of pasture plants after mowing to different heights in two different habitats is shown in Fig. 11.3. Annual production after the 5-cm mowing was higher than that following the 3-cm mowing in the flat valley bottom. On the other hand, annual production following the 3-cm mowing was higher than that

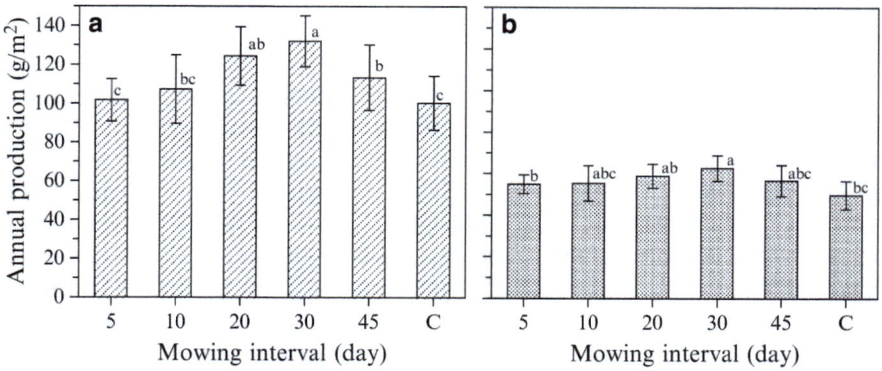

Fig. 11.2 Relationship between the annual production of pasture plants and mowing interval in days. *C* indicates that no mowing was conducted before the last mowing in late August. (**a**) Flat valley bottom. (**b**) Upper slope. *a*, *b*, and *c* on each *bar* show statistical significance ($P<0.01$)

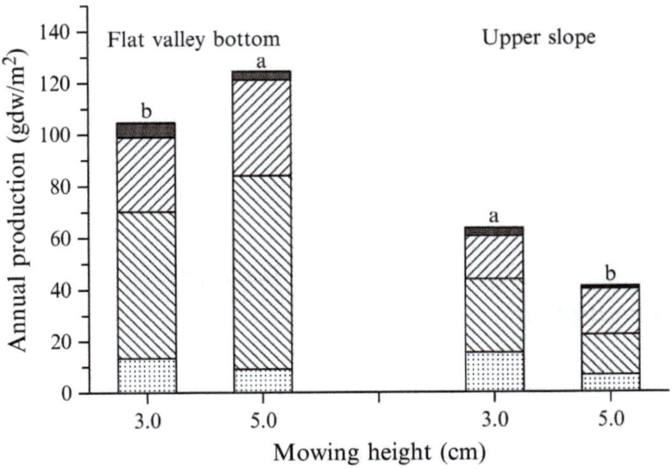

Fig. 11.3 Annual production of pasture plants mowed to 3- and 5-cm heights monthly in the flat valley bottom and on the upper slope. Each *bar* shows the biomass at each monthly mowing from June (*bottom*) to August (*top*)

after the 5-cm mowing on the upper slope. Thus, the 3-cm mowing, which simulated grazing by sheep and goats, and the 5-cm mowing, which simulated cattle grazing, produced opposite results in terms of the optimal annual plant production depending on the location.

11.3.3 Effects of Different Combinations of Mowing Intervals and Heights on Annual Production

The annual production of pasture plants following a combination of mowing regimes at 0- and 3-cm heights and 1-month and half-month intervals are shown in Fig. 11.4. In 2006, the first year of the experiment (Fig. 11.4a), annual production was lowest in the control plot, whereas it was highest following the 0-cm mowing at both 1-month and half-month intervals. Although the 1-month interval resulted in fewer mowings than the half-month interval, the annual production was higher following the 1-month mowing interval. On the other hand, in 2009, the fourth year of the experiment (Fig. 11.4b), annual production tended to be higher following the 3-cm mowing for both the 1-month and half-month intervals. The annual production following the 0-cm mowing at half-month intervals was the lowest and was lower than that of the control.

11.3.4 Effects of Different Combinations of Mowing Intervals and Heights on Species Diversity

The species diversity of pasture plants undergoing a combination of mowing regimes at 3- and 0-cm heights and at 1-month and half-month intervals are shown in Fig. 11.4. In 2006, the first year of the experiment (Fig. 11.4a), species diversity did not differ among treatments. In 2009, the fourth year of the experiment (Fig. 11.4b), species diversity was highest following the 3-cm mowing treatments, whereas it was lowest after mowing to 0 cm, with the control treatment falling in the middle. After 4 years of experiments, when underground biomass of pasture plants was

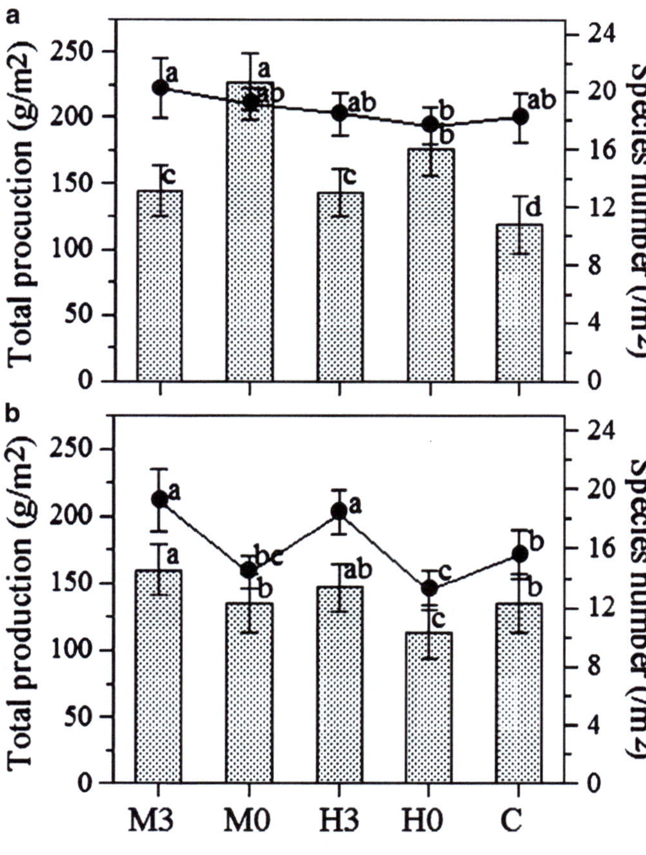

Fig. 11.4 Annual production and species diversity of pasture plants at different heights, intervals, and years in the flat valley: the first year mowing, 2006 (**a**); the fourth year of yearly mowing intensity, 2009 (**b**). *M3*, monthly mowing to 3-cm plant height; *M0*, monthly mowing to 0-cm plant height; *H3*, mowing at half-month interval to 3-cm plant height; *H0*, mowing at half-month interval to 0-cm plant height; *C*, a single mowing event in August to 3-cm plant height. *a*, *b*, *c*, and *d* on *bar* and *line* show statistical significance ($P<0.01$)

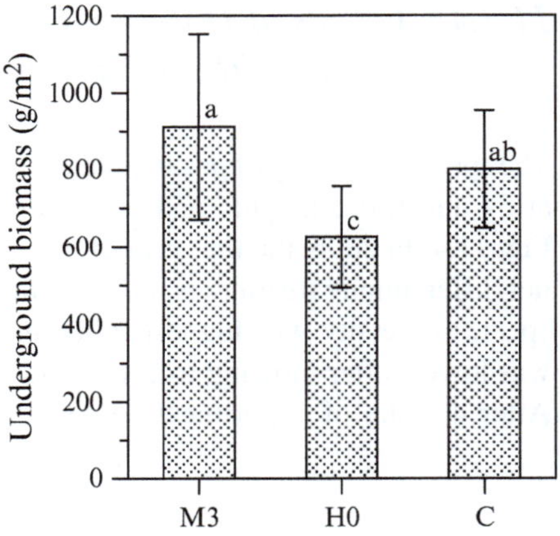

Fig. 11.5 Underground biomass of pasture plants after 4 years of experiment. *M3*, *H0*, and *C* are same as in Fig. 11.4. *a*, *b*, and *c* on *bar* and *line* show statistical significance ($P<0.01$)

compared among the 3-cm mowing height and 1-month mowing intervals, the 0-cm mowing height and half-month mowing intervals, and the control, it was significantly lowest for the 0-cm mowing height and half-month mowing intervals, and it tended to be higher in the 3-cm mowing height and 1-month mowing intervals compared with the control (Fig. 11.5).

11.4 Discussion

11.4.1 Optimization of Plant Production by Mowing

Some previous reports have supported the grazing optimization hypothesis (e.g., McNaughton 1979; Hilbert et al. 1981; Dyer et al. 1986), whereas others have not (e.g., Belsky 1986; Belsky et al. 1993; Milchunas and Lauenroth 1993; Pucheta et al. 1998). Our mowing experiments supported the optimization of annual production of pasture plants by livestock grazing. Specifically, mowing pasture plants at 1-month intervals resulted in the highest annual production among different test intervals, including no mowing as a control (Fig. 11.2). However, the intense 5-day mowing interval did not improve annual production when compared with the control (Fig. 11.2). Additionally, the intense mowing regime entailing a 0-cm mowing height at half-month intervals increased annual production in the first year but lowered it thereafter, whereas the intermediate mowing regime, which entailed a 3-cm mowing height at 1-month intervals, resulted in consistently improved plant production (Fig. 11.4). This observation suggests that intermediate grazing pressure optimizes the annual production of pasture plants, whereas intense grazing pressure has a deleterious effect. These results appeared to depend on the habitat productivity (Fig. 11.3) and the years of perturbation (Fig. 11.4). Pasture plants showed good growth of aboveground parts in the first year under intense grazing pressure (Fig. 11.4) because they could compensate well for strong grazing of aboveground biomass during the first year of intense grazing pressure. This compensatory growth might be the result of translocation from underground storage, because the underground parts of pasture plants declined markedly by the fourth year of continuous mowing. Thus, long-term overgrazing may drastically reduce the annual production of pasture plants.

In terms of the mechanisms behind grazing optimization, it has been proposed that nutrient cycling caused by grazing improves plant production (Bryant et al. 1983; Georgiadis et al. 1989; Holland and Detling 1990; de Mazancourt et al. 1998, 1999; Leriche et al. 2001; Yamauchi and Yamamura 2004). However, our mowing experiments demonstrated optimization of pasture production by grazing without nutrient cycling, as our mowing experiments were carried out in exclosures lacking livestock. The optimized pasture production may have been the result of improved photosynthetic activity of the remaining plants from the rejuvenation of leaves after grazing. Hayashi et al. (2007) demonstrated this mechanism by computer simulation. Photosynthetic rate declines with leaf senescence (Mondal et al. 1978). After grazing by livestock, plant regrowth produces new leaves with high photosynthetic rates, although time is required for this regrowth. In the Poaceae family, *Carex* and *Allium* possess a growing point at the base of the leaf, such that they can grow radial leaves with no time lag even if the tops of the leaves have been grazed. Therefore, these pasture plants exhibit an advantageous trait against the grazing of large herbivores.

Compared between mowing height and frequency, pasture plants showed high compensatory growth after mowing by soil surface mowing in 2006, the first year. However, this compensatory growth became very low in 2009, the fourth year, which means that soil surface mowing has a higher overgrazing effect by livestock than frequent mowing at the 3-cm height. High production of pasture plants during times of overgrazing by livestock is possible at the first year by their compensatory growth, although good compensation by the pasture plants does not continue for many years because of overconsumption of underground storage by compensatory growth. This theory is proved by the fact that underground biomass of pasture plants after 4 years of experiments was the lowest in the 0-cm mowing height and half-month mowing intervals (Fig. 11.5). Annual production of pasture plants at 3-cm mowing height and 1-month mowing interval was higher for 4 years of experiments than that in the control. It might be possible that underground biomass became larger with the 3-cm mowing height and 1-month mowing interval regime than that in the control area because of higher annual production, although the difference was not statistically significant (Fig. 11.5).

We observed that grazing optimization differed between the flat valley bottom and the upper slope. In the flat valley bottom, mowing to a 5-cm height resulted in a higher annual production of pasture plants compared with the 3-cm mowing despite a more intense mowing frequency, whereas on the upper slope, the 5-cm mowing yielded a lower annual production compared with the 3-cm mowing (Fig. 11.3). This difference may be a reflection of the differences in soil moisture between the two habitats. On the flat valley bottom, where the soil is moist, many plants can grow to heights greater than 15 cm in 1 month. Conversely, on the upper slope, where the soil is relatively dry, most plants reach a maximum height of less than 10 cm in 1 month. On the flat valley bottom, which experiences a high growth rate, the relatively greater mass of remaining pasture plants after mowing to a 5-cm height was more effective for enabling the regrowth of pasture plants than mowing the plants to a 3-cm height. On the other hand, on the upper slope, which experiences a low growth rate, mowing to a height of 5 cm was less effective, perhaps because insufficient plant mass had been cut to induce regrowth. Thus, it is more effective for pasture and livestock production to graze cattle in a flat valley bottom, whereas sheep and goats should graze on the upper slopes. Indeed, this tallies with traditional Mongolian knowledge; to lead cattle to the valley but sheep and goats to a slope. From livestock production, the pasture that regrows relatively tall after cattle grazing is better for cattle growth than the pasture that regrows relatively short after grazing by sheep and goats, because bite size linearly increases with increase of sward surface height in temperate grasses (Forbes 1988).

11.4.2 Enhancement of Species Diversity by Mowing

Fujita et al. (2002) demonstrated the intermediate disturbance hypothesis of plant species diversity in a Mongolian pasture, which is well known in tropical forests and intertidal zones (Levin and Paine 1974; Connell 1978; Pacala and Crawley 1992). Livestock grazing or mowing by humans may enhance the species diversity of pasture plants. However, we previously reported that species diversity increased in the flat valley bottom but declined on the upper slope after livestock grazing (Fujita et al. 2009). A variety of patterns for the effects of grazing pressures by large herbivores on plant species richness have been reported, including positive effects (Pucheta et al. 1998; Collins et al. 2002; Bakker et al. 2003; Pykälä 2004; Frank 2005), negative effects (Wilsey and Polley 2003; Guo 2004; Hendricks et al. 2005), or little effect (Stohlgren et al. 1999; Tracy and Sanderson 2000). Proulx and Mazumder (1998) proposed the hypothesis that plant species richness increases with increased grazing pressure in nutrient-rich environments, whereas it decreases in nutrient-poor environments. This hypothesis is in accordance with our previous results (Fujita et al. 2009).

Our current results showed that mowing in the first year had little effect on the species diversity of pasture plants, whereas after the fourth year of continuous yearly mowing, the species diversity of pasture plants for the 0-cm height mowing and no mowing decreased compared to plots of the first year of mowing, whereas species diversity for the 3-cm height mowing remained constant (Fig. 11.4). The intense mowing regime, whereby plants were cut to 0 cm in height at half-month intervals, reduced the species diversity to less than that observed in the no-mowing control plot. This result suggests that intermediate grazing pressure enhances species diversity, whereas intense grazing pressure decreases it. Our previous data also demonstrated that whether grazing pressure is optimal or too intense is dependent on habitat fertility, because the same grazing pressure by livestock had opposite affects on two habitats with different soil moisture levels (Fujita et al. 2009). In Mongolia, along with grazing pressure, the maximum species diversity of pasture plants was found in the most heavily grazed plots in the forest-steppe zone, in the least grazed plots in the steppe zone, and not in the grazed plots in the desert-steppe zone (Fernandez-Gimenez and Allen-Diaz 1999). These results are understandable in that the grazing pressure that optimizes species diversity of pasture plants increases with the rise of habitat fertility. Grazing optimization may also change with the disturbance period, as plant responses change depending on underground storage, even under the same grazing pressure. Therefore, it is understandable that a variety of patterns for the effects of grazing pressure by large herbivores on plant species richness have been described by previous studies. If overgrazing is too intense, one season is enough to drastically decrease the species diversity of pasture plants in locations close to nomad residences and sheds for livestock where sheep and goats are put together every day (see Fujita et al., this volume, Chap. 13). However, sometimes intense grazing pressure, such as mowing at half-month intervals to a 0-cm plant height, may require several years to considerably reduce the species diversity of a pasture.

Acknowledgments We thank Professor N. Yamamura (Research Institute for Humanity and Nature, Japan), Academician C. Dugarjav (Institute of Botany, Mongolian Academy of Science), and Dr. J. Tsogtbaatar (Institute of Geoecology, Mongolian Academy of Science) for their kind assistance and advice. This work was financially supported in part by RIHN research project "Collapse and Restoration of Ecosystem Networks with Human Activities."

References

Bakker C, Blair JM, Knapp AK (2003) Does resource availability, resource heterogeneity or species turnover mediate changes in plant species richness in grazed grasslands? Oecologia (Berl) 137:385–391.

Belsky AJ (1986) Does herbivory benefits plants? A review of the evidence. Am Nat 127: 870–892.

Belsky AJ, Carson WP, Jensen CL, Fox GA (1993) Overcompensation by plants: herbivore optimization or red herring? Evol Ecol 7: 109–121.

Bryant JP, Chapin FS, Klein DR (1983) Carbon nutrient balance of boreal plants in relation to vertebrate herbivory. Oikos 40: 357–368.

Bryant, FC, Kothmann, MM, Merrill, LB (1979) Diet of sheep, Angora goats, Spanish goats and white-tailed deer under excellent range conditions. *J Range Manag* 32: 412–417.

Collins SL, Glenn SM, Briggs JM (2002) Effect of local and regional processes on plant species richness in tall grass prairie. Oikos 99: 571–579.

Connell JH (1978) Diversity in tropical rain forests and coral reefs. Science 199: 1302–1310.

Côtè, SD, Rooney, TP, Tremblay, J-P, Dussault, C, Walker, DM (2004) Ecological impacts of deer overabundance. Annu Rev Ecol Evol Syst 35: 113–147.

De Mazancourt, C, Loreau, M, Abbadie, L (1998) Grazing optimization and nutrient cycling: when do herbivores enhance plant production? Ecology 79: 2242–2252.

De Mazancourt, C, Loreau, M, Abbadie, L (1999) Grazing optimization and nutrient cycling: potential impact of large herbivores in a savanna system. Ecol Appl 9: 784–797.

Dyer MI, DeAngelis DL, and Post WM (1986) A model of herbivore feedback on plant productivity. Math Biosci 79: 171–184.

Fernandez-Gimenez, ME, Allen-Diaz, B (1999) Testing a non-equilibrium model of rangeland vegetation of dynamics in Mongolia. J Appl Ecol 36: 871–885.

Fleischner, TL (1994) Ecological costs of livestock grazing in western North America. Conserv Biol 8: 629–944.

Forbes, TDA (1988) Researching the plant-animal interface: the investigation of ingestive behavior in grazing animals. J Anim Sci 66: 2369–2379.

Frank DA (2005) The interactive effects of grazing ungulates and aboveground production on grassland diversity. Oecologia (Berl) 143: 629–634.

Fujita N, Amartuvshin N, Uchida T, Wada E (2002) Biodiversity and sustainability of Mongolian herbaceous plants subjected to nomadic grazing. In: Sustainable watershed with emphasis on lake 3cosystems (eds Fujita N, et al.). DIWPA Series 3. Nauka-Center, Novosibirsk, pp 101–107.

Fujita, N, Amartuvshin, N, Yamada, Y., Matsui, K., Sakai, S, and Yamamura, N (2009) Positive and negative effects of livestock grazing on plant diversity of Mongolian nomadic pasturelands along a slope with soil moisture gradient. Grassland Sci 55: 126–134.

Georgiadis NJ, Ruess RW, McNaughton SJ, Western D (1989) Ecological conditions that determine when grazing stimulates grass production. Oecologia (Berl) 81: 316–322.

Guo Q (2004) Slow recovery in desert perennial vegetation following prolonged human disturbance. J Veg Sci 15: 757–762.

Hayashi M, Fujita N, Yamauchi A (2007) Theory of grazing optimization in which herbivory improves photosynthetic ability. J Theor Biol 245: 630–637.

Hendricks HH, Bond WJ, Midgley JJ, Novellie PA (2005) Plant species richness and composition a long livestock grazing intensity gradients in a Namaqualand (South Africa) protected area. Plant Ecol 176: 19–33.

Hilbert DW, Swift DM, Detling JK, and Dyer MI (1981) Relative growth rates and the grazing optimization hypothesis. Oecologia (Berl) 51: 14–18.

Holland EA, Detling JK (1990) Plant response to herbivory and belowground nitrogen cycling. Ecology 71: 1040–1049.

Leriche, H, Le Roux, X, Gignoux, J, Tuzet, A, Fritz, H, Abbandie, L, Loreau, M (2001) Which functional processes control the short-term effect of grazing on net primary production in grasslands? Oecologia (Berl) 129: 114–214.

Levin SA, Paine RT (1974) Disturbance, patch formation, and community structure. Proc Natl Acad Sci USA 71: 2744–2747.

Lu, CD (1988) Grazing behavior and diet selection of goats. Small Ruminant Res 1: 205–216.

McNaughton, SJ (1979) Grazing as an optimization process: grass–ungulate relationships in the Serengeti. Am Nat 113: 691–703.

Milchunas DG, Lauenroth WK (1993) Quantitative effects of grazing on vegetation and soils over a global range of environments. Ecol Monogr 63: 327–366.

Mondal, MH, Brun, WA, Brenner, ML (1978) Effects of sink removal on photosynthesis and senescence in leaves of soybean (*Glycine max* L.) plants. Plant Physiol 61: 394–397.

Mysterud, A., 2006. The concept of overgrazing and its role in management of large herbivores. Wildl Biol 12, 129–141.

Olff, H, Ritchie, ME (1998) Effects of herbivores on grassland plant diversity. *Trends Ecol Evol* 13: 261–265.

Pacala SW, Crawley MJ (1992) Herbivores and plant diversity. Am Nat 140: 243–260.

Paige, KN, Whitham, TG (1987) Overcompensation in response to mammalian herbivory: the advantage of being eaten. Am Nat 129: 407–416.

Proulx M, Mazumder A (1998) Reversal of grazing impact on plant species richness in nutrient-poor vs. nutrient-rich ecosystems. Ecology 79: 2581–2592.

Pucheta E, Cabido M, Díaz S, Funes G (1998) Floristic composition, biomass, and aboveground net plant production in grazed and protected sites in a mountain grassland of central Argentina. Acta Oecol 19: 97–405.
Pykälä J (2004) cattle grazing increases plant species richness of most species trait groups in mesic semi-natural grasslands. Plant Ecol 175: 217–226.
Squires, VR (1982) Dietary overlap between sheep, cattle, and goats when grazing in common. J Range Manag 35: 116–119.
Stohlgren, TJ, Schell, LD, Heavel, BV (1999) How grazing and soil quality affect native and exotic plant diversity in Rocky Mountain grasslands. Ecol Appl 9: 45–64.
Tracy BF, Sanderson MA (2000) Patterns of plant species richness in pasture lands of the northeast United States. *Plant Ecol* 149: 169–180.
Wilsey B J, Polley HW (2003) effects of seed additions and grazing history on diversity and productivity of subhumid grasslands. *Ecology* 84: 920–931.
Yamauchi, A, Yamamura, N (2004) Herbivory promotes plant production and reproduction in nutrient-poor conditions: effects of plant adaptive phenology. Am Nat 163:138*153.

Chapter 12
Use of Steppe Vegetation by Nomadic Pastoralists in Mongolia

Urianhai Galzuud Nachinshonhor

Keywords Mongolia • Steppe ecosystem • Aboveground productivity • Nomadic pastoralism • Sustainability

12.1 Introduction

About a quarter of global terrestrial ecosystems are occupied by herbaceous vegetation (Shantz 1954), such as steppe, prairie, savanna, and pampas. Almost all these herbaceous ecosystems are subject to a strong human influence. In the severe continental climates on the Mongolian plateau, steppe ecosystems cover more than 70% of the land (Yunatov 1976; Zhang 1990), and nomadic pastoralism developed here in ancient times (Bazargur et al. 1989; Tumurjav and Erdenetsogt 1999; Sodnoi 2009).

Several studies have described the nomadic pastoralism regime in the Mongolian plateau in terms of geographic, meteorological, economic, and anthropological aspects (Simukov 1935; Batnasan 1972; Myagmarjav 1974; Bazargur et al. 1989; Humphrey and Sneath 1999). These studies proposed that the productivity of steppe vegetation is the main driving factor of nomadic pastoralism, but concrete evidence is still lacking. There are also several studies on Mongolian steppe ecosystem in terms of vegetation ecology (Yunatov 1968; Ulzykhutag 1985; Hilbig 1995), but how the ecosystem is influenced by nomadic pastoralism is still unclear.

Recently, aspects of the relationship between the robustness of the steppe ecosystem and nomadic pastoralism in Mongolia have been reported (Fernandez-Gimenez and Allen-Diaz 1999; Fernandez-Gimenez 2000; Kakinuma et al. 2008;

U.G. Nachinshonhor (✉)
Okayama University Graduate School of Environmental Science, Okayama, Japan
e-mail: nachin@cc.oakayama-u.ac.jp

N. Yamamura et al. (eds.), *The Mongolian Ecosystem Network: Environmental Issues Under Climate and Social Changes*, Ecological Research Monographs,
DOI 10.1007/978-4-431-54052-6_12,

Sasaki et al. 2009; Wesche et al. 2010), but these studies were conducted over short experimental periods and thus lacked sensitivity. The present study attempted to clarify the relationship between the sustainability of steppe ecosystem and nomadic pastoralism using 12 years of vegetation data obtained in a livestock exclosure and in control sites and 7 years of nomadic migration information from Bayan-Unjuul County, Tuv Prefecture, Mongolia.

12.2 Material and Methods

12.2.1 Study Site

The study site (46°49′47.46″ N, 105°47′35.34″ E; elevation, 1,354 m) is located 30 km southwest of Bayan-Unjuul County, the center of Tuv Prefecture in Mongolia. Between 1998 and 2009, mean annual temperature was 0.6°C and mean annual precipitation was 154.3 mm (Institute of Metrology and Hydrology of Mongolia). The precipitation coefficient of variation (CV) was 30.7% in this period, corresponding to a nonequilibrium system according to Ellis and Galvin (1994), where the precipitation CV of the system is near to or exceeds 30%.

The vegetation is dominated by monocotyledons (*Agropyron cristatum* and *Stipa krylovii*), perennial dicotyledons (*Artemisia adamsii*), annual dicotyledons (*Chenopodium* spp.), and shrubs (*Caragana* spp.). As in almost all Mongolian steppe areas, the natural vegetation in Bayan-Unjuul is used as rangeland to pasture camels, horses, cattle, sheep, and goats in a traditional nomadic pastoral system, where livestock are moved in both time and space (Nachinshonhor et al. 2003).

12.2.2 Experimental Method and Data Collection

In June 1999, we built a 1-ha (100 × 100 m) exclosure of wooden poles and iron wire to prevent livestock grazing (Exclosure site), and the area outside the exclosure was used as the Control site. Large wild ungulates are found in the Control site, but their effects on natural vegetation are assumed to be very small in comparison with the effects of livestock. Aboveground plant biomass is maximal in late August, and we adopted the aboveground biomass in the Exclosure site in late August as an index of annual plant production. The difference in aboveground biomass between Exclosure and Control sites was considered as the grazing effect of livestock.

In late August of every year from 1999 to 2010, we sampled plant aboveground parts from 1 × 1 m areas, with four replicates, in both the Exclosure and Control sites. Species were identified, and dry mass assessed. We identified aboveground parts of shrubs as new branches and leaves, and old branches were grouped separately. The taxonomic nomenclature follows Yunatov (1968).

Table 12.1 Effects of year and grazing on aboveground biomass and species diversity (Shannon–Wiener index) of community and functional groups

Variable	Factor	*df*	*F*	*P*
Aboveground biomass				
Community	Years	11	10.35	***
	Exclosure	1	10.47	*
	Years × exclusion	23	16.95	***
Monocotyledon	Years	11	12.26	***
	Exclosure	1	18.24	***
	Years × exclusion	23	14.96	***
Perennial dicotyledon	Years	11	12.26	***
	Exclosure	1	3.99	*
	Years × exclusion	23	3.25	***
Pioneer dicotyledon	Years	11	15.2	***
	Exclosure	1	0.76	ns
	Years × exclusion	23	28.68	***
Shrub	Years	11	2.5	ns
	Exclosure	1	7.11	**
	Years × exclusion	23	3.43	***
Species diversity (Shannon–Wiener index)				
Community	Years	11	9.18	***
	Exclosure	1	0.25	*
	Years × exclusion	23	5.06	***

ns not significant
*$P<0.05$; **$P<0.01$; ***$P<0.001$

In late August of 2002 to 2006, 2008, and 2010, interviews about nomadic pastoralism were carried out with local peoples who pasture livestock around the study site. Considering the answers, a period from September of the previous year to August of the current year was assumed as a nomadic fiscal year. Interview questions included (a) species and numbers of livestock managed in the previous nomadic physical year; and (b) distance traveled for livestock pasturing in the previous nomadic physical year. Livestock statistical information was provided by the National Statistical Office of Mongolia and the administration of Bayan-Unjuul County.

12.2.3 Data Analysis

According to the growth form and life history, plant species were classified into the following functional groups: (1) monocotyledons; (2) perennial dicotyledons; (3) annual dicotyledons; and (4) shrubs (Table 12.1). The palatability of species (Table 12.2) was categorized according to Yunatov (1968), Ulzykhutag (1985), Dashzeveg (1986), Ma (1994), Jigjidsuren and Douglas (2003), Eastern Oregon

Table 12.2 Numbers of species of different palatability among functional groups

	Palatability (number)			
	High	Moderate	Low	Unknown
Monocotyledon	9	1	–	–
Perennial dicotyledon	2	7	2	4
Annual dicotyledon	1	6	1	–
Shrub	2	–	1	–
Total	14	14	4	4

Sources: Yunatov (1968), Ulzykhutag (1985), Dashzeveg (1986), Jigjidsuren and Douglas (2003), Ma (1994), Eastern Oregon Agricultural Research Center (2005), Batjargal (2007)

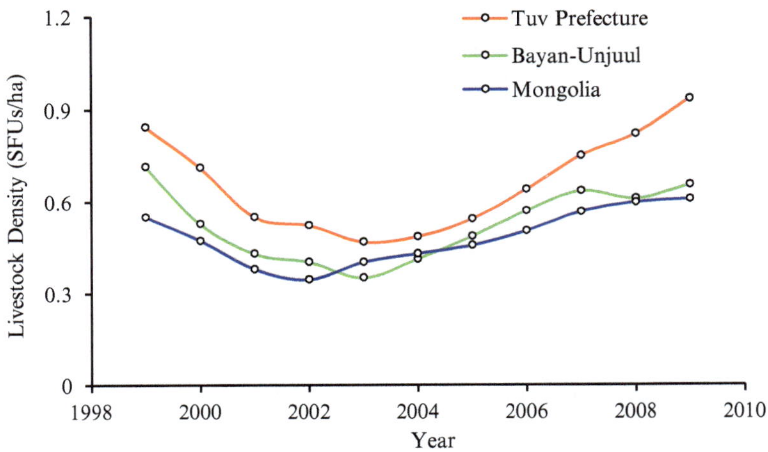

Fig. 12.1 Changes of livestock (sheep forage units, *SFUs*) intensity of Mongolia, Tuv Prefecture and Bayan-Unjuul County from1999 to 2009. Data from National Statistical Office of Mongolia (2004a, b, 2006, 2008, 2010), Administration of Bayan-Unjuul County

Agricultural Research Center (2005), and Batjargal (2007). Species diversity was evaluated using the Shannon–Wiener index. Analysis of variance (ANOVA) was applied to test for significant differences in aboveground biomass and Shannon–Wiener index among years and between Exclosure and Control sites.

Livestock densities were converted to sheep forage units (SFU) according to Tserendash (2000): 1 camel = 5.7 SFU, 1 horse = 6.6 SFU, 1 cattle = 6 SFU, 1 sheep = 1 SFU, and 1 goat = 0.9 SFU. During the study period, a severe cold disaster, called a "Zud," caused extensive loss of livestock from winter 2002 to spring 2003 in Bayan-Unjuul (Fig. 12.1).

12.3 Results

12.3.1 Variation of Aboveground Biomass Among Years and Effects of Livestock

Aboveground biomass of the community and each functional group, except shrubs, showed significant annual variation (Table 12.1). Exclosure treatment had the strongest effect on the monocotyledon group, with relatively weak effects on shrubs and perennial species. There was no significant difference in dicotyledons between Exclosure and Control sites (Table 12.1), suggesting that grazing pressure was most concentrated on the dicotyledon species. The interaction of year and exclusion showed significant effects on aboveground biomass of the community and all functional groups.

12.3.2 Relationship Between Aboveground Production and Grazing Intensity

In 2000, 2005, 2008, 2009, and 2010, years that had comparatively high production, the aboveground biomass decreased significantly in the Control sites (Fig. 12.2). Aboveground biomass of monocotyledons and annual dicotyledons between Exclosure and Control sites was strongly correlated ($R^2=0.7032$, $R^2=0.6106$,

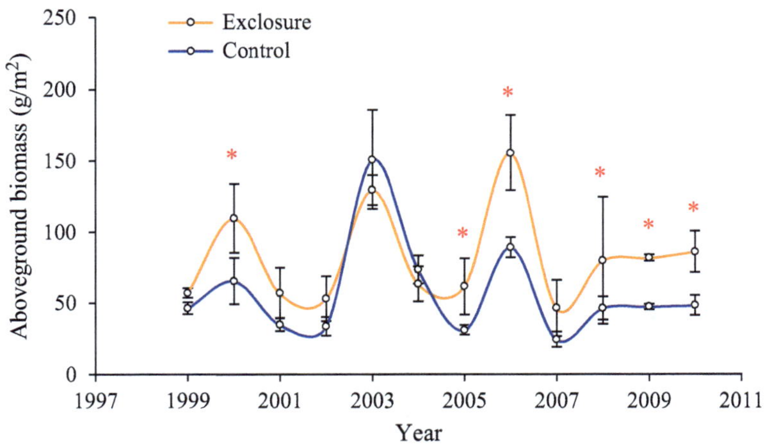

Fig. 12.2 Variation in aboveground biomass according to year and condition. *Dots* (mean ± SD, $n=4$) indicate values for Exclosure and Control sites. When the year × exclusion interaction was statistically significant, the repeated-measures analysis of variance (ANOVA) was followed by post hoc contrast tests for within-year comparisons. *Asterisk*: Statistical significance within each year at $P<0.05$. Data for 1999–2009 are from Nachinshonhor et al. (2009)

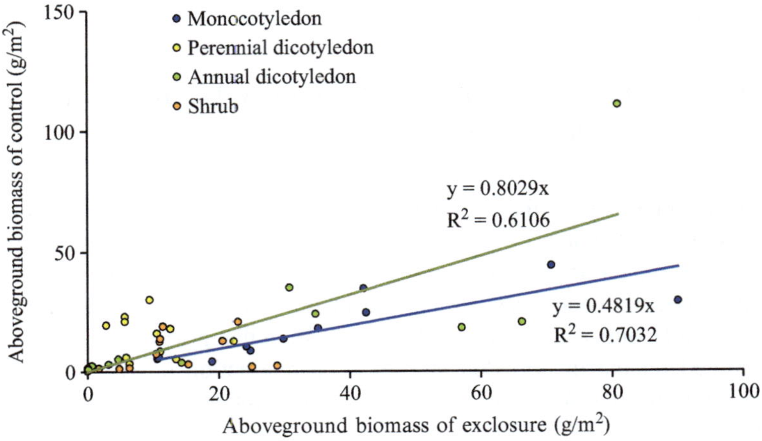

Fig. 12.3 Comparison of aboveground biomass of functional groups between Exclosure and Control sites. Regression lines show relationships for monocotyledon (*bottom*) and annual dicotyledon (*top*) groups. Data for 1999–2009 are from Nachinshonhor et al. (2009)

respectively), but for perennial dicotyledons and shrubs, there was no significant correlation (Fig. 12.3).

12.3.3 Variation in Species Diversity (Shannon–Wiener Index) Among Years and Grazing Pressure

The Shannon–Wiener index showed a significant annual variation (Table 12.1; $P<0.0001$) and a weak but significant effect of exclosure (Table 12.1; $P<0.05$). In terms of annual changes, species diversity significantly increased more in the Exclosure than in the Control site only in 2009 (Fig. 12.4; $P<0.05$).

12.3.4 Relationship Among Nomadic Pastoral Mobility, Number of Livestock, and Aboveground Production

The interview results showed a trend indicating that annual migration distance increased with increasing numbers of livestock, with significant correlations found in four of seven cases (Fig. 12.5). Average annual migration distance differed significantly among years ($P<0.0001$) and was correlated with aboveground production (Fig. 12.6; $R^2=0.4636$). Although this correlation was not significant, there was a significant difference between the longest migration, in 2003–2004, and the shortest migration, in 2007–2008 ($P<0.001$).

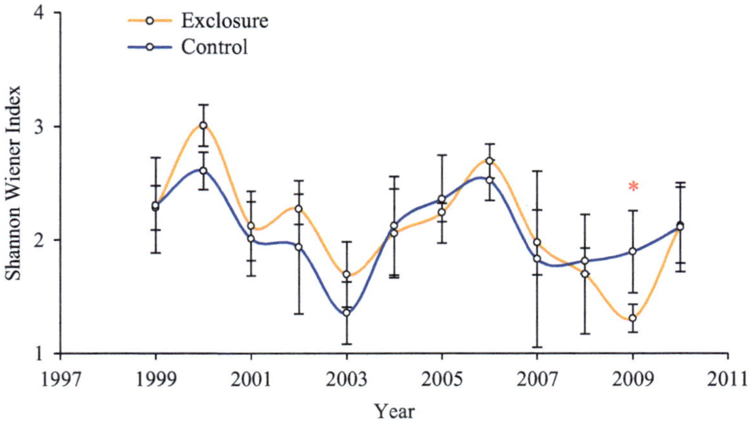

Fig. 12.4 Variation in aboveground Shannon–Wiener index according to year and treatment. *Dots* (mean ± SD, $n=4$) indicate values for Exclosure and Control sites. When the year × exclusion interaction was statistically significant, the repeated-measures ANOVA was followed by post hoc contrast tests for within-year comparisons. *Asterisk*: Statistical significance within each year at $P<0.05$. Data for 1999–2009 are from Nachinshonhor et al. (2009)

12.4 Discussion

12.4.1 Relationship Between Nomadic Livestock and Aboveground Production of Steppe Vegetation

The aboveground biomass in the Control site was strongly affected by livestock in the relatively high productivity years of 2000, 2006, and 2008–2010 (Fig. 12.2) as a result of grazing and trampling by livestock. This observation suggests that in years of limited production caused by adverse abiotic conditions such as drought events, the biotic effects of livestock were reduced. As a result, the vegetation escaped the double effects of abiotic and biotic damage. This natural control of grassland, with more feed in vegetation-rich areas, is based on the premise of free access to space and pastoral mobility. In this sense, the common use of land is important to Mongolian traditional nomadic pastoralism. The opposite case, limited mobility, which leads to degradation of natural vegetation, had been reported from Inner Mongolia (Wang 2006; Zhao et al. 2006; MunkhDalai et al. 2007), where traditional nomadic pastoral livestock management virtually stopped with the introduction of land policies dictating semi-privatization.

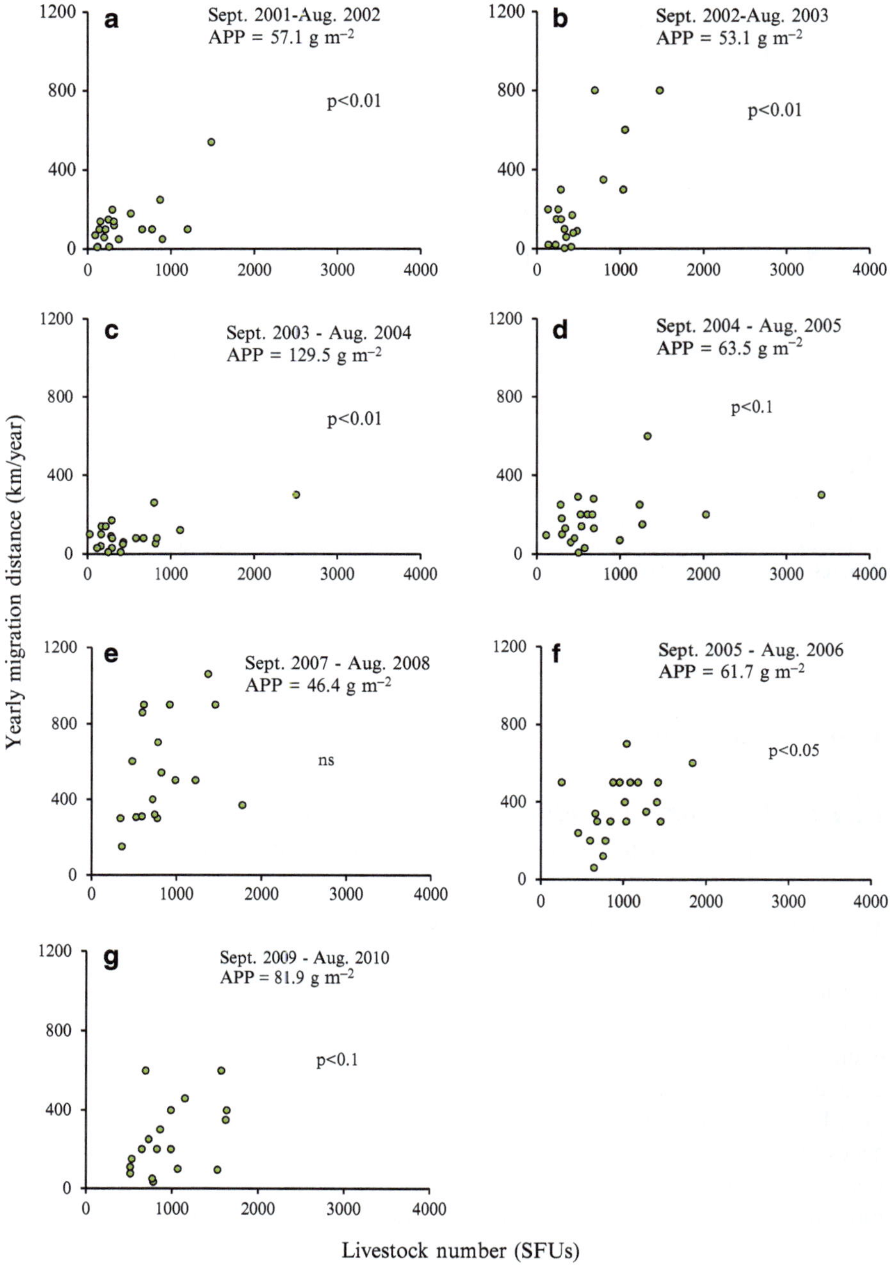

Fig. 12.5 Relationship between livestock numbers (SFUs) and yearly average migration distances was significantly different during the study period ($P<0.001$). Data for 2001–2009 are from Nachinshonhor et al. (2009)

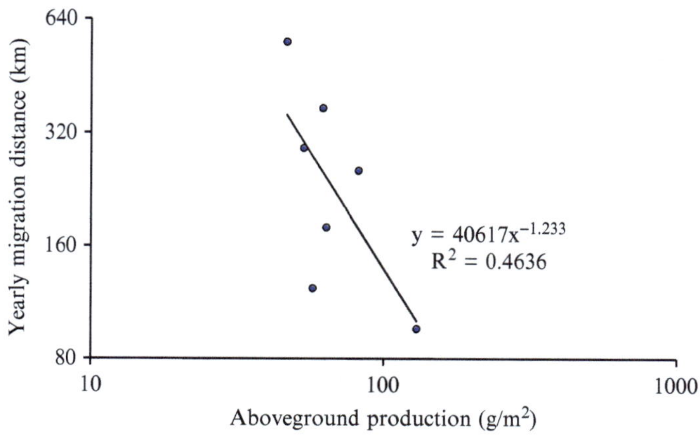

Fig. 12.6 Relationship between aboveground production of the plant community and average yearly migration distance for nomadic pastoralists (log scale: 2002, $n=22$; 2003, $n=20$; 2004, $n=20$; 2005, $n=22$; 2006, $n=22$; 2008, $n=23$; 2010, $n=18$. Data for 2001–2009 are from Nachinshonhor et al. (2009)

12.4.2 Effects of Nomadic Pastoral Herding on Species Composition of the Plant Community

Ulzykhutag and Dariimaa (1977) showed that graminoids, such as xerophilous species, and shrubs dominate this region and also noted the important role of these plants for the local nomadic pastoral livestock industry. All monocotyledons in our site were palatable to livestock (Table 12.2), especially *A. cristatum*, *S. krylovii*, and *Allium* spp. Local nomadic people call these plants "Nariin Nogoo" or "Nariin Övs" in Mongolian, which means "fine plants" (G.U. Nachinshonhor, personal communication). The same names were reported by Kakinuma et al. (2008) from the Mongolian forest steppe. Compared with monocotyledons and shrubs, perennial dicotyledons showed a slight but significant decrease in the Control site, possibly because they were dominated by unpalatable species such as *A. adamsii* (Table 12.2; Eastern Oregon Agricultural Research Center 2005). Palatable species such as *Artemisia frigida* and *Convolvulus ammanii* made only a weak contribution to the community.

12.4.3 Relationship Between Species Diversity and Nomadic Pastoral Herding

We observed that in 2003, 2008, and 2009, relatively wet years following heavy drought events, some pioneer species such as *Chenopodium* spp. increased greatly (data not shown), as also reported by Yunatov (1976). The mass-generating pioneer species had reduced species diversity in these years (Fig. 12.4). Although species diversity was lowest in 2009, the Control site had a significantly higher Shannon–Wiener index than did the Exclosure site. In contrast to these results, a study in Inner Mongolia showed that species diversity significantly decreased after heavy grazing (Zhang et al. 2004. Our result showed that grazing can increase species diversity (Shannon–Wiener index; Table 12.1, Fig. 12.4), suggesting that a switch in grazing affects the vegetation composition and is related to management regime.

12.4.4 Some Lessons from Interviews on the Sustainability of the Steppe Ecosystem and Nomadic Pastoralism

After two successive years of heavy drought (2001 and 2002), a disastrous winter Zud occurred during winter 2002 to spring 2003 in Bayan-Unjuul. As a result, by the end of 2003, the livestock numbers (SFUs) had decreased by about 12.5% compared with the end of 2002, and the density dropped to the lowest value measured during the study period (Fig. 12.1; National Statistical Office of Mongolia 2004a, b, 2006, 2008, 2010). Despite high production in 2003, the decrease in livestock density reduced the grazing pressure in 2003 (Fig. 12.2). The reduction in livestock numbers and the high vegetation production may have led to relatively less mobility from September 2003 to August 2004 (Fig. 12.5c).

Interview results showed that grazing intensity was regulated by migration distance, which was adapted to vegetation production (Fig. 12.6). Although this relationship was not significant, the average migration distance was significantly different between the longest and shortest migration years during the study period, and the production of plant communities between these years was also significantly different ($P<0.05$), which suggests that annual migration patterns were adapted to the productivity of the steppe communities.

Noboru et al. (2009) showed that, in relatively arid conditions, continuous grazing pressure over long periods not only reduces the quantity but also changes the quality of plant communities. Although a clear decrease in species diversity after livestock grazing was not shown, our study supports the idea that grazing intensity regulation will contribute to the sustainability of steppe vegetation. Nomadic herders know that they cannot change nature (weather), but they can move livestock to avoid difficult conditions, and this mode seems to be the most basic principle of nomadic pastoralism. Interview results further suggested that the benefits of nomadic pastoralism lie not only in obtaining food for livestock but also in assuring the health of livestock through their movement. Feedback systems between steppe vegetation and nomadic pastoralism contribute to sustainability of both human society and the natural environment of the nomadic pastoralist Mongolian region.

Acknowledgments I thank Dr. T. Hirose, Dr. L. Jargalsaikhan, Sh. Dugarjav, Tserendash, Tsooj, Tuvshintogtokh, and the many people who invested their time and energy to make this research possible. This research was supported by the Social Welfare Corporation NANKO AIRINKAI,

Institute of Botany of the Mongolian Academy of Science, Tohoku University Graduate School of Science, Tohoku University Study Center for North East Asia, the Japan Science Society, Tokyo Metropolitan University, Meiji University, the Toyota Foundation, Niigata University, the Arid Land Research Center of Tottori University, the Japan Society for the Promotion of Science, the National Museum of Ethnology, and the Research Institute for Humanity and Science.

References

Batjargal N (2007) Color reference book of some forage plants. Ulaanbaatar, Mongolia

Batnasan G (1972) Negdelchdiin nuudel, suurishhiin zarim asuudald. Some issues on migration, settlement of nomadic herders. Mongolian Academy of Sciences, Institute of History, Ulaanbaatar, Mongolia

Bazargur D., Shiirev-Adya S., Chinbat G. (1989) Mobility of nomadic peoples in the People's Republic of Mongolia. National Publishing House, Ulaanbaatar, Mongolia

Dashzeveg L. (1986) A study of forage plant palatability. Ulaanbaatar, Mongolia

Eastern Oregon Agricultural Research Center (2005) Palatability of Mongolian rangeland plants. Circular of Information No. 3. Eastern Oregon Agricultural Research Center, Union, OR http://ir-dev.library.oregonstate.edu/bitstream/handle/1957/3087/Palatability%20of%20Mongolian%20Rangeland%20Plants.pdf?sequence=1. Accessed 7 Feb. 2011

Ellis J, Galvin K.A. (1994) Climate patterns and land use practices in the dry zones of Africa. Bioscience 44(5): 340–348

Fernandez-Gimenez M.E. and Allen-Diaz B.H. (1999) Testing a non-equilibrium model of rangeland vegetation dynamics in Mongolia. J Appl Ecol 36: 871–885

Fernandez-Gimenez M. E. (2000) The role of Mongolian nomadic pastoralist's ecological knowledge in rangeland manegement. Ecol Appl 10 (5): 1318–1326

Humphrey C. and Sneath D. (1999) The end of nomadism? Society, state and the enlironment in inner Asia. Duke University Press, Durham, NC

Hilbig W. (1995) The vegetation of Mongolia. SPB Academic Publishing, Amsterdam

Jigjidsuren S., Douglas A. J. (2003) Forage plants in Mongolia. Ulaanbaatar, Mongolia

Kakinuma K., Ozaki T., Takatsuki S. and Chuluun J. (2008) How pastoralists in Mongolia perceive vegetation changes caused by grazing. Nomadic Peoples 12 (2): 67–74

Ma Y. (1994) Flora of Inner Mongolia, vol 5, 2nd edn. Huhhot, China

MunkhDalai A.Z., Borjigin E., and Zhang H. (2007) Mongolian nomadic culture and ecological culture: On the ecological reconstruction in the agro-pastoral mosaic zone in Northern China. Ecol Econ 62: 19–26

Myagmarjav (1974) The basic moving issues for herdsman. Ulaanbaatar, Mongolia

Nachinshonhor G.U., Tserendash S., Jargalsaikhan L., Tsooj Sh., Tsendeekhuu Ts., Dugarjav C., Hirose T. (2003) Productivity of plant communities and the type of animal husbandry in the Mongolian grassland. Bull Jpn Assoc Mongolia Study 33: 29–37

Nachinshonhor G.U., Jargalsaikhan L., Tserendash S., Tsooj Sh., Tsendeekhuu Ts., Dugarjav Ch., Tadaki Hirose (2009) Functions of pastoral nomadism on the sustainability of Mongolian steppe vegetation. Proceedings of "Special Issue Dedicated to the 35th Anniversary of the Institute of Botany," Mongolian Academy of Sciences, 21: 151–157. Ulaanbaatar, Mongolia

National Statistical Office of Mongolia (2004a) Mongolian statistical yearbook 2003. Ulaanbaatar, Mongolia

National Statistical Office of Mongolia (2004b) Mongolia in a market system statistical yearbook 1989–2002. Ulaanbaatar, Mongolia

National Statistical Office of Mongolia (2006) Mongolian statistical yearbook 2005. Ulaanbaatar, Mongolia

National Statistical Office of Mongolia (2008) Mongolian statistical yearbook 2007. Ulaanbaatar, Mongolia

National Statistical Office of Mongolia (2010) Mongolian statistical yearbook 2009. Ulaanbaatar, Mongolia

Noboru F., Amartuvshin N., Yamada Y., Matsui K., Sakai S., Yamamura N. (2009) Positive and negative effects of livestock grazing on plant diversity of Mongolian nomadic pasturelands along a slope with soil moisture gradient. Grassland Sci 55(3): 126–134

Sasaki T., Okayasu T., Ohkuro T., Shirato Y., Jamsran U., Takeuchi, U. (2009) Rainfall variability may modify the effects of long-term exclosure on vegetation in Mandalgobi, Mongolia. J Arid Environ 73: 949–954

Shantz H. L. (1954) The place of grasslands in the Earth's cover. Ecology. 35: 143–145

Simukov A. (1935) Materiali po kochevomy bitu naseleniya MNR (Materials on the nomadic life of the population of Mongolia). Sovremini Mongolii 6:13, Ulaanbaatar Mongolia

Sodnoi T. (2009) Ecology of Mongolian livestock. Ulaanbaatar, Mongolia

Tserendash S. (2000) Ecology and quality assessment for the capacity of Mongolian pasture land. Report of the Science and Technology Project 1999–2000. Ministry of Education, Culture and Science, Research Institute of Animal Husbandry, Ulaanbaatar, Mongolia

Tumurjav M., Erdenetsogt N. (1999) Nomads of Mongolia. Ulaanbaatar, Mongolia

Ulzykhutag N., Dariimaa S. (1977) Some features of plant communities in Unjuul Sum. Bulletin of Botanical Institute, Mongolian Academy of Sciences 1977(3): 131–150

Ulzykhutag N. (1985) Reference book of forage plants in the Mongolian People's Republic. Ulaanbaatar, Mongolia

Wang J. (2006) Agricultural ecology and traditional Mongolian society. Shandong People's Publication, Jinan, China

Wesche K., Ronnenberg K., Retzer V., and Miehe G. (2010) Effects of large herbivore exclusion on southern Mongolian desert steppes. Acta Oecol 36: 234–241

Yunatov A.A. (1968) Forage plants in pastures of the Mongolian People Republic. Ulaanbaatar, Mongolia

Yunatov A.A. (1976) Main features of vegetation cover of the Mongolian People's Republic. Ulaanbaatar, Mongolia

Zhang T.-H, Zhao H.-L., Li S.-G., Zhou R.-L. (2004) Grassland changes under grazing stress in Horqin sandy grassland in Inner Mongolia, China. New Zealand J Agric Res 47: 307–312

Zhang Z. (1990) Grassland resource of Inner Mongolia. Huhhot, China

Zhao M., Han G., and Hong M. (2006) Grassland resource and its situation in Inner Mongolia, China. Bull Fac Agric Niigata Univ 58(2): 129–132

Chapter 13
Vegetation Interactions for the Better Understanding of a Mongolian Ecosystem Network

Noboru Fujita, Narantsetsegiin Amartuvshin, and Erdenegerel Ariunbold

Keywords Forest and steppe distribution • Grazing tolerance • Precipitation and pasture production • Shrub and herb production • Soil alkalization

13.1 Introduction

An ecosystem network is composed of inorganic, organic, and human nodes, which interact directly and indirectly. Ecologically, important nodes in Mongolia include the atmosphere, vegetation, stock farming (livestock and herders), and a social system (Fig. 13.1). To better understand the Mongolian ecosystem network, we considered important interactions between adjacent nodes. In this network, vegetation interactions have important implications: the causes of forest mosaics, the effects of precipitation and temperature on primary production in pastures, the relationship between grazing-tolerant plant domination in pastures and soil alkalization, the recovery of edible pasture plant species diversity after disturbance removal, herbs and shrub leaves as a livestock food source, taproot depth of shrubs, the cause of shrub species distribution and that of shrub absence in pastureland, the relationship between shrub degradation and wind erosion, and the different grazing effects of goats and sheep on shrubs and herbs. We examined these subjects because they are important in elucidating the Mongolian ecosystem network. The nomenclature follows Grubov (2001), except for *Leymus chinensis*, which is based on Gubanov (1996).

N. Fujita (✉)
Research Institute for Humanity and Nature, Kyoto, Japan
e-mail: fujita@chikyu.ac.jp

N. Amartuvshin
Institute of Botany, Mongolian Academy of Sciences, Ulaanbaatar, Mongolia

E. Ariunbold
Institute of Geoecology, Mongolian Academy of Sciences, Ulaanbaatar, Mongolia

N. Yamamura et al. (eds.), *The Mongolian Ecosystem Network: Environmental Issues Under Climate and Social Changes*, Ecological Research Monographs,
DOI 10.1007/978-4-431-54052-6_13,

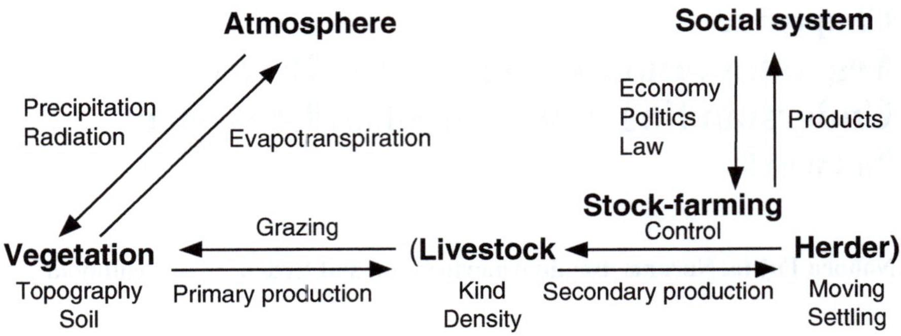

Fig. 13.1 The main nodes, actions, and flows of inorganic elements and organisms in the Mongolian ecosystem network

We used Tukey's honestly significant difference test to compare means. The statistical analysis was performed using JMP for Macintosh, ver. 5.1.2 (SAS Institute, Cary, NC, USA).

13.2 Forest Presence Versus Slope Orientation and Location on the Slope

Forest is distributed discontinuously in the forest-steppe zone. We investigated forest distribution on the slopes of ridges on Bogd Khan Mountain located south of Ulan Bator from field surveys on the site with auxiliary use of satellite photographs by Landsat. For each slope, we determined the slope direction and checked for the presence of forest, distinguishing between the upper and lower slopes.

Figure 13.2 shows forest frequency on upper and lower slopes with different slope directions. On the upper slope, forest was distributed on all north-facing, northeast-facing, and west-facing slopes, whereas it was limited on other slope directions. On south-facing slopes, it exhibited the least frequency. On lower slopes, forest frequency decreased in all slope directions.

What is the reason why the frequency of forest existence differed among different slope directions? Forest distribution in the forest-steppe zone may depend on soil moisture and human activity. Atmospheric processes that affect soil moisture include precipitation, which supplies water to the soil, and solar radiation, which supplies light energy for photosynthesis and removes water from the soil directly via heat energy and indirectly via evapotranspiration. Slope direction affects the quantity of precipitation and radiation that reaches the ground. In the Northern Hemisphere, radiation is greater on south-facing slopes than on north-facing slopes, and this difference increases with latitude. Precipitation is greater on windward slopes than leeward slopes between high mountain ranges. For example, the forests around Lake Baikal are richer on the north-facing slope than the south-facing slope

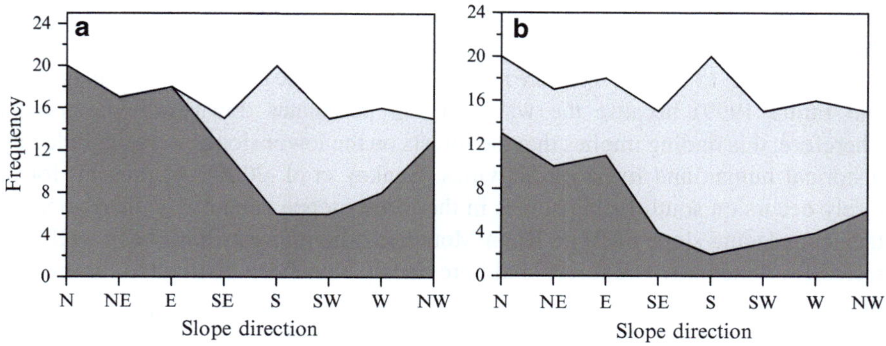

Fig. 13.2 Frequency of the presence and absence of forest on the upper and lower slopes facing different directions on Bogd Khan Mountain, south of Ulan Bator: upper (**a**) and lower (**b**) slopes. *Shaded area*, forested area; *light area*, no forest

on the east and west shores and are distributed more on the leeward east shore slope than the windward west shore slope, possibly because of the prevailing northerly winds (Fujita 1997). In Mongolia, the prevailing wind blows from the northwest, and between the north-facing and south-facing slopes of large mountains such as Khangai and Altai, this difference is observed. However, among closely spaced slopes of the same mountain facing different directions, differences in precipitation must be small. Then, differences in the presence of forests between the north-facing and south-facing slopes in the Bogdkhan Mountains is caused by other factors. Between adjacent north-facing and south-facing slopes with the same precipitation in Gachuurt, soil moisture after precipitation clearly remains higher on the north-facing slope (see Ishii and Fujita, this volume, Chap. 6). The deciduous conifer *Larix sibirica* Ledeb. dominates the east side of the Bogd Khan Mountain, whereas the evergreen conifers *Picea obovata* Ledeb. and *Pinus sibirica* Du Tour are abundant on the west side. These evergreen conifers grow in habitats with higher soil moisture than those occupied by *L. sibirica* (Fujita 1997). There is a possibility that precipitation is greater on the west side on the large scale of the Bogd Khan Mountains, as the prevailing wind blows from the northwest.

Forest occurred on the upper east-facing slopes, whereas it was restricted on the upper west-facing slopes. Differences in solar radiation and precipitation do not explain this difference. Insolation does not differ on east-facing slopes in the morning and west-facing slopes in the afternoon. It can be explained by differences in evapotranspiration rates on the east-facing and west-facing slopes. Although solar radiation does not differ, air temperature and soil surface temperature are likely higher on the west-facing slope before dusk than on the east-facing slopes after dawn. Therefore, evapotranspiration may be greater on the west-facing slopes.

Several possible explanations exist as to why forests occur mainly on the north-facing slopes in the forest-steppe zone. The distribution of forest and steppe, differing between slope directions, was thought to be natural (Gunin et al. 1999; Dulamsuren et al. 2005). Conversely, Hilbig (2000) concluded that it was the result of centuries of human activity, including logging and pasture use. When comparing the upper and lower slopes, forest is dominantly distributed on the upper slope in all directions (Fig. 13.2). Soil moisture is usually higher on the lower slopes (Yanagisawa and Fujita 1999) because the water in soil percolates downward via gravity. Therefore, this finding implies that the forests on the lower slopes were degraded by historical human and livestock activities (Sankey et al. 2006). At present, forest rarely occurs on south-facing slopes in the forest-steppe zone but is distributed on the south-facing slope of Bogd Khan Mountain where forests have been well protected from the past. The forest could previously have been distributed throughout the forest-steppe zone, including the flat valley and south-facing slope. Permafrost occurs beneath the north-facing forests, but not beneath the south-facing pastures (Ishikawa et al. 2005). Permafrost can contribute to soil moisture, but the degradation of forests could have damaged the permafrost, leading to its disappearance on south-facing pastures.

13.3 Pasture Production Versus Precipitation and Air Temperature

Primary production increases with the mean annual precipitation in Inner Mongolia, China, and Mongolia (Li et al. 2005; Bai et al. 2008). To investigate the effects of precipitation and air temperature on pasture plant production in different vegetational zones in Mongolia, we compared pasture production with precipitation and air temperature data from nearby meteorological stations at Gachuurt, east of Ulan Bator, Mandalgobi, and Dalanzadgad located in forest-steppe, steppe, and desert-steppe zones, respectively. At each site of Gachuurt (48°01′27″ N, 107°10′49″ E), Mandalgobi (45°43′44″ N, 106°16′16″ E), and Dalanzadgad (43°34′55″ N, 104°25′40″ E), we mowed pasture plants to 3 cm twice monthly in five 1-m^2 quadrats in an exclosure excluding livestock from May to September during 2006–2009. Annual precipitation of the three sites from 2006 to 2009 and their normal years are listed in Table 13.1. During the 4 years, the annual precipitation was never more than that of the normal year in Mandalgobi. We selected areas covered with herb vegetation without main shrub patches for the quadrats. After mowing, we dried the samples at 80°C and measured the dry weight as a measure of production during the half-month. We calculated the total precipitation and mean air temperature during the 15 days before each mowing. To determine the spatial variation of precipitation in the short term, we measured precipitation continuously at 16 points arranged in a netlike fashion at 5-km distances in Bayan-Unjuul from May to September in 2009. The northeast corner of the mesh was 46°58′19″ N, 105°52′44″ E. Despite human and bird disturbances, we were able to obtain perfect data from 13 points.

The relationships between total precipitation and mean air temperature and pasture plant production during a half-month for each measurement at all sites from 2006 to 2009 are shown in Fig. 13.3. The relationships between annual precipitation and annual production and pasture plant biomass in August at all sites are shown in

Table 13.1 Annual precipitation (mm) from 2006 to 2009. Measured durations of normal year are Ulan Bator (1969–2007), Mandalgobi (1944–2007), and Dalanzadgad (1937–2007)

	Gachuurt	Mandalgobi	Dalanzadgad
Normal	293.2	170.6	130.8
2006	264.1	98.6	89.0
2007	266.8	70.5	142.6
2008	352.0	107.0	135.0
2009	204.0	84.6	47.8

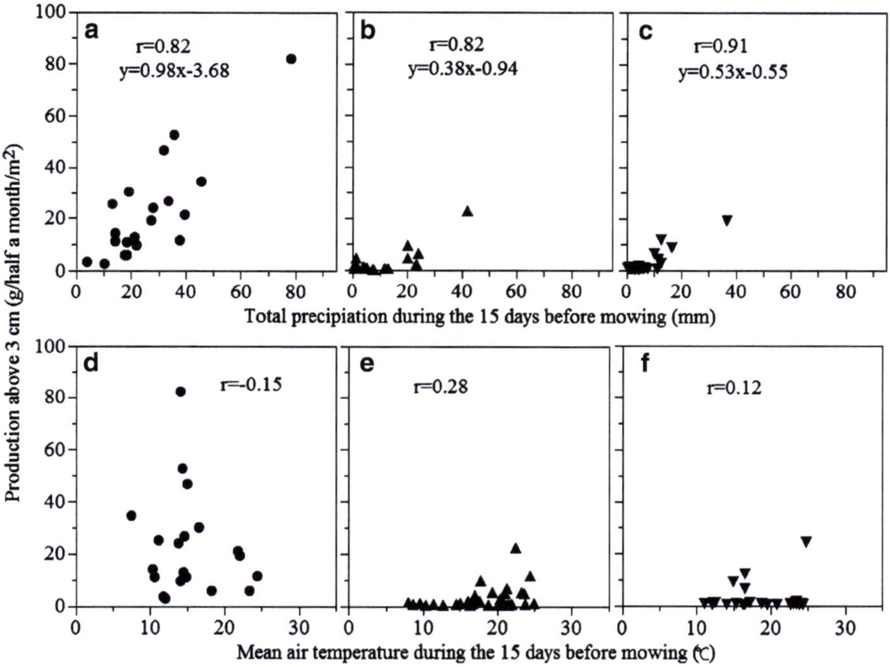

Fig. 13.3 Relationship between total precipitation and average air temperature and pasture plant growth over a half-month during the growing season from 2006 to 2009 in Gachuurt (**a**, **d**), Mandalgobi (**b**, **e**), and Dalanzadgad (**c**, **f**). *r* correlation coefficient, *y* regression line

Fig. 13.4. A significant correlation was observed between precipitation and pasture production over a half-month at all sites without reaching a ceiling of production, but air temperature was not correlated with pasture production. In the dry steppe and desert-steppe climates and the moist forest-steppe climate also, precipitation was the main factor determining pasture production during a half-month. Good pasture production, more than 20 g/half-month/m^2 resulting from high precipitation was very

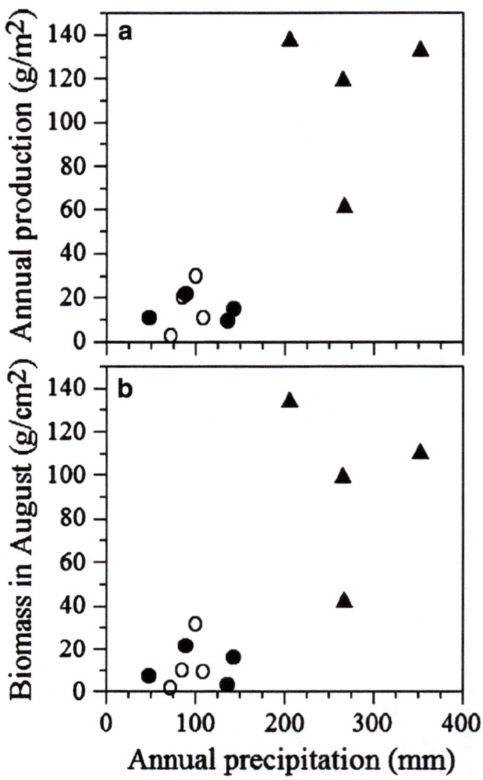

Fig. 13.4 Relationships among annual precipitation, annual production, and pasture plant biomass growth in August from 2006 to 2009. *Black triangles*, Gachuurt; *white-filled circles*, Mandalgobi; *black circles*, Dalanzadgad

rare, occurring only once during the 4 years from 2006 to 2009 in both Mandalgobi and Dalanzadgad. Total precipitation less than 10 mm for the half-month before mowing had little effect on pasture production at Mandalgobi and Dalanzadgad because it could not maintain soil moisture resulting from rapid water loss to the dry atmosphere. It was reported that the threshold precipitation becomes 25 mm in the Mojave Desert (Beatley 1974). At Dalanzadgad, a threshold of 10 mm is not clear compared with Mandalgobi, because small shrubs, such as *Youngia tenuicaulis* (Babc. et Stebbins) Czer. and *Ajania achilleoides* (Turcz.) Tzvel., are mixed more with herbs at Dalanzadgad than at Mandalgobi. These shrubs can produce some leaves in the growing season under drought conditions. At Gachuurt, even low precipitation had an effect, possibly because evapotranspiration was low as a result of low air temperature and wind velocity. A short, heavy rain may be more beneficial for pasture production than prolonged, low-intensity rain. During the growing season when the air temperature exceeded the threshold value of plant growth, air temperature was independent of pasture production with little precipitation.

Annual precipitation did not correlate with the annual or August pasture plant biomass production at any site from 2006 to 2009, but a correlation was observed when compared through all sites (Fig. 13.4). In a regional comparison along a large

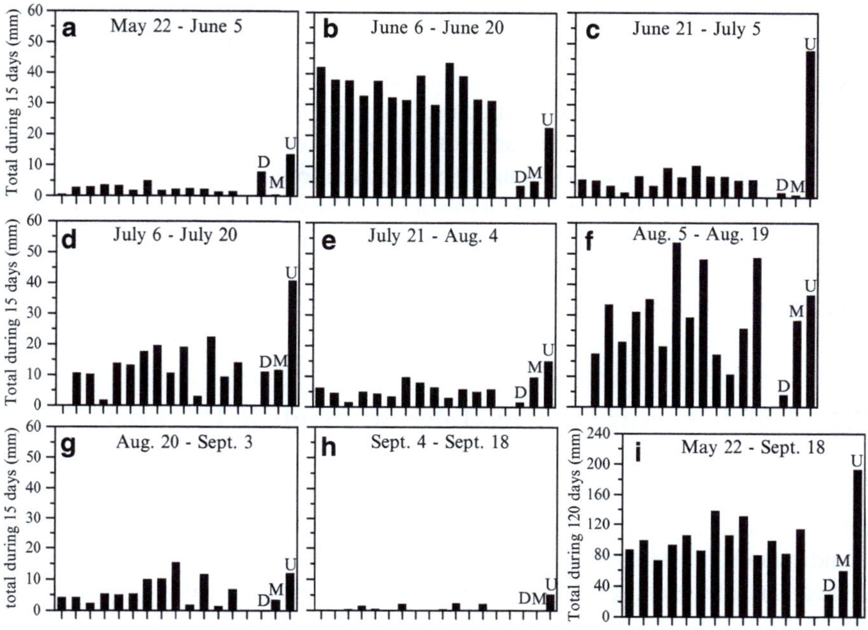

Fig. 13.5 Temporal changes of total precipitation during 15 days in 2009 at 13 points in Bayan-Unjuul and at Ulan Bator (*U*), Mandalgobi (*M*), and Dalanzadgad (*D*). Duration is given in each figure. Total precipitation throughout these durations is shown in (**f**)

climatic gradient, annual precipitation correlated with annual production. However, total annual precipitation cannot explain the yearly variation of annual production at each site because a heavy rainfall at a limited, critical time, for example, 10 mm during a half-month, is necessary for plant production, especially for the growth of herbs, which absorb water from the surface soil (see Satoh et al., this volume, Chap. 5). The effective precipitation timing for plant growth depends on the quantity of precipitation, weather after precipitation, specific vegetation, and soil texture. We used total precipitation during the 15 days before mowing, but this term was expedient, not absolute. The effect of precipitation on soil moisture after rainfall continues longer with increased mass of precipitation.

Temporal and spatial variation of precipitation at short range in Bayan-Unjuul in summer 2009 is shown in Fig. 13.5. Temporal variation in this area was very large, and precipitation was below the threshold (10 mm during 15 days) for more than half of this period. Spatial variation changed according to the time. The spatial variation among points was large in middle July (Fig. 13.5d), early August (Fig. 13.5f), and late August (Fig. 13.5g), and especially high and low differences between the threshold during their terms were serious among points for pasture production. All points showed high precipitation in middle June (Fig. 13.5b) and did not reach the threshold in late May (Fig. 13.5a), late July (Fig. 13.5e), and early

September (Fig. 13.5h). Temporal variations at Ulan Bator, Mandalgobi, and Dalanzadgad and spatial variations in a wide area from Ulan Bator to Dalanzadgad were large in 2009. As a result of seasonal and annual precipitation variation, pasture production over a half-month fluctuated greatly in all three sites (Fig. 13.3). This fluctuation shows the importance of abiotic factors such as precipitation for pasture production, even at Gachuurt, located in the forest-steppe zone. The instability of pasture carrying capacity makes nomadic livestock farming difficult unless herds can be moved from one pasture to another. Measures to overcome low-production seasons include moving to relatively higher production pastures and keeping livestock density below the carrying capacity of the pasture.

13.4 Edible Versus Grazing-Tolerant Plants in Pastures

Livestock prefer to graze certain edible plants and avoid other plants; therefore, the grazing-tolerant plants often remain without being eaten after livestock grazing. To understand the effects of the remaining grazing-tolerant plants on the recovery of edible plants after grazing is stopped, we conducted a selective mowing experiment. The study site was the same as that described in Fujita et al. (2009) in Gachuurt, east of Ulan Bator. In autumn 2003, five 4-m^2 exclosures were assembled at two sites, one on the upper slope, at an altitude of 1,615 m, and one in the flat valley bottom, at an altitude of 1,548 m. We made the exclosures from pillars and rectangular lumber to avoid creating shade. We used four mowing treatments to compare the annual production of edible plants among treatments: mowing edible and grazing-tolerant plants to 3 cm, mowing only edible plants to 3 cm, mowing edible and grazing-tolerant plants to 5 cm, and mowing only edible plants to 5 cm. We arbitrarily placed a 1-m^2 quadrat for each treatment in each exclosure. In total, we used five 1-m^2 quadrats for the duplicate of each mowing treatment in the five exclosures at each site. We mowed monthly from early June to early September 2004. After each mowing, we transported the mowed material to the laboratory where we dried the samples at 80°C for 2 days and measured the dry weight. The annual production for each treatment was the sum of the biomass after mowing four times.

The annual production of edible plants for each treatment at the two sites is shown in Fig. 13.6. When the treatments with only edible plants cut were compared to the treatments with all plants mown, the annual production of edible plants was higher in the 5-cm mowing of both types of plants in the flat valley bottom and in the 3-cm mowing of both plants on the upper slope. No significant difference was observed between the two other treatments. In the flat valley bottom, the unmown grazing-tolerant plants likely put the edible plants at a disadvantage, as the edible plants remained low after mowing because of light competition with grazing-tolerant plants. Conversely, on the upper slope, the annual production of edible plants increased in the quadrat with only edible plants mown to 3 cm where grazing-tolerant plants remained without mowing. Little competition for light likely occurred because of the lower height and mass of pasture plants in the quadrats; however, this does not

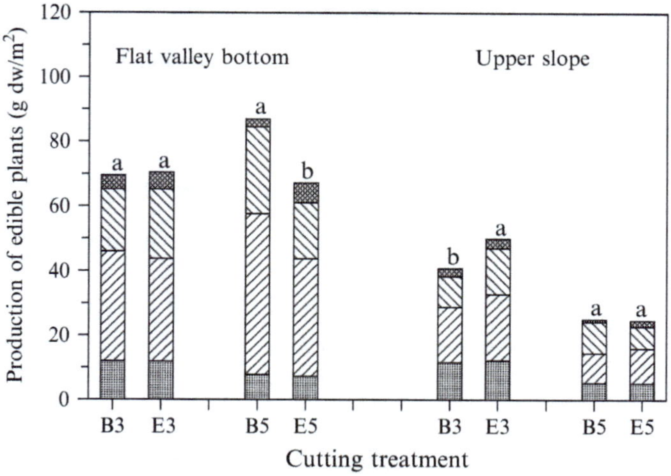

Fig. 13.6 Annual production of edible pasture plants after mowing to 3 or 5 cm with or without mowing the grazing-tolerant plants at the bottom of the flat valley and on the upper slope. *B3* mowing to 3 cm, including the grazing-tolerant plants, *E3* mowing only edible plants to 3 cm, *B5* mowing to 5 cm, including the grazing-tolerant plants, *E5* mowing only edible plants to 5 cm. Each *bar with a different pattern* in the annual production shows the monthly mowing in early June, July, August, and September from the base. *a*, *b* statistical significance at $P<0.01$

fully explain the reversed trend in annual production. One explanation is the difference in soil moisture. When measured in mid-July 2004, the soil moisture was 16.1% and 13.6% for the quadrats mown to 3 cm on the upper slope with grazing-tolerant plants remaining and quadrats with all plants mown, respectively. Soil moisture was higher in the quadrat in which the grazing-tolerant plants were unmown. Although rich vegetation transports more water from the ground to the atmosphere via evapotranspiration than does poor vegetation (Baldocchi et al. 2004), the surface soil temperature could have been lower in the quadrat with grazing-tolerant plants as a result of greater ground coverage after mowing. Moreover, edible plants mixed with unmown grazing-tolerant plants were not disadvantaged in terms of production in the drier habitat.

13.5 Grazing-Tolerant Plants and Soil pH

Several types of grazing-tolerant plants dominate Mongolian pastureland, and these plants have physical, chemical, and habitual defenses against livestock grazing. Grazing-tolerant plants with physical and chemical defenses can grow tall because livestock actively avoid grazing them. Grazing-tolerant plants with habitual defenses remain low to avoid livestock grazing. Overgrazing allows these plants to dominate pastures and increases soil pH (Hiernaux et al. 1999). We placed 1-m^2 quadrats at

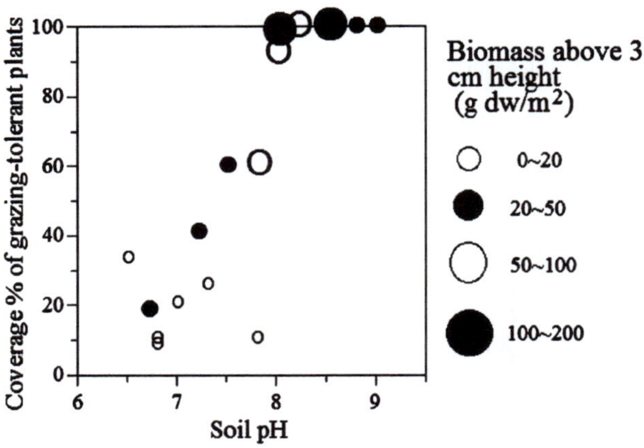

Fig. 13.7 Relationship between soil pH and weight percentage of grazing-tolerant plants versus all pasture plants in 1 m^2 using biomass greater than 3 cm

different pastureland sites in the forest-steppe zone around Ulan Bator in July 2005. We measured the surface soil pH and plant species coverage in each quadrat to determine the coverage domination between grazing-tolerant plants and all plants and mowed all pasture plants to 3 cm. After mowing, we dried the plants at 80°C for 2 days and measured their dry weight. To determine surface soil pH, we mixed surface soil with deionized water in a small bottle and measured the pH of the clear top layer of water. We used the average pH of three points in each quadrat. We calculated the total coverage separating edible and grazing-tolerant plants and compared them.

Figure 13.7 shows the relationship between soil pH and percent coverage of grazing-tolerant plants versus total coverage with the biomass of plants above 3 cm. A strong positive correlation was observed between the percent coverage of grazing-tolerant plants and soil pH, indicating that domination by grazing-tolerant plants in pasture coincides with soil alkalization. In northeast Thailand and southern Queensland, Australia, with high precipitation, land degradation caused soil acidification (Noble et al. 2000), but, in Mongolia, with little precipitation, pasture degradation is accompanied by soil alkalization and domination by grazing-tolerant plants. Plant biomass in good pasture dominated by edible plants for livestock becomes low as a result of grazing by livestock, whereas in degraded pasture it becomes relatively high because of the domination of grazing-tolerant plants that are little grazed by livestock. This difference makes it difficult to judge the degree of pasture degradation from the biomass of pasture plants, especially in the forest-steppe zone.

13.6 Pasture Plant Species Diversity Recovery After Disturbance Versus Soil pH

Plant species diversity declines greatly with heavy overgrazing in pastures and plowing in agricultural land. When the disturbance is removed, such as when herders move their livestock or abandon cultivation, plant species diversity should begin to recover. To determine the effects of disturbance history on the diversity recovery of plant species, we compared recovery after short- and long-term disturbances.

For the short-term disturbance, in 2001, we selected pasture near a winter shed moved annually in west Ih-Tamir, North Khangai Province. For the long-term disturbance, in 2005, we used agricultural land abandoned after more than 20 years of cultivation in the west of Ulan Bator. In the pasture, the herder yearly moved the winter shed to one side, so we could track its location up to 4 years previously. For the agricultural land, we determined the year of abandonment of each area by questioning people. We found sites abandoned for 1, 4, and 8 years in the west of Ulan Bator and for 20 years in the south of Ulan Bator. These sites had been farmed for more than 20 years before being abandoned. We set five 1-m^2 quadrats at and around each site and measured the number of pasture plants species and the surface soil pH in each quadrat.

Figure 13.8 shows the average number of pasture plant species from the five quadrats at each site for the length of abandonment and the surface soil pH, including the number of pasture plant species and surface soil pH in the surrounding pasture. Pasture plant species diversity recovered quickly where soil was not alkalized over 4 years after moving the site of the winter shed used for one winter, while it recovered slowly as the high soil alkalinity decreased after the agricultural land was abandoned. After deforestation, 10–15 years were necessary for soil pH to recover from alkalinization (Sangha et al. 2005). Soil alkalization progress depends on disturbance intensity and duration. Pasture plant species diversity recovers quickly after short-time disturbances because little soil alkalization occurs and also possibly

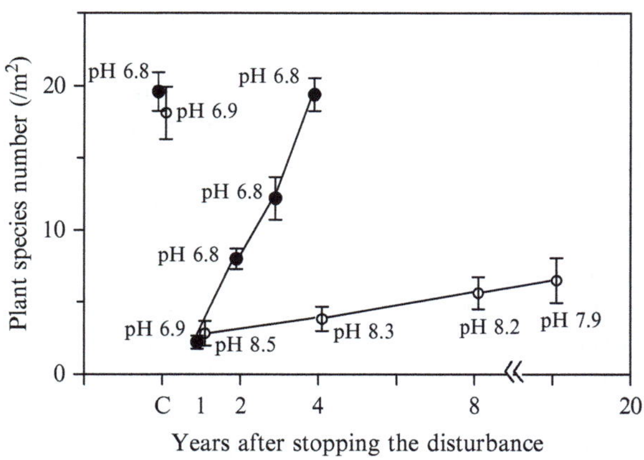

Fig. 13.8 Yearly recovery of plant species diversity and soil pH after removing the disturbance. *Black circles*, previous site of winter shed; *white-filled circles*, abandoned plowed land; *C* (*x*-axis), pastures around winter shed and abandoned plowed land

because root structures survive. Conversely, the slow recovery of pasture plants after a long-time disturbance occurs because high soil alkalinity prevents the invasion of pasture plants, especially edible plants. Soil alkalization causes a regime shift or positive feedback. In abandoned plowed land with very alkaline surface soil, the specialized grazing-tolerant species *Artemisia macrocephala* Jacq. dominates exclusively. This species is biennial, and its flowers cause pollinosis.

Once pasture soil becomes strongly alkaline, it cannot be used for breeding livestock for years because soil pH and edible plant diversity recover slowly. Short disturbances are inevitable with nomadism because livestock, especially sheep and goats, must be rounded up nightly near the residence. However, the seasonal and annual migration of traditional nomads avoids the soil alkalization associated with long-term disturbances.

13.7 Herb Versus Shrub Leaf Production

In the steppe and desert-steppe zones, both shrubs and herbs often grow in pastures. Dominant shrubs form patches of different sizes and densities. We measured the production and grazed mass of herbs and shrub leaves near Mandalgobi in 2009 to determine shrub edibility and their contribution to the livestock diet. We set ten 1-m^2 quadrats in the pasture exclosure (45°49′12″ N, 106°17′43″ E) to measure the seasonal production and grazed herb quantity. We mowed pasture plants to 3 cm at half-month intervals in five quadrats to determine herb production. We measured the height of herbs at ten points outside the exclosure monthly and mowed pasture plants at the mean height outside the exclosure in the other five quadrats in the exclosure to determine the grazed herb quantity outside the exclosure. Inside and outside the exclosure, *Caragana microphylla* (Pall.) Lam. was the dominant shrub, comprising about 20% of the canopy coverage. We used it for shrub measurement, although *Caragana leucophloea* Pojark. was mixed sparsely with it. We collected five duplicates of new leaves from a 20-cm^2 area of leaf canopy of each shrub patch in early June, July, and August in the exclosure. Then, we measured the dry weight per area of leaf canopy to determine shrub leaf production. We measured the leaf canopy size of five shrub patches inside and outside the exclosure by approximating ellipses in early May, June, July, and August. We calculated the monthly grazed mass of shrub leaves from the sum of the decreased leaf canopy outside the exclosure and the increased leaf canopy inside the exclosure. Some leaves at the patch canopy center had been grazed by livestock; however, this amount was not included in the measured mass because of the added difficulty in measurement. We measured the leaf canopy size of all shrub patches in the 1-ha exclosure and calculated the area percentage of shrub leaf canopy to land in early July and found that the canopy covered about 20% of the land. Based on the area percentage of the leaf canopy, we calculated the production and grazed mass of shrub leaves for 1 month.

Daily precipitation, herb production over a half-month, grazed herb mass, shrub leaf production, and monthly grazed shrub leaf mass are shown in Fig. 13.9. Daily

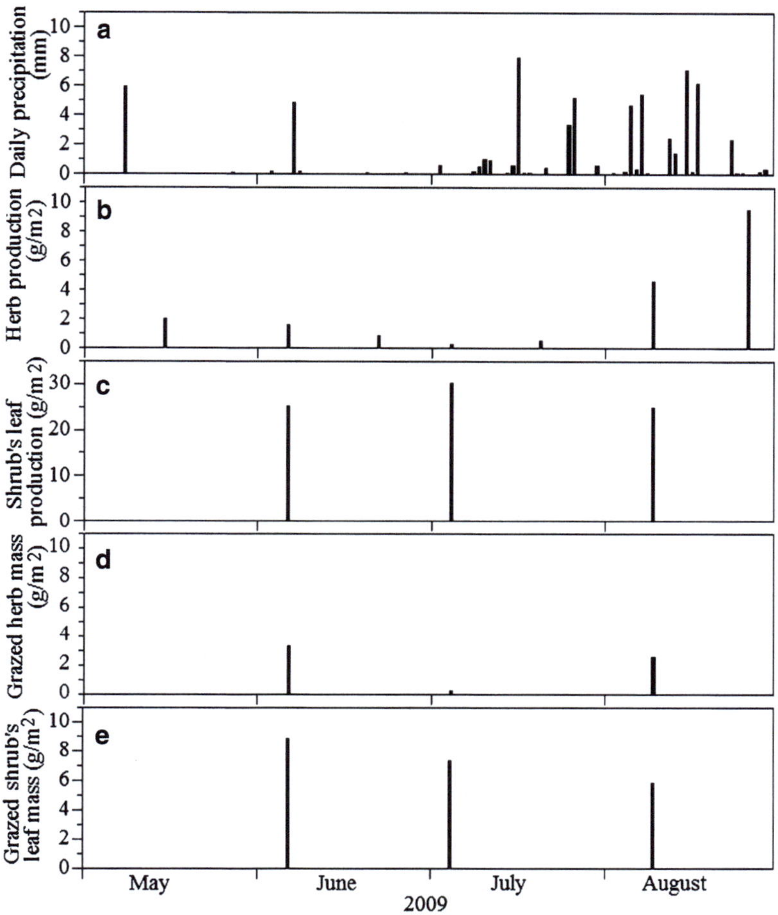

Fig. 13.9 Precipitation, and production and grazed mass of herbs and shrubs, during the growing season in Mandalgobi from May to August 2009: daily precipitation (**a**); herb production higher than 3 cm over a half-month (**b**); *Caragana microphylla* leaf production during 1 month (**c**); herb mass grazed by livestock during 1 month (**d**); *Caragana microphylla* leaf mass grazed by livestock over 1 month (**e**)

precipitation did not exceed 10 mm throughout the season in 2009. Precipitation was rare for the first half of the season; the total amount during a 15-day period was less than 10 mm. Herb production was low until July. After mid-July, the total precipitation for a 15-day period increased to more than 15 mm, despite daily precipitation of less than 10 mm, and herb production was high in August. Conversely, *C. microphylla* leaf production was high throughout the season. The grazed herb mass was also lower than grazed *C. microphylla* leaves.

In pasture with high shrub density, shrubs produced leaves better than herbs, and the shrub leaves were a good food source for livestock. *Caragana* belongs to

Fabaceae, which fix nitrogen gas; therefore, *Caragana* leaves and flowers are nutritious with good protein and fiber content (Jigjidsuren and Johnson 2003). If there is little precipitation in spring, herbs do not create leaves in spring, even if the temperature becomes warm enough for leafing. On the other hand, *Caragana* can produce leaves in spring regardless of precipitation, as reported by Yamada et al. (2009). Furthermore, when drought continues for a long time during the growing season, the aboveground parts that herbs have produced after heavy rains wither, but *Caragana* still keeps its leaves. Therefore, *Caragana* leaves are a valuable food for livestock in spring and the dry season. *Caragana* sprouts leaves in the dry season because it can absorb water deep in the soil. In *C. microphylla* habitat in Dalanzadgad, the surface soil is sandy and the deep soil layer holds water that percolates down during heavy rains (see Satoh et al., this volume, Chap. 5). Water use by shrubs and herbs at different soil depths has been reported on other steppes (Sala et al. 1989).

13.8 Distribution and Root Depth of Shrub Versus Soil-Particle Distribution

In central Mongolia, three shrub types dominate steppe and desert-steppe pastures (see Fujita and Amartuvshin, this volume, Chap. 3). In the steppe between Ulan Bator and Mandalgobi, *C. microphylla* is dominant. In the desert steppe between Mandalgobi and Dalanzadgad, two shrub types occur: *Caragana stenophylla* Pojark., and a mixture of *Kalidium foliatum* (Pall.) Moq. and *Reaumuria soongorica* (Pall.) Maxim. Two soil types occur in these zones: sandy and silty or clayey. *C. stenophylla* appeared to occur mainly in the sandy soil and *R. soongorica* also in the clayey soil (Grubov 2001). We collected 100 ml surface soil, removing fine roots, at a depth of 10 cm at five sites in the habitat of each shrub type (*C. microphylla*: 47°03′43″ N, 105°44′44″ E; 45°49′12″ N, 106°17′43″ E; 46°21′10″ N, 106°27′26″ E; 47°05′30″ N, 106°35′48″ E; 47°12″30″ N, 106°38′14″ E; *C. stenophylla*: 45°13′33″ N, 105°49′18″ E; 44°52′28″ N, 105°37′52″; 44°52′42″ N, 105°27′57″ E; 44°12′08″ N, 105°04′26″ E; 43°34′55″ N, 104°25′40″ E; mixture of *K. foliatum* and *R. soongorica*: 45°12′06″ N, 105°47′01″ E; 44°18′41″ N, 105°13′02″ E; 44°11′58″ N, 105°04′19″ E; 43°38′51″ N, 104°32′32″ E; 43°46′29″ N, 104°30′04″) in middle October, 2010. At the same time, we collected a taproot of each shrub species at 10-cm-depth intervals. At Mandalgobi (45°49′12″ N, 106°17′43″ E), we collected *C. leucophloea* Pojark. At Huld (44°52′42″ N, 105°27′57″ E), we gathered *C. stenophylla*. At the north of Dalanzadgad (43°38′51″ N, 104°32′32″ E), we collected *K. foliatum*, *R. soongorica*, *Kalidium gracile* Fenzl, and *Anabasis brevifolia* C.A. Mey. After drying the soil and the taproot at 90°C and 70°C, respectively, for 3 days, we sorted the soil and measured the dry weight of the taproot. We used 2-, 1-, 0.5-, 0.25-, 0.106-, and 0.053-mm-diameter sieves for soil sorting with a shaking apparatus for 15 min. We weighed the soil in successive sieves and determined the coarse sand, fine sand, and silt content of the soil from the particle diameters of 2–0.25, 0.25–0.053, and <0.053 mm, respectively.

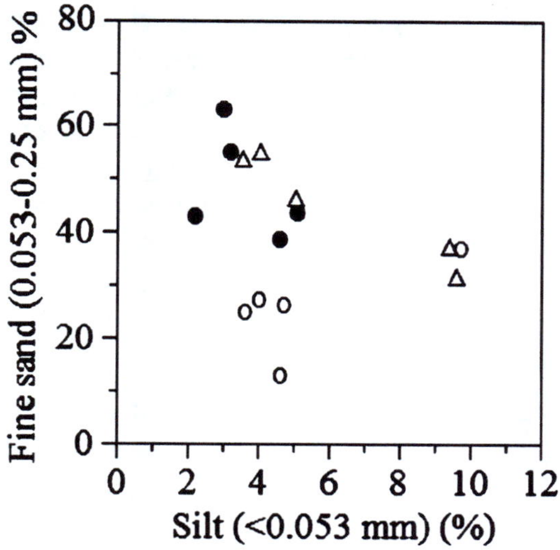

Fig. 13.10 Weight percentage of silt with diameter <0.053 mm and fine sand with diameter 0.053–0.25 mm at depth of 10 cm. Remainder is weight percentage of coarse sand with diameter 0.25–2 mm. *White triangle*, *Caragana microphylla* site; *white circle*, *Caragana stenophylla* site; *black circle*, *Kalidium-Reaumuria* site

The weight percentages of fine sand and silt at each site are shown in Fig. 13.10. The soil composition of the *C. stenophylla* sites was >38% fine sand and <5% silt (fine sand type), whereas the soil composition from the mixed *K. foliatum* and *R. soongorica* sites was <24% fine sand and <5% silt (coarse sand type) or >9% silt (silt type). *C. microphylla* had soil composition of the fine sand type and silt type. *C. microphylla* that grows in the soil of the fine sand type can absorb water from deep soil with the deep taproot for a long time, while herbs with shallow roots absorb water from shallow soil only just after precipitation in Mandalgobi with the fine sand type (see Satoh et al., this volume, Chap. 5). *C. stenophylla* and *C. leucophloea*, which occur in the soil of the fine sand type, have a deep taproot similar to *C. microphylla*, whereas *K. foliatum*, *R. soongorica*, *K. gracile*, and *A. brevifolia* that grow on the silty soil type have a shallow tap root (Fig. 13.10). *C. stenophylla* extends a taproot horizontally at 40-cm depth to penetrate deeply through the hard calcium carbonate layer with heavier root weight at the 40–50 cm depth (Fig. 13.11a). *C. stenophylla* and *C. leucophloea* might be able to absorb water from the deep soil under soil conditions similar to *C. microphylla*. On the other hand, *K. foliatum*, *R. soongorica*, *K. gracile*, and *A. brevifolia* seem not to try to obtain water from the deep soil. Near Mandalgobi, very little percolation of precipitation occurred to soil depths greater than 20 cm in 2001 (Yamanaka et al. 2007) However, after much precipitation, water percolates to the deep soil layer, and shrubs that have deep taproots, such as *Caragana* sp., suck water from the deep soil layer (see Satoh et al., this volume, Chap. 5). On the other hand, shrubs that have shallow taproots, such as *K. foliatum*, *R. soongorica*, *K. gracile*, and *A. brevifolia*, and herbs absorb water only from the shallow soil layer. To maintain much precipitation in the deep soil layer for a long time, it is necessary for much precipitation to penetrate deeply into the soil, with little being drawn up or down in the soil column.

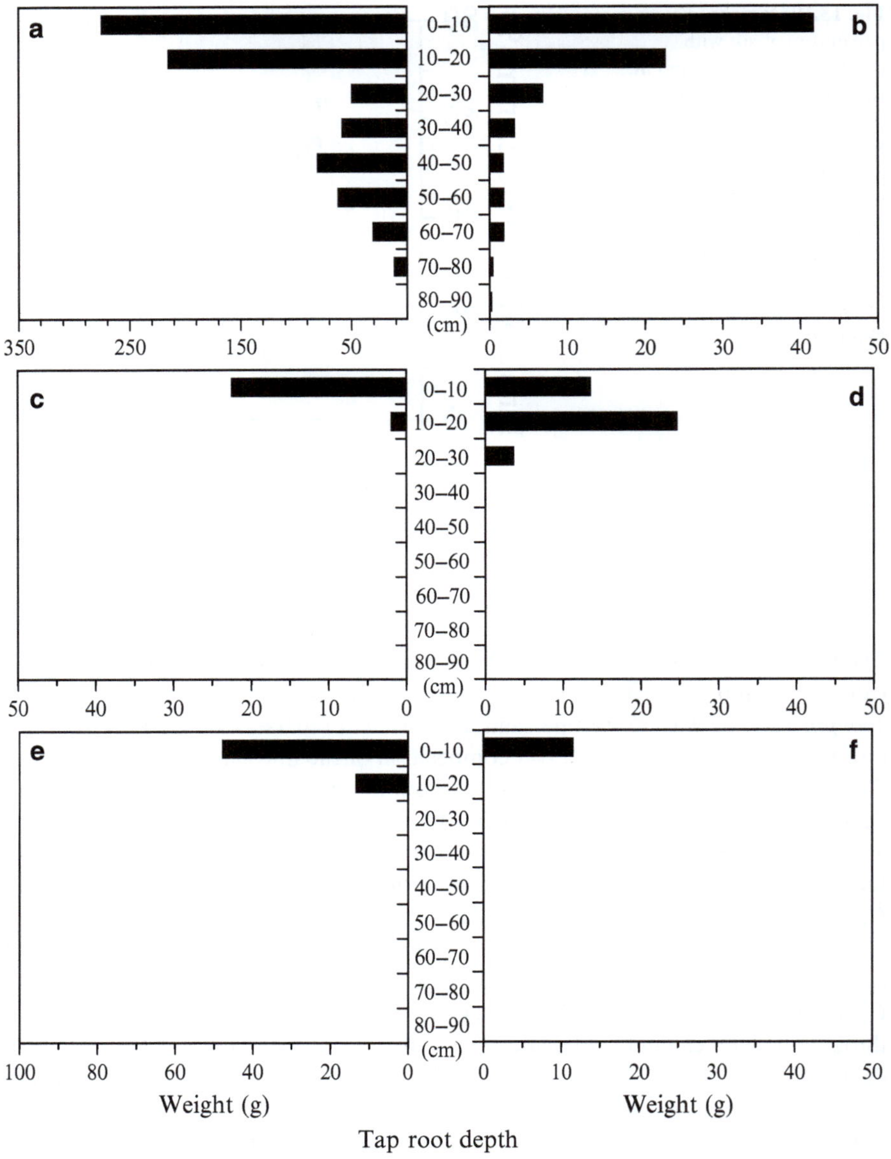

Fig. 13.11 Vertical taproot distribution of shrubs at intervals of 10-cm depth in the soil for *Caragana stenophylla* (**a**); *Caragana leucophloea* (**b**); *Kalidium foliatum* (**c**); *Reaumuria soongorica* (**d**); *Kalidium gracile* (**e**); *Anabasis brevifolia* (**f**)

The difference in the soil-particle distribution in the surface soil layer might be related to water percolation to the deep soil and water suction up to the soil surface. Shrubs with a shallow taproot must have drought tolerance because they are exposed to drought after when there is little precipitation, which is supported by the fact that *K. foliatum* and *A. brevifolia* are C4 shrubs (Pyankov et al. 2000).

13.9 Pasture Domination by Shrubs Versus the Effects of Human Activity

Shrub density varied in the pastures. To determine the distribution of shrub density and its relationship to human activity, we observed shrub density by driving a car along a road in August 2008. We classified shrub density into four levels: high, low, poor, and none, roughly equal to shrub patch coverages of >10%, 1–10%, <1%, and 0%, respectively. As we drove, when we observed a change in the shrub density level, we recorded the point using a global positioning system (GPS). We also recorded positions of herders' summer residences and winter sheds by GPS, noting the species present and the shrub density level. We calculated the distances between successive density change points. From Ulan Bator to Mandalgobi in the steppe zone, the main shrub was *C. microphylla*. From Mandalgobi to Dalanzadgad, the dominant species were *C. stenophylla*, *K. foliatum*, and *R. soongorica*, in different soil habitats (Fig. 13.10), with some other shrub species. *C. stenophylla* grew in sandy soil, while *K. foliatum* and *R. soongorica* were distributed, often together, in silty soil. We observed shrub density separately between *C. stenophylla* and the mixed *K. foliatum* and *R. soongorica* in the desert-steppe zone. The shrub type for areas where no shrubs grew was based on the shrub types growing on either side of the area. In all cases, the shrub types on both sides coincided.

The total distance for each shrub type level is shown in Fig. 13.12. The distribution patterns of the three shrub types differed. For *C. microphylla*, high density accounted for one-third of the distance from Ulan Bator to Mandalgobi, which was not observed in *C. stenophylla*. High-density patches accounted for more than half the distance of the mixed *K. foliatum* and *R. soongorica*. For all shrub types, shrub density decreased as the density of herders' summer residences and winter sheds increased. It is a question whether, with overgrazing by livestock, shrub density has increased or, conversely, has decreased. The density change of shrubs under overgrazing conditions depends on the palatability of shrubs (see Fujita and Amartuvshin, this volume, Chap. 3). These shrubs are all edible to greater or lesser degrees. Then, shrubs might have grown densely throughout the pastures in the past, but the low shrub densities observed could have been caused by degradation over long-term nomadic use. For example, *Caragana*, found in lower densities than *K. foliatum* and *R. soongorica*, is a better livestock food source. Comparing the two *Caragana* species, most *C. stenophylla* was strongly degraded and *C. microphylla* was partially degraded. In the desert-steppe zone where *C. stenophylla* grows, pasture production must be small because of lower precipitation compared with the steppe zone. Therefore, *C. stenophylla* had been degraded more often and more severely than *C. microphylla* in the steppe zone. *C. stenophylla* is a dry type of shrub with thin leaves and spines, similar to *C. leucophloea*. *C. microphylla* and *C. leucophloea* often grow together, and *C. leucophloea* is grazed more heavily by livestock. In addition to being a good food source, its weaker spines, a physical grazing-tolerance defense, is another reason for heavy *C. stenophylla* grazing by livestock. In mid-July 2009, Huld was very dry and *C. stenophylla* lost all their leaves. With *Caragana*

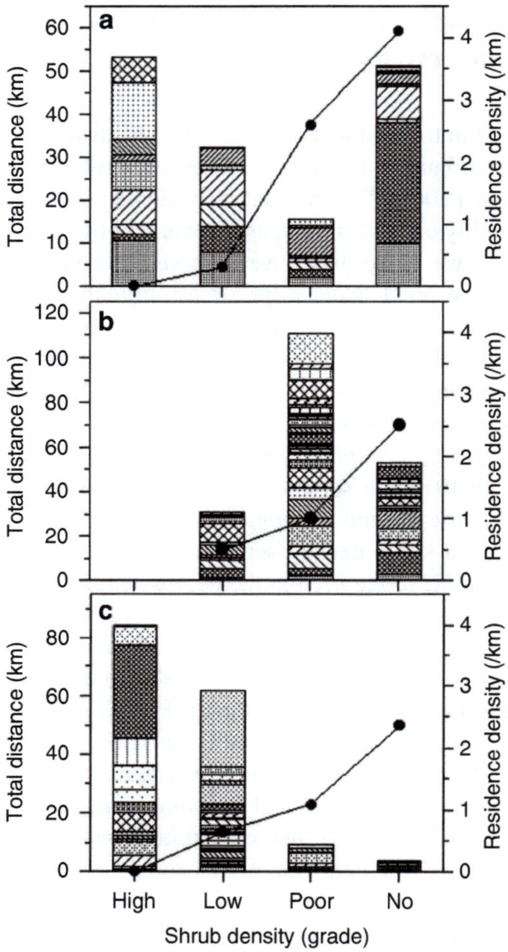

Fig. 13.12 Distribution of shrub density levels and density of summer residences and winter sheds at each level in August 2008. *Bar*, total distance of each shrub density level; *line*, density of nomad summer residences and winter sheds at each shrub density level. (**a**) *Caragana microphylla* between Ulan Bator and Mandalgobi; (**b**) *Caragana stenophylla* between Mandalgobi and Dalanzadgad; (**c**) mixed *Kalidium foliatum* and *Reaumuria soongorica* between Mandalgobi and Dalanzadgad. *High* shrub patch with greater than about 10% land coverage, *Low* about 1–10% land coverage, *Poor* coverage <1%, *No* no coverage. The patterned sections in the bars indicate the continuous units of each level

degradation, herb species also change. For example, *Allium mongolicum* Rgl. and *Stipa glareosa* P. Smirn. change to *Allium polyrrhizum* Turcz. ex Rgl. and *Stipa gobica* Roshev., respectively. According to Jigjidsuren and Johnson (2003), *A. mongolicum* has higher fiber content than *A. polyrrhizum*, and *S. glareosa* is highly nutritious. Therefore, *Caragana* degradation is accompanied by decline of nutrient value in herbs.

Decreasing shrub density is a major problem in the steppe and desert-steppe zones for sustainable nomadism. In semiarid and arid zones similar to the desert-steppe zone in Mongolia, it was reported that abiotic factors such as precipitation have stronger effects on vegetation than biotic factors such as herbivore grazing (Ellis and Swift 1988; Fernandez-Gimenez and Allen-Diaz 1999; Illius and O'Connor 1999; Wesche and Retzer 2005). Climate exerts short-term effects on vegetation, whereas herbivore grazing has long-term effects (Fuhlendorf et al. 2001). Actually, species diversity, coverage, biomass, and production of pasture plants change largely depending on the spatial, seasonal, and yearly fluctuation of precipitation (Figs. 13.3 and 13.4). However, pasture degradation is not caused by fluctuation in precipitation, because pasture plants have drought tolerance. Pasture vegetation damaged by a drought year nearly recovered after a following normal precipitation year (Shinoda et al. 2010). On the other hand, effects of herbivore grazing depend on the pressure and duration of grazing. Long-term hard grazing by livestock near the summer residences of nomads and the winter sheds that have been used through many years has brought pasture degradation such as shrub extinction to the steppe and desert-steppe zones in Mongolia.

Shrub preservation and recovery is an urgent issue in Mongolia, and *C. stenophylla* is especially at risk. For some shrubs, regeneration is sufficient to induce pasture succession (Vieira et al. 1994). In Inner Mongolia and China, *Caragana* can be grown by seed-spraying of pastures. However, we have not observed natural *C. microphylla* and *C. stenophylla* regeneration in these pastures. For *Caragana* to recover, individuals might have to be cultivated and grown to full size. Furthermore, a problem exists with planting fast-growing trees such as *Populus* in the steppe and desert-steppe zones where they do not normally grow, because they waste soil water, advancing drought, and are useless for nomads. *Caragana* species are best planted in the sandy soil of the steppe and desert-steppe zones.

13.10 Wind Erosion Versus Shrub Patch Degradation

Wind erosion is a problem in Mongolia, especially in the spring, because the prevailing westerlies carry soil eastward, sometimes as far as Japan. Shrub patches accumulate airborne soil and form mounds. When the shrub patches degrade, the soil mounds are eroded. To determine the wind erosion occurring on and around shrub patches at different levels of degradation, we measured the vertical change in soil surface through one windy season on and around normal and degraded shrub patches in the *C. microphylla* sites near Mandalgobi and the *C. stenophylla* sites in Huld. The average shrub patch coverage was almost 21% at the *C. microphylla* site and 5% at the *C. stenophylla* site. In early October 2009, we inserted seven thin iron rods in the soil so that their tops were flush with the soil surface, one on top, two on the shoulders, and two on the edges of the soil mound, as well as two in the flat ground 1 m from the edges along the axis of the prevailing wind (Fig. 13.13). We used shrub patches >1.5 m in diameter along the major axis of the prevailing wind. We distinguished normal and degraded patches as having leaf canopy areas of >80%

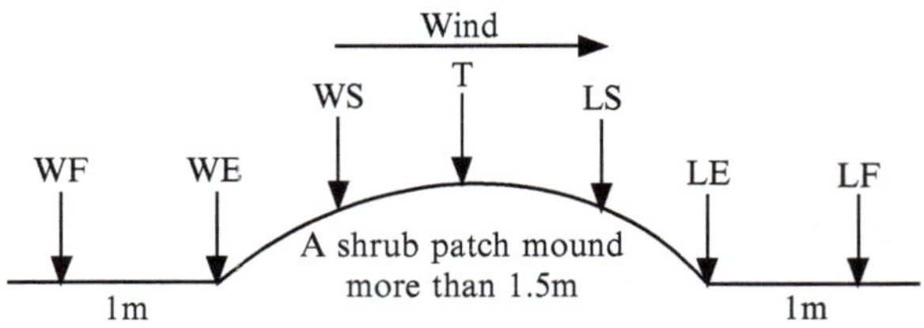

Fig. 13.13 Seven iron rods on and near a shrub patch soil mound on the axis of the prevailing wind inserted at the top of the mound (*T*), windward mound shoulder (*WS*), leeward mound shoulder (*LS*), windward mound edge (*WE*), leeward mound edge (*LE*), windward flat pasture 1 m from the windward mound edge (*WF*), and leeward flat pasture 1 m from the leeward mound edge (*LF*)

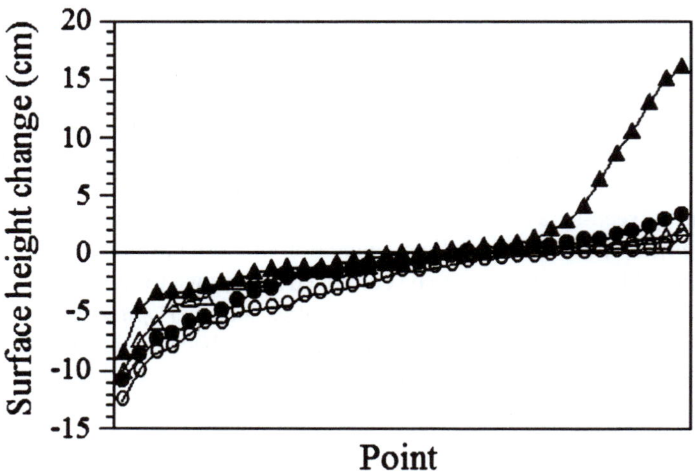

Fig. 13.14 Increase and decrease in soil surface around normal and degraded *Caragana* patches between October 2009 and August 2011. *Triangles*, *C. microphylla* in Mandalgobi; *circles*, *C. stenophylla* in Huld. *Black figures*, normal patches; *white figures*, degraded patches

and <20% of the patch area, respectively. We used five patches for each group at both sites. We measured the vertical distance up or down from the soil surface to the iron rod tip in mid-July 2010 and mid-August 2011. We judged soil accumulation and erosion by the position from the top of each iron rod.

Figure 13.14 shows the changes in soil surface along the prevailing wind axis in normal and degraded shrub patches from early October 2009 to mid-August 2011. The soil erosion was greater at the *C. stenophylla* site in Huld than the *C. microphylla* site in Mandalgobi, and the mound soil erosion of the normal shrub patch was lower than that of the degraded patch at both locations. The average changes were +1.2 and −1.3 cm on and around the normal and degraded patches, respectively, at the *C. microphylla* site in Mandalgobi and −1.7 and −2.8 cm on and around the normal and degraded patches, respectively, at the *C. stenophylla* site in Huld. In Huld, in mid-July 2010, the soil erosion was low as a whole and the soil was on average accumulated on and around the normal shrub patches, but, in mid-August 2011, soil erosion had occurred even on and around the normal shrub patch. In Huld, shrub degradation has already progressed because the present coverage of shrub patch was only 5%. The mound soil of a normal shrub patch that has been isolated can be eroded. Therefore, the degradation of shrub patches promotes wind erosion in pastureland and increases sand fall to the surroundings. Vertical fluctuations of the soil surface differed at different mound positions. Generally, the soil is blown from the windward mound and the top, and the soil is accumulated in the leeward side.

13.11 Goat Grazing Versus Sheep Grazing

To compare grazing effects on vegetation between sheep and goats, we performed raising experiments in the exclosure at two different vegetation sites in Bayan-Unjuul. One is a shrub site. In the pasture (47°01′55″ N, 105°48″08″ E) where *C. microphylla* grew dominantly with rare *C. leucophloea*, we constructed 24 fenced areas, 10 m square, in early October 2009. In the herb layer, perennials such as *Stipa krylovii* Roshev. and *Artemisia frigida* Willd. occurred exclusively. After measuring the area of *Caragana* leaf canopy in each fenced area and the weight of each sheep and goat, from 5 August 2010, we raised sheep and goats in each fenced area in four combinations; that is, one sheep, two sheep, one goat, and two goats, with water supplied using a plastic container. Averaged weights of sheep and goats before the experiment were 43.5 and 35.2 kg, respectively. All combinations had five repetitions. We measured the leaf weight of *C. microphylla* with a 20-cm^2 quadrat and the aboveground biomass of herbs within 1 m^2 using five repetitions in the remaining fenced areas. After 4 days, on 9 August 2010, we measured the weight of each sheep and goat and the remaining coverage of shrub leaf canopy and herbs in each area to determine the weight reduction of each sheep and goat and the mass grazed by sheep and goats in each fenced area for 4 days. We calculated the mass of *Caragana* leaves and herbs grazed from their coverage reduction and dry weight per area. We raised sheep and goats inside fences for two more days, up to 11 August 2010, to examine the effects of heavy grazing pressure by sheep and goats on *Caragana* and herb growth in the next year. We released sheep and goats from the fences on 11 August 2010, but we retained the fences to prevent livestock grazing until the next year. On 16 August 2011, we measured the area of *Caragana* leaf canopy and the coverage of herbs in each fenced area.

The second site is a herb site without shrubs. In the pasture (47°02′27″ N, 105°55′58″ E) where no shrubs were distributed, we constructed 20 fences of 200 m^2 (10 × 20) in early October 2010. In August 2011, annuals and perennials occurred wholly within fences. The most dominant species are *Chenopodium* annuals such as *C. accuminatum* Willd., *C. aristatum* L., and *C. urbicum* L., and dominant perennials

Fig. 13.15 Photographs inside fence after 4 days of experiment: strongly grazed *Caragana* in a fenced enclosure for goats (**a**), and remaining *Caragana* in a fenced enclosure for sheep (**b**)

were *S. krylovii*, *A. frigida*, and *Cleistogenes songorica* (Roshev.) Ohwi. After measuring the coverage percentages of *S. krylovii*, *A. frigida*, and total *Chenopodium* species in each fence and the weight of each livestock, from 13 August 2011, we raised sheep and goats for 3 days in each fenced area in three combinations, that is, two sheep, two goats, and one sheep and one goat, with water supply using a plastic container. Averaged weights of sheep and goats before the experiment were 42.1 and 35.3 kg, respectively. All combinations had five repetitions. On 16 August 2011, we measured the weight of each sheep and goat inside each fence to know the weight reduction of each sheep and goat for 3 days.

Diet preference was clearly different between sheep and goats in the shrub site. Sheep preferred herbs over *Caragana* leaves, whereas goats preferred *Caragana* leaves over herbs (Fig. 13.15). Averaged biomass per area for *C. microphylla* leaves and herbs before the experiment was 265.3 and 17.0 g/m^2, respectively, thus the biomass was more than ten times larger in *Caragana* leaves than herbs (Fig. 13.16a). Grazed coverage of *Caragana* leaves was significantly greater by goats than by sheep, and the mass of grazed herbs was significantly less for goats than sheep (Fig. 13.16b). Weight reduction did not correlated with coverage of *Caragana* leaves within a fence in sheep, but negative correlation between weight and coverage was found to be significant ($P<0.01$) in goats (Fig. 13.16c). This negative correlation was derived from a difference in the mass of *Caragana* leaves and herbs actually grazed (Fig. 13.16d). The difference in diet preference between sheep and goats was recognized in an early stage of livestock hunger, because sheep and goats had eaten up both *Caragana* leaves and herbs under starving hunger conditions after 6 days of the experiment. Growth recovery of *Caragana* and herbs in the next year was not different between sheep and goats (Fig. 13.17). The growth recovery was very good in *Caragana* even after complete defoliation and partial browsing of twigs after strong grazing by goats or sheep (Fig. 13.17a), but recovery was not as good in herbs (Fig. 13.17b), perhaps because of partial grazing of the underground parts of

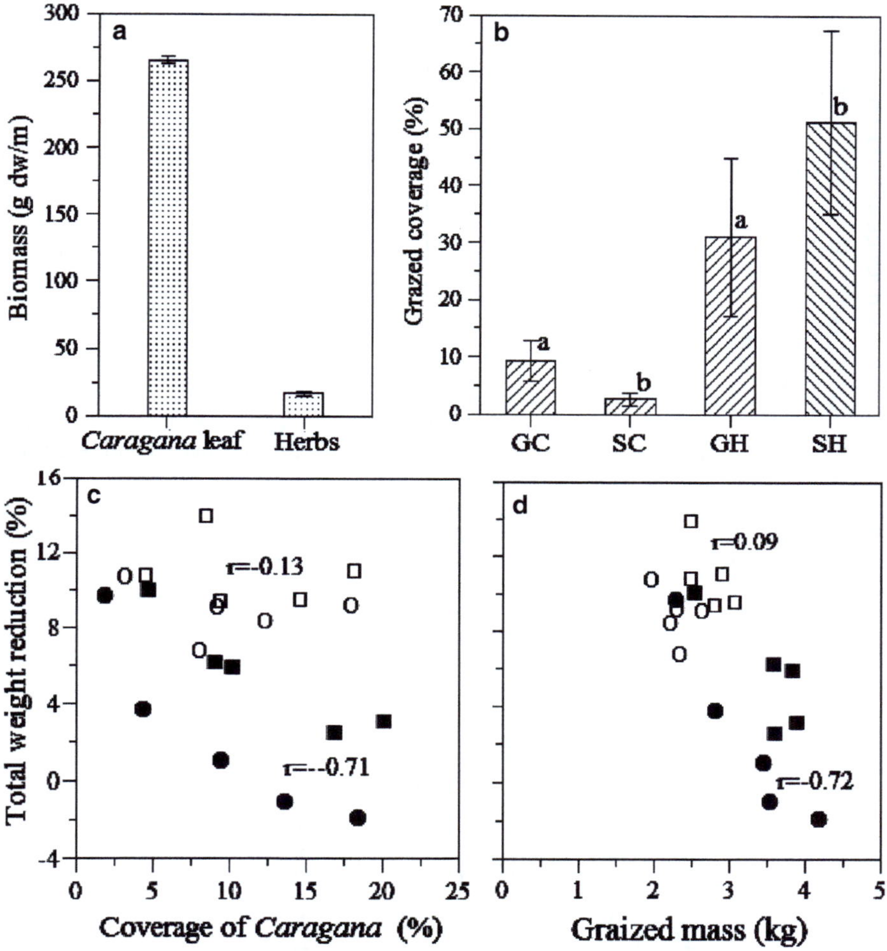

Fig. 13.16 Averages with standard deviations for biomass of *Caragana* leaves and herbs per 1 m^2 before the experiment (**a**), averages with standard deviations for grazed coverage percentages of *Caragana* leaves and herbs per one sheep or goat after 4 days (**b**), relationships between *Caragana* leaf coverage before experiment and total weight reduction of sheep or goats during 4 days inside each fence (**c**), and relationship between total grazed mass of both *Caragana* leaves and herbs by sheep or goats and total weight reduction percentage of sheep or goats during 4 days inside each fence (**d**). *GC* average with standard deviation for grazed coverage of *Caragana* leaves by goats inside fence, *SC* average with standard deviation for grazed coverage of *Caragana* leaves by sheep inside the fence, *GH* average with standard deviation for grazed coverage of herbs by goats inside the fence, *SH* average with standard deviation for grazed coverage of herbs by sheep inside the fence. *a*, *b* statistical significance at $P<0.01$. *White circles* one sheep, *white squares* two sheep, *black circles* one goat, *black square* two goats. *r* correlation coefficient for sheep or goats

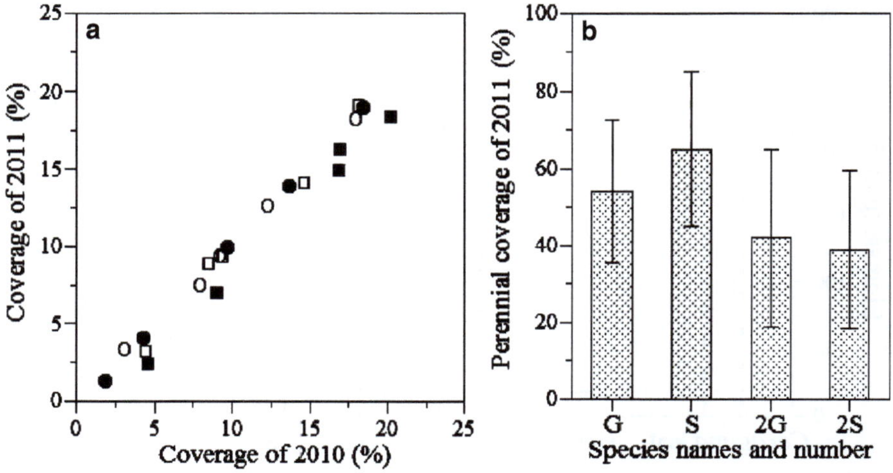

Fig. 13.17 Relationship of *Caragana* leaf coverage inside each fenced area between 2010 and 2011 (**a**) and averaged perennial coverage with standard deviation inside fence areas after the experiment for one and two sheep or goats (**b**). Symbols as in Fig. 13.16. *a, b* statistical significance at $P<0.01$

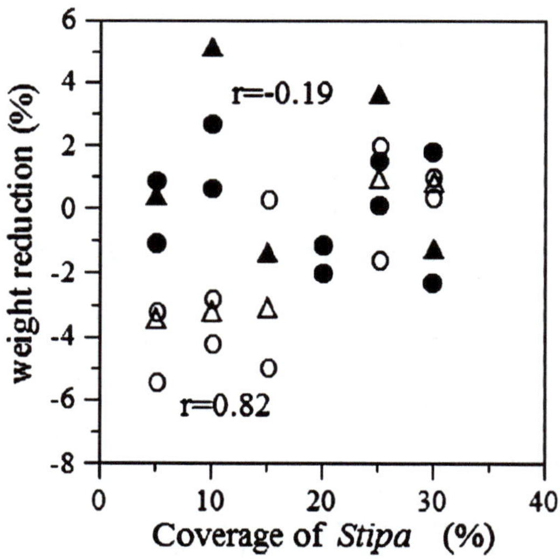

Fig. 13.18 Relationship between coverage of *Stipa krylovii* before experiment in each fenced area and weight reduction of each sheep and goat during 3 days. *White circle* one sheep within a two-sheep fence, *white triangle* one sheep in a sheep–goat-mixed fenced area, *black circle* one goat in a two-goats fence, *black triangle* one goat in a sheep–goat-mixed fenced area. *r* correlation coefficient for sheep or goats

the herbs. Diet preference for herbs alone differed between sheep and goats at the herb site. Weight reduction of goats was not seen during the 3 experimental days, although the weight of sheep decreased in negative correlation with coverage of *S. krylovii* after three experimental days (Fig. 13.18). At this time, diets were not deficient in each fenced area, because the aboveground biomass of herbs remained at more than one-third from that recorded at the start of experiment. Goats grazed various herbs without weight reduction for 3 days inside the fence, but sheep preferred fewer herbs other than *S. krylovii* in spite of weight reduction. The diet preference of sheep for *S. krylovii* only did not change when they coexisted with goats.

There are many reports about differences in diet preference between sheep and goats. Goats consumed more shrubs (Magadlela et al. 1995; Rogosic et al. 2006) and aculeate herbs (Holst et al. 2004) than sheep, although goats ignored some shrubs (Wilson et al. 1975). Sheep and goats preferred better grasses and forbs, respectively (Animut et al. 2005). Sheep grazed deeply but goats feed shallowly in the sward horizon (Gong et al. 1996). Our results do not contradict these reports in general. As for the comparison between grasses and forbs for diet, sheep preferred *S. krylovii* alone for grasses and did not eat *Cleistogenes songolica*, and goats preferred both grasses and forbs. Reaction to hunger from insufficient diet differed between goats and sheep. Goats maintained their weight by grazing every available plant until their weight was reduced by shortage in their diet, while sheep loose their weight easily after grazing all favorite diets available. Although both goats and sheep finally exhaust all plants in response to extreme hunger, goats might avoid weight reduction by grazing less preferred diets whereas sheep might choose weight reduction rather than grazing a lesser diet before hunger become strong.

Based on difference in diet preference, grazing effects on pasture differ between goats and sheep. In early stages of overgrazing, goats might bring about shrub degradation, whereas sheep might cause herb degradation. Pasture species of shrubs and herbs have some degree of grazing tolerance. *Caragana* species could recover well from complete but temporary leaf grazing (Fig. 13.17a). A serious problem is caused in pasture degradation when overgrazing continues for a long time or year after year.

13.12 Conclusions

Pasture production is controlled by precipitation, although temperature is a necessary condition. Sufficient annual herb production in pastureland requires a certain dose of precipitation during a limited timeframe for continued production in the dry regions. For example, at least 10 mm over 15 days is required for useful pasture plant production. Unpredictable, spatial, seasonal, and yearly fluctuation of precipitation causes great spatial and temporal fluctuation of pasture production, although drought damage to a pasture is recoverable by normal precipitation. The free moving of livestock and frequent and seasonal migration of livestock camps in traditional nomadism is a good adaptation to this fluctuation. Migration also helps to avoid pasture degradation and soil alkalization caused by continuous overgrazing, which occurs even in the forest-steppe zone, and irreversibly changes the vegetation through negative feedback.

Moderate livestock grazing pressure facilitates pasture plant production through a positive feedback effect, but overgrazing has a negative feedback effect. Herders, who control livestock density and migration, affect pasture production indirectly, while pasture production affects herders indirectly by influencing livestock production. For example, herders lead cattle to flat valley bottoms and sheep and goats to the upper slopes.

Shrub patches in steppe and desert-steppe pastures are very important for Mongolian nomadism. *Caragana* ssp., including *C. microphylla* in the steppe and *C. stenophylla* in the desert steppe, fix nitrogen from the air and can grow leaves not only after adequate precipitation but also throughout the growing season by absorbing water from deep in the soil. They provide good fodder for livestock, especially during the dry and winter seasons. *K. foliatum* and *R. soongorica* have high drought tolerance and become food sources for livestock in the desert-steppe zone. For this reason, the shrubs have become strongly degraded in the steppe and desert-steppe zones. If no efforts are made to save these shrubs, future extinction is inevitable. Moreover, the degradation of *Caragana* is leading to a crisis for sustainable nomadism. Shrub conservation and recovery is an urgent concern in Mongolia.

Goats prefer *Caragana* leaves, while sheep prefer herbs, especially *Stipa*. In the first stage of hunger, goats graze various plants without weight reduction, but sheep keep to their diet preferences in spite of weight reduction. Under conditions of great hunger, both goats and sheep degrade pasture severely by browsing shrubs and grazing the underground parts of herbs.

Acknowledgments This work was supported financially in part by the RIHN research project "Collapse and Restoration of Ecosystem Networks with Human Activities." We are grateful to Professor N. Yamamura and Dr. E. Kusano for their helpful suggestions on the draft of this manuscript.

References

Animut G, Goetsch AL, Aiken GE, Puchala R, Detweiler G, Krehbiel CR, Merker RC, Sahlu T, Dawson LI, Johnson ZB, Gipson TA (2005) Performance and forage selectivity by sheep and goats co-grazing grass/forb pastures at three stocking rates. Small Ruminant Research 59: 203–215.

Bai Y, Wu J, Xing Q, Pan Q, Huang J, Yang D, Han X (2008) Primary production and rain use efficiency across a precipitation gradient of the Mongolian Plateau. *Ecology* 89: 2140–2153.

Baldocchi DD, Xu L, Kiang N (2004) How plant functional-type, weather, seasonal drought, and soil physical properties alter water and energy fluxes of an oak–grass savanna and an annual grassland. *Agric For Meteorol* 123: 13–39.

Beatley JC (1974) Phenological events and their environmental triggers in Mojave desert ecosystems. *Ecology* 55: 856–863.

Dulamsuren C, Hauck M, Mühlenberg M (2005) Ground vegetation in the Mongolian taiga forest–steppe ecotone does not offer evidence for the human origin of grasslands. *Appl Veg Sci* 8: 149–154.

Ellis, JE, Swift, DM (1988) Stability of African pastoral ecosystems: alternate paradigms and implications for development. *J Range Manag* 41: 450–459.

Fernandez-Gimenez, ME, Allen-Diaz, B (1999) Testing a non-equilibrium model of rangeland vegetation dynamics in Mongolia. *J Appl Ecol* 36: 871–885.

Fujita N (1997) Forest vegetation around Lake Baikal. In: *New Scope on Boreal Ecosystems in East Siberia* (Wada A, Timoshkin OA, Fujita N, Tanida K, eds). DIWPA Series, vol 2. Scientific Publishing Center of the UIGGM SB RAS, Novosibirsk, pp 88–98.

Fujita N, Amartuvshin N, Yamada Y, Matsui K., Sakai S, Yamamura N (2009) Positive and negative effects of livestock grazing on plant diversity of Mongolian nomadic pasturelands along a slope with soil moisture gradient. *Grassland Sci* 55: 126–134.

Fuhlendorf, SD, Briske, DD, Smeins, FE (2001) Herbaceous vegetation change in variable rangeland environments: the relative contribution of grazing and climatic variability. *Appl Veg Sci* 4: 177–188.

Gong Y, Hodgson J, Lambert MG, Gordon IL (1996) Effects of contrasting sward heights within forage species in short-term ingestive behavior of sheep and goats grazing grasses and legumes. *N Z J Agric Res* 39: 75–82.

Grubov VI (2001) *Key to the Vascular Plants of Mongolia, vols I & II*. Science Publishers, Enfield, NH, pp. 1–817.

Gubanov IA (1996) *Conspectus of Flora of Outer Mongolia*. Valang Publishers, Moscow, pp. 1–132.

Gunin PD, Vostokova EA, Dorofeyuk NJ, Tarasov PE, Black CC (1999) *Vegetation Dynamics of Mongolia*. Kluwer, Dordrecht, The Netherlands.

Hiernaux P, Bielders CL, Valentin C, Bationo A, Fernández-Rivera S (1999) Effects of livestock grazing on physical and chemical properties of sandy soils in Sahelian rangelands. *J Arid Environ* 49: 231–245.

Hilbig W (2000) Forest distribution and retreat in the forest steppe ecotone of Mongolia. *Marburger Geographische Schriften* 135: 171–187.

Holst PJ, Allan CJ, Campbell MH, Gilmour AR (2004) Grazing of pasture weeds by goats and sheep. 1. Nodding thistle (*Carduus nutans*). *Aust J Exp Agric* 44: 553–557.

Illius, AW, O'Connor, TG (1999) On the relevance of nonequilibrium concepts to arid and semi-arid grazing systems. *Ecol Appl* 9: 798–813.

Ishikawa M, Sharkhuu N, Zhang Y, Kadota T, Ohta T (2005) Ground thermal and moisture conditions at the southern boundary of discontinuous permafrost, Mongolia. *Permafrost and Periglacial Processes* 16: 209–216.

Jigjidsuren S, Johnson DA (2003) *Forage Plants of Mongolia*. Admon, Ulan Bator.

Li SG, Asanuma J, Eugster W, Kotani A, Liu JJ, Urano T, Okikawa T, Davaa G, Oyunbaatar D, Sugita M (2005) Net ecosystem carbon dioxide exchange over grazed steppe in central Mongolia. *Global Change Biol* 11: 1941–1955.

Magadlela AM, Dabaan ME, Bryan WB, Prigge EC, Skousen JG, D'Souza GE, Arbogast BL, Flores G (1995) Brush clearing on hill land pasture with sheep and goats. *J Agric Sci* 174: 1–8.

Noble, AD, Gillman, GP, Ruaysoongnern, S (2000) A cation exchange index for assessing degradation of acid soil by further acidification under permanent agriculture in the tropics. *Eur J Soil Sci* 51: 233–243.

Pyankov, VI, Gunin, PD, Tsoog, S, Black, CC (2000) C4 plants in the vegetation of Mongolia: their natural occurrence and geographical distribution in relation to climate. *Oecologia* (Berl) 123: 15–31.

Rogosic J, Pfister JA, Provenza FD, Grvesa D (2006) Sheep and goat preference for and nutritional value of Mediterranean maquis shrubs. *Small Ruminant Res* 64: 169–179.

Sala, OE, Golluscio, RA, Rauenroth, WK, Soriano, A (1989) Resource portioning between shrubs and grassed in the Patagonian steppe. *Oecologia* (Berl) 81: 501–505.

Sangha KK, Midmore DJ, Rolfe J, Jalota RK (2005) Tradeoffs between pasture production and plant diversity and soil health attributes of pasture systems of central Queensland, Australia. *Agric Ecosyst Environ* 111: 93–103.

Sankey TT, Montagne C, Graumlich L, Lawrence R, Nielsen J (2006) Lower forest–grassland ecotones and 20th century livestock herbivory effects in northern Mongolia. *For Ecol Manag* 233: 36–44.

Shinoda, M, Nchinshonhor, GU, Nemoto, M (2010) Impact of drought on vegetation dynamics of the Mongolian steppe: a field experiment. J Arid Environ 74 : 63–69.

Vieira ICG, Uhl C, Nepstad D (1994) The rule of the shrub *Cordia multispicata* Cham. as a 'succession facilitator' in an abandoned pasture, Paragominas, Amazonia. *Vegetation* 115: 91–99.

Wesche, K, Retzer, V (2005) Is degradation a major problem in semi-desert environments of the Gobi region in southern Mongolia. *Erforsch Biol Ress Mongolei* 9: 133–146.

Wilson AD, Leigh FH, Hindley NL, Mulham WE (1975) comparison of the diets of goats and sheep on a *Casuarina cristata–Heterodendrum oleifolium* woodland community in western New South Wales. *Aus J Exp Agric Anim Husb* 15: 45–53.

Yamada, Y, Yamaguchi, Y, Undarmaa, J, Hirobe, M, Yoshikawa, K (2009) Environmental factors controlling leaf emergence in Caragana microphylla, a deciduous shrub of the Mongolian steppe. *J Arid Land Stud* 19: 137–140.

Yamanaka, T, Kaihotsu, I, Oyunbaatar, D, Ganbold, T. (2007) Summertime soil hydrological cycle and surface energy balance on the Mongolian steppe. *J Arid Environ* 69: 65–79.

Yanagisawa N, Fujita N (1999) Different distribution patterns of woody species on a slope in relation to vertical root distribution and dynamics of soil moisture profiles. *Ecol Res* 14: 165–177.

Part III
Socioeconomic Activity and the Ecosystem Network

Chapter 14
Pastoral Mobility and Pastureland Possession in Mongolia

Akira Kamimura

Keywords Herder group • Land law • Mongolia • Pastoral mobility • Pastureland possession

14.1 Introduction

Pastoralists in Mongolia have adopted a number of models of European- or Western-style modernization or development since the middle of the last century: in the Socialist period, collectivization to establish herding collectives, which may be recognized as a Socialist version of development; in the early 1990s, during the transition to a market economy, a market-oriented model based on private property, with the dissolution of the herding collectives and privatization of livestock, although pastureland remains for common use; and currently, since the late 1990s, "Community-based natural resource management" (CBNRM), a development model for common-pool resources management.

The situation in the past two decades alleged to require new management institutions in the pastoral sector in Mongolia could be summarized as follows. Mongolia has been undergoing two processes impacting the pastoral economy and society: integration to the global economy or the transition to a market economy on a national scale since the early 1990s, and global warming on a global scale, the impact of which has been growing during the past decade. These processes have concentrated human and livestock population into areas less arid and closer to a large city or market such as Ulaanbaatar. The problem is that most of those movements are unidirectional: herders in favorable areas always accept those from outside their areas, but rarely share benefits of reciprocal ethics (even though adverse weather,

A. Kamimura (✉)
Tokyo University of Foreign Studies, Tokyo, Japan
e-mail: kamimura.akira@tufs.ac.jp

N. Yamamura et al. (eds.), *The Mongolian Ecosystem Network: Environmental Issues Under Climate and Social Changes*, Ecological Research Monographs,
DOI 10.1007/978-4-431-54052-6_14,

heavy snowfall, or cold weather, happened, they could not escape to southern areas because there is little forage available), which reciprocation has long been a main aspect of institutions in managing pastureland use. This change appears to damage the resilience of pastoral society and makes it difficult for customary institutions to function effectively if not in the whole country.

CBNRM as an approach to common property management emerged from empirical critiques of Hardin's "tragedy of the commons," which predicts the degradation of communal pastures for reasons of the lack of institutions to control resource use and calls for privatization. Many large-scale development projects implemented by the international donor community in the pastoral sector in arid Africa in the 1970s and 1980s, and also the allotment of pastures to individual households carried out in the 1980s in Inner Mongolia, were inspired by Hardin's thesis, which had negative effects not only on pastureland ecology but also on the pastoral economy (Fratkin 1997). The economic advice that the Asian Development Bank (ADB) gave the Government of Mongolia in the early 1990s had the same logic. Since the publication of Hardin's article there have been reported from around the world many cases in which a common-pool resource was successfully managed, contrary to his prediction.

"Since 1999, Mongolia has become a de facto testing-ground for community-based rangeland management, with the establishment of over 2000 'herder groups' or 'pasture user groups,' facilitated by over 12 different donor and NGO-sponsored programs" (Fernandez-Gimenez et al. 2008: 3; Mau and Chantsalkham 2006). Donors funding those programs, for example, the United Nations Development Programme (UNDP) and the Swiss Agency for Development and Cooperation (SDC), also initiated legislation to allocate a pasture possession right to herder groups or pasture user groups (PUGs). A draft of a law on pastureland was prepared by a working group consisting of staff of those programs, some Parliament members, and officials of Mongolian ministries. The draft was presented to Parliament on 5 October 2010, although it was yet not passed at the time of this writing. In some project sites, *sum* governors contract with PUGs to allow for pastureland possession rights beyond the stipulation of the current Land Law, in anticipation of the future implementation of a Pastureland Law.

Despite the high expectations, "the practical implementation of community based natural resource management (CBNRM) initiatives, however, has frequently fallen short of expectations" (Leach et al. 1999: 226). In Mongolia, too, difficulties in implementation of CBNRM programs have been reported recently (for example, Upton 2005, 2008; Bedunah & Schmidt 2004). These difficulties or failures predict efficacy of the proposed Pastureland Law.

This chapter considers how the proposed legislation to provide for pasture possession rights to PUGs or herder groups could contribute to the sustainability[1] of

[1] The definition of sustainability here is "maintaining the productive and adaptive capacity of ecosystems while providing for the well-being of human communities" (Fernandez-Gimenez and Swift 2003: 821).

pastoralism in Mongolia. First, I give a brief review on the background of the legislation for land tenure in Mongolia; second, I examine some articles of draft versions of pastureland law that would have impact on pastoral management; and then consider several issues critical to the subject, drawing on empirical data collected in a field survey in Mongolia.

14.2 Legislation Background

Since the beginning of the 1990s, attempts have been made to introduce a new pastureland tenure system or to formalize institutions to regulate pastureland use. Although the focus has shifted from privatization to ecological sustainability in the last decade, the logic and narrative of the rationalization seem to have remained unchanged.

When the transition to a market economy became a main national issue in the early 1990s, the private ownership of land was supposed to be an essential prerequisite for the market economy, although few Mongolian people shared such an assumption. In 1991, following the collapse of the Soviet Union and the dissolution of the Council for Mutual Economic Assistance (CMEA), the Mongolian government was urged to initiate a number of economic reforms designed to create a market economy. Those reforms were conducted actually in response to advice and recommendations of international development agencies such as the International Monetary Fund (IMF), the World Bank, the ADB, and other donor agencies. Legislation was soon enacted to privatize state assets. The new Constitution, which allows the privatization of state-owned land, but excepted pastureland, was promulgated in 1992. However, debate at Parliament over a new Land Law that would allow the private ownership of land led to nation-wide controversy.

Sneath (2003) suggests that although the Constitution excludes pastureland from privatization, some international agencies such as ADB required the Government to establish the private ownership of pastureland, citing a study report submitted to ADB: "Currently, there is no private ownership of land. As a consequence, land tenure insecurity causes disincentives to invest in land improvements. … The Land Law … aims to provide positive incentives to herders, farmers and others to maximize production and to protect land from damage or degradation" (p. 443).

ADB implemented the Agriculture Sector Program (ASP) designed to support the transition of Mongolian agriculture from a centrally planned system to a market-based economy in 1994–1998. As a policy reform measure to promote markets in land-use rights, ADB required passage of a Land Law. Parliament passed the law on 11 November 1994 after 3 years of debate. The Land Law allows long-term (up to 60 years and renewable for another 40 years) lease rights for possession of state land for residential and commercial use, and for cropland, but not for pastureland. The ASP complete report criticized that the Land Law contained ambiguities and inconsistencies, and thus the law was amended three times: in 1997, 1999, and 2002. The 1997 amendment allowed for possession of lands under winter and spring campsites, and the 2002 amendment for transference of leases of cropland and use of the leases as collateral, but not ownership of land. Other legislation was also issued in 1997, including the Land Fee Payment Law, the Law on Registration of Immovable Property, and the Geodesy and Cartography Law (ADB 2002: 4–5).

It was only in 2002 that the legislation for ownership of land for residential and commercial purposes was enacted in Mongolia. A new Land Privatization Law (a law on allocation of land to Mongolian citizens for ownership) passed Parliament and became effective on March 1, 2003. Reflecting citizens' lack of interest, the law was amended twice in 2005 and 2008 to prolong the term of privatization, for 2 years until May 1, 2008, and for another 3 years until 2013, respectively.

The ADB report (2002) concludes as follows:

In the early 1990s, the complexity of land ownership issues in Mongolia was probably underestimated. Since Mongolia has no history of private land ownership, and herders have the constitutional right to move their animals freely, opposition to private ownership of land, especially pastureland, has been very strong. Because of the difficulties, the Land Law had been under debate for 3 years and was passed in 1994 only after removal of its most controversial clauses relating to private ownership and pasture-use rights. In retrospect, the ASP could have set a more moderate but realistic target and focused on cropland only (p. 5).

Since the mid-1990s, as negative effects of the reforms that had been carried out in the early 1990s clearly appeared, poverty, equity, and environment became more stressed than before. Swift (1995) argued that "unambiguous and secure tenure is an essential prerequisite for poor households to produce more and that an enactment of a new Land Law would set the framework for future land-allocation" (p. 113). On the other hand, he wrote: "Grazing-land title is not, under Mongolian conditions, needed as security for credit (herds can play this role), and title should not be used as a basis for creation of a market in grazing land, since this would encourage the privatisation and sale of the highest productivity pieces of land, without which the rest of the grazing system cannot be productive or sustainable" (p. 115), and he advocated group-leases of pastures.

During and after serial *zuds* in 1999–2001, the vulnerability of Mongolian pastoralism to natural disasters and the need of modernization or transition from extensive herding to intensive livestock farming were emphasized. In 2001, "Nambaryn Enkhbayar [ex-President and then a prime minister: A.K.] expressed unequivocally his vision of development and modernity for Mongolian pastoralists. … the livestock management would shift to intensive, large farming units, involving only 10 per cent of the population" (Marin 2008: 78). Although his view provoked criticizes from both inside and outside the country (Rossabi 2005), this has given the impression to many Mongolian pastoralists and citizens that the transition to intensive livestock farming by business entities, and thus the private ownership of pastures in the future, is one of the established national policies.

The background of legislation for pastureland possession as just described should have been reflected in the opinions of drafters and advocators of a new pastureland law. As Fernandez-Gimenez et al. (2008) reviewed several articles written by some of them, I highlight here three main points in their views calling for pastureland possession: stress on the need for incentives to invest in pastures, to which responsibility to protect pastures is inversely parallel; orientation toward intensive livestock husbandry run by formal business entities; and preference for emancipation from the responsibility to protect pastureland and regulate its use from the government. The "Introduction of the Proposed Pastureland Law to Parliament"[2] also presents these points as rationale of the legislation.

14.3 Proposed Legislation

The proposed Pastureland Law was prepared not only to create consistency in pastureland tenure with other large pastoral nations (e.g., Australia, United States, Iceland, China, and Argentina) (Hannam 2008: 110), but also with reference to the Grassland Law of China.

Fernandez-Gimenez et al. (2008) reviewed pastureland possession provisions of the July 20, 2007 draft version. Although a number of amendments have been added to new versions, I would like to mention some of them related to this paper on the versions of March 27, 2008 (A version) and the newest one of May 13, 2010 (B version).

The Draft Law provides for a possession right of winter and spring pastures, and areas for managing intensified livestock husbandry purposes to PUGs or herder groups,[3] and to legal entities, respectively (Article 4.1.3 in A version; Article 4.1.3 and Article 9.3 in B version).[4] In the A version, the area for managing intensified livestock husbandry purposes is limited to pastureland near cities (Article 13.3); however, in the newer B version this article has been deleted. In the B version, Article 3.1.9 provides that the size of a PUG will be not less than ten households (five households in the Govi areas). A PUG or herder group will be a legal entity, an "unregistered union" or "partnership" formed in accordance with Article 481.1 of the Civil Code (Article 3.1.9), which "shall participate in the civil legal relationships through appointed representative or one of the members" (Civil Code, Article 481.2).

While Article 20 of the previous version of July 20, 2007 provided that a Pastureland Possession Certificate will be both transferred and put as mortgage to

[2] On the website of Parliament: http://www.parliament.mn/new/law/project/index/page/5.

[3] "Нөхөрөл" ("friendship" or "partnership") in the Mongolian original. Unofficial English version on the Green Gold website translates this word as "community" and "herders' friendship and groups" in A Articles 4.1.3 and 13.1 respectively. The newest B version (May 13, 2010) uses "малчдын бүлэг," which is a translation of a "herder group" in English.

[4] The previous version of July 20, 2007 allowed for possession of pastures only by PUGs.

others, the revised article of A and B versions allows only for transfer. In the newest B version, mortgage, sale, and donation of a certificate are explicitly forbidden (Article 14.1.1); this should help to decrease the possibility of possession by entities other than herder communities.

The amendment to insert a new Article 18.2.4 in the A version that establishes an obligation to make an assessment of impact on the environment before fencing possessed pastures may prevent inducing "the dilemma of fences" in Inner Mongolia (Taylor 2006: 379). Fencing pastures leads to overuse and accelerated degradation of unfenced pastures. In the B version, there are no articles about fencing pastures, but Article 12.3.2 allows for the right of a pastureland possessor to demarcate, put signs on the boundary of, and take measures to keep others' livestock from entering the possessed pasture.

The B version (May 13, 2010) provides for the demarcation of pastureland of *otor*[5] (Article 6.1) and passage (Article 6.2) for common use. Pastureland for *otor* should be more than 10% of the total pastureland (Article 9.5). Article 12.3.3 provides that a part of the pastureland possessed by a PUG will be used by others temporarily and by contract with a majority approval of all the PUG members.

Although these amendments, as well as Article 9.11, which provides for regulation in the case of natural disaster, may help to maintain the flexibility of pastoralism, the proposed law is composed within a frame of the exclusive property rights system so as to protect pastureland by clarifying the responsibility of pastoralists as possessors of pastureland, or, in other words, to ascribe the degradation of pastureland to local herders. The A version of July 20, 2007 provides that "if no rehabilitation work has been done on pastureland in possession, and/or pastureland condition was the same as before possession or deteriorated … pastureland possession contract can be changed into another one prior to expiration" (Article 22.2). In the newest B version, even if these articles are deleted, Article 12.4.2 still stipulates that a possessor should keep the condition of possessed pastureland from getting worse, with Article 14.1.4 prohibiting from herding a number of animals exceeding the carrying capacity of the pastureland.

14.4 Household Surveys

We conducted household surveys[6] to identify factors that affect pastoral household mobility. Three *sums* were selected from different ecological zones (Table 14.1). We selected one *bag* from each *sum* that is not a *sum* center *bag* or one too much

[5] *Otor* is an opportunistic move for a specific purpose (for example, to fatten livestock in autumn for the winter, or to escape from an adverse natural condition, drought, or *zud*), sometimes taking only a specific part of livestock, usually without a *ger* but with a tent, by a specific part of the household.

[6] The survey is based on the survey conducted in the summer of 2007 as part of a CALE, Nagoya University research project (see Fernandez-Gimenez et al. 2008).

Table 14.1 Characteristics of the sites

Study site and time	O'goomor *bag* (Erdene sum) To'v *aimag* Sept. 2008	Tseel *bag* (Bayan-O'njuul *sum*) To'v *aimag* Sept. 2009	Hondot *bag* (Hanhongor sum) O'mnogovi *aimag* Feb. 2009
Ecological zone	Forest steppe	Steppe	Desert steppe
Mean annual precipitation	<300 mm	<200 mm	<200 mm
Land area (1,000 ha)	212.0 (803.0)	164.6 (479.0)	n/a
Human population (at the end of the previous year)	522 (3,607)	965 (2,324)	524 (2,422)
Herding households (at the end of the previous year)	183 (1,105)	259 (618)	136 (721)
Head of livestock (at the end of the previous year)	28,330 (126,085)	69,551 (179,229)	31,210 (149,283)

influenced by other sectors such as mining. The households to survey were selected with systematic sampling from tax lists in order of the total number of owned livestock (Table 14.2). Because those lists include households permanently resident in the *sum* center and those artificially split off in an attempt to get more aid supply and avoid tax, we omitted such households.

The mobility of a herding household is significantly correlated only with total livestock number (see Table 14.3), however, in contrast to our assumption, with no other expected factors such as attitudes toward pastureland tenure (i.e., a herder who has the opinion that pastures for all seasons should be owned by individual households might actually move more than a herder who thinks that pastures should be used publicly), age of household head, or even labor power available within the household, in all three sites. There is also a tendency to increase the degree of correlation if using the total number of small livestock and/or that herded by a *hot ail* (pastoral residential group).[7] This finding agrees with herders' allegations that they move so as to satisfy the needs of livestock or forage. Even in less-arid areas, the more livestock a herder wants to keep, the more mobility is needed.

It is to be noted here that in more arid Hanhongor *sum* herders appear to move less and less according to the number of livestock than in the two other sites, mainly because of the smaller number of livestock owned or herded by a household as well as reflecting the "Gobi Type" movement system (Humphrey & Sneath 1999). They move around a well, usually allocated to the household, spending half a year on their winter campsite and the rest of year tracking fodder availability: they call all the moves out of their winter campsite "*otor*." The fact that herders in Hanhongor use

[7] The number of livestock herded by *hot ail* is as at the time of survey, but does not reflect the previous condition: whether the household had been part of a *hot ail*, and the number of livestock the *hot ail* or single household had herded.

Table 14.2 Average household of the sites

Study site	O'goomor *bag*, Erdene *sum*, To'v *aimag* $N=168$, $n=20$	Tseel *bag*, Bayan-O'njuul *sum*, To'v *aimag* $N=203$, $n=52$	Hondot *bag*, Hanhongor sum, O'mnogovi *aimag* $N=122$, $n=52$
Number of *hot ail* member households	2.05	2.08	1.66
Head of livestock owned (SFU)	447.38	595.53	402.49
Sheep (head)	123.5	178.3	74.5
Goats (head)	67.7	148.2	191.7
Cattle (head)	23.3	15.5	1.2
Horses (head)	17.6	26.6	11.2
Camels (head)	0.0	1.0	13.9
Head of livestock herded by hot ail (SFU)	887.15	1,086.21	556.15
Avg. moves per year	3.32	3.55	3.05
Avg. distance per year (km)	48.54 (of 32 months)	58.05 (of 47 months)	50.32 (of 44 months)
Frequent use of customary camps in the last 10 years (%)	30.43	20.31	12.85
Livestock income 2008 (2007 Erdene) (Tg)	2,380,976	3,707,662	2,761,296
Sheep	581,210	1,137,399	326,053
Goats	886,071	1,736,590	1,894,932
Cattle	724,267	456,846	6,604
Horses	189,429	322,212	65,443
Camels	0	54,615	468,264
Livestock expenditure 2008 (2007 Erdene) (Tg)	834,714.29	390,296.15	225,256.60

SFU: Sheep Forage Unit. Sheep are estimated as 1 SFU, goats as 0.9, cattle as 6, horses as 7, camels as 5. *Tg*: Tugrug

their own customary camp less frequently than in the other two sites (see Table 14.2) reflects the opportunistic characteristic of their movements whereas the movements in the other two sites are rather transhumant.

Surveys with formal or informal interviews (for example Table 14.4) were also conducted in O'ndorshireet *sum*, To'v *aimag*[8] in 2008 and 2009, and in Duut *sum*, Hovd *aimag* in 2008 and 2010, where the SDC Green Gold project has been implemented since 2008.

[8] Jamsranjav (2009) is a report on the GG project implemented in the *sum* in 2008.

Table 14.3 Correlation between mobility and livestock income, and number of livestock

	Erdene N=168, n=20			Bayan-O'njuul N=203, n=52			Hanhongor N=122, n=52		
r=	Total moves	Total distance	Total livestock income	Total moves	Total distance	Total livestock income	Total moves	Total distance	Total livestock income
Total livestock owned (SFU)	0.58	0.54	0.68	0.46	0.44	0.70	0.31	0.31	0.69
Total livestock herded by *hot ail* (SFU)	0.64	0.70	0.43[a]	0.48	0.48	0.44	0.34	0.25	0.41

[a]Not significant

Table 14.4 Perception of pastureland degradation by pastoralists

	Erdene	Bayan-Unjuul	Hanhongor
Percent of respondents who perceive degradation of pasture	90.5%, $n=19$	90.6%, $n=48$	69.8%, $n=37$
Percent of the above respondents who allege climate change as a main cause of degradation	78.9%, $n=15$	70.8%, $n=34$	91.9%, $n=34$

14.5 Discussion

14.5.1 Mobility: Paradigm of Equilibrium Versus Nonequilibrium Model

In recent literature on the Mongolian rangeland ecosystem, the paradigm of equilibrial and non-equilibrial models has become one of the topics often discussed. This paradigm, however, may have implications for attitudes toward the legislation for pastureland tenure as well as grazing management (Mearns 1994), although it is considered as a continuum rather than a clear dichotomy. In an equilibrial ecosystem, grazing pressure explains the condition of the vegetation, and conventional range management techniques, such as maintaining appropriate average stocking rates, are deemed to be most suitable for sustainable grassland management, whereas in a non-equilibrial ecosystem abiotic factors such as rainfall explain the vegetation conditions with livestock populations and vegetation dynamics being only loosely related, and opportunistic means of "tracking" fodder availability, usually through mobility, are typical pastoral management adaptations to the spatial and temporal variability predicted under this ecosystem (Bruce and Mearns 2002: 39). Thus, if most pastureland in Mongolia would be deemed to conform to the equilibrium model, the legislation of more secure and formal rights to pastures with control of stocking rates might well be adopted (Table 14.5).

Although nonequilibrium dynamics are expected to occur where the coefficient of variation (CV) of annual precipitation is greater than 33% and mean annual precipitation is less than 250 mm in Mongolia (Ellis & Chuluun 1993; Ellis 1995), researchers differently estimate the area under nonequilibrium dynamics in Mongolia: Swift (1995) estimates this to be nearly half the country; a 2003 World Bank report concludes as "Mongolia's rangelands would be expected to conform to the characteristics of 'non-equilibrial' as opposed to 'equilibrial' grazing systems" (p. 3); Wesche and Retzer (2005), at least "the entire Gobi part of Mongolia, i.e., some 40% of the country" (p. 134); and Sneath (2003), "a third or more of Mongolia" (p. 444).

Fernandez-Gimenez and Allen-Diaz (1999), although they estimated the steppe to be under the equilibrium grazing system, suggest that "the steppe and mountain-steppe are also strongly influenced by inter-annual variation, a characteristic of non-equilibrium

Table 14.5 Implications of paradigm of equilibrium versus nonequilibrium ecosystem for grazing management and legislation policy

Grazing system	Equilibrium	Nonequilibrium
Annual precipitation	CV<33% Mean annual precipitation >250 mm	CV>33% Mean annual precipitation <250 mm
Main factor to vegetation condition	Grazing pressure (human-induced degradation)	Rainfall (climate-induced degradation)
Herd management	Carrying capacity (control of stocking rates)—no or less mobility	Opportunistic means of "tracking" fodder availability—more mobility
Land tenure	Secure tenure Formal property (ownership) rights	Flexible access to large areas Communal usufruct—customary institutions
Spatial and social boundary	Clearly delineated	Ambiguous, multilayered

rangelands" and that "ecosystems usually possess characteristics of both non-equilibrium and equilibrium systems" (p. 882). Wesche and Retzer (2005) also argued that "Evidence for non-equilibrium conditions depends on the variables analyzed" (p. 143). This disparity shows that it is meaningless to demarcate areas corresponding to equilibrium or nonequilibrium grazing systems with two different herd management types.

Yet, arguments over the paradigm do not seem to be reflected in the design of CBNRM projects and the proposed pastureland law, which have adopted the control of stocking rate to balance with the carrying capacity of a pasture as a means to protect the condition of pastureland, whether under an equilibrium or nonequilibrium system. The limitation of general stocking rate is known to be hard to implement, and in Mongolia too I have known no project sites where it is actually in operation.[9] Instead, it translated into "mobility" or deferred rotation grazing by seasons, in particular, deferred utilization of winter and spring pastures in summer and autumn, not only in the Govi areas (Schmidt 2006: 20) but also in other areas. Carrying capacity as presented by governments or projects estimates grassland productivity too low to match the perception of herders because of flaws in its calculation: for example, compensatory regrowth of grazed and browsed plants is ignored (Behnke & Scoones 1993: 19). In most areas in Mongolia, and all the more in less-arid areas corresponding to equilibrium ecosystems, except in some regions such as Selenge aimag, mobility is the most effective way to spread grazing pressure and preserve the condition of pasturelands because the impact of herding on grassland becomes greater as a herd stay longer in one place.

[9] The Millennium Challenge Corporation (MCC) Peri-Urban Rangeland Project aims to form herder groups to run a semi-intensified form of livestock farming. Every herder group will consist of 2–6 households with up to 900 sheep units of livestock. No other projects define the limit of the size of livestock.

If we focus on herd management, we should also take into account risk management to escape from sudden severe natural conditions: droughts and *zud*, the latter of which may be caused by other environment factors than precipitation and grazing pressure, such as heavy snowfall and/or severe cold temperature. In reality, the herd management in Mongolia is based on the combination of transhumance to move regularly between seasonally suitable, usually customary camps, and opportunistic movements to escape from weather changes and to track fodder availability. "Fluctuating size and composition of herds, climatic variations and availability of drinking water all combine to influence a family's decision on when and where to move stock" (Mau & Chantsalkham 2006: 12). Mobility and flexibility, corresponding to nonequilibrium ecosystems, are still a main strategy in most areas, and it would be more intensified so as to manage uncertainty: not only interannual variation of precipitation and sudden adverse weather conditions but also the socioeconomic condition of the household and the society, even though the frequency and total distance of moves mainly depends on the size of herd: the number of the livestock that the household are herding, as our survey shows.

14.5.2 "Community:" Size and Boundaries

Leach et al. (1999) criticize that CBNRM approaches rest on not only an assumption of a distinct, relatively stable, local environment: in other words an equilibrium ecosystem, but also an equilibrium model of a distinct community (p. 228), which is rooted in much early social theory: "Early sociology and anthropology conceived of society as a bounded object or closed social system" (p. 229). The proposed pastureland law organizes a distinct "community" as a legal entity, and CBNRM projects implemented in Mongolia wish it to also work as a business enterprise, the interests and activities of which are, however, not necessarily compatible with those of a PUG.[10]

It is open to question to adopt some design principles in common-pool resource (CPR) literature into the pastoral sector in Mongolia with no reference to the nature of the resource system: this is a "blueprint" approach. Small size and clearly defined boundaries of group membership are commonly supposed to be critical conditions to sustainable management of CPRs (Agrawal 2002), and the first one of Ostrom's eight design principles, "clearly defined boundaries" (Ostrom 1990:90, 2009), is often referred to in donor documents (for example, Jamsranjav 2009). According to these principles, PUGs or herder groups have been organized, the size of which is usually two to three times smaller than a *bag*. Certainly, the smaller a natural resource user group and its target resource system are, the lower the transaction cost is. Moreover, when their boundaries are well defined, it will be easy to define to

[10] NBNRM projects often linked environmental issues to livelihood improvement or poverty alleviation of herder households by optional incomes such as selling felt craft. There is an assumption that poverty is likely to cause overexploitation of natural resources.

whom the fruits of investment into or the responsibility of degradation of pastures belong. The proposed pastureland law is based on the same assumptions, with articles requiring herders to include the number of group members (albeit in the newest draft version only the name of group) and a map showing the area, location, and boundaries in a contract with the *sum* government, and to control the stocking rate.

However, in arid and semiarid rangeland ecology, including Mongolian pastureland, the resources "may be multiple, overlapping, and vary across space and time," and "user groups may vary in size or composition with respect to different resources within an area, and with respect to a single resource over time" (Fernandez-Gimenez 2002: 51), which are characteristics of nonequilibrium ecosystems and communities. Thus, to whom the fruits or responsibility of investment into a pasture belong is not clear, as already noted. Our household survey also shows that a majority of herders perceive climate change as a main cause of pastureland degradation whether they live in arid or less-arid areas (see Table 14.4).

Some herders in Bayan-O'njuur answered "it is a *bag*" when asked what the expanse of "your community"[11] is. They said, "It might possibly be a *sum* in some cases, but it's too big. In a *bag* everyone knows every household, if not everyone else. In our *sum* there are many I don't know." When asked "what right to be allocated to wells in a future pastureland law," a number of herders in every site answered "they should be 'managed' ('*ezemshih*,' which is usually translated as 'to possess') by one person and 'used commonly' ('*niiteer ashiglah*' usually translated as 'to use publicly' or open access)". They explained that "*niiteer ashiglah*" does not mean "open access" as usually understood and translated, but "common use," and that outsiders who want to use resources have to negotiate with local herders or the *sum* government anyway. This description shows that we may take a *bag* as a community in most cases, although it was in the collective era that *bags* were established with seasonal nomadic moves nominally restricted within *bag* territories (Mearns 1993, 2004).

Fragmentation of a resource system into parts, in particular a pasture into smaller parts than a *bag*, is likely to impede the effective performance of resources management. In summer 2009 in O'ndorshireet *sum*, one of the SDC Green Gold project sites, the project coordinator told me that when they planned to conduct *otor* in less-used pastures allocated to a PUG in the *sum*, members of the PUG severely objected, claiming the land was their own. An old herder, one of the PUG leaders who came to a project meeting, complained that he was receiving persistent demands from members of another PUG to abandon his spring camp situated in the territory of the PUG even though he had used it for more than 20 years. In autumn 2010 a female PUG member in Bayan-O'njuul sum told us, "We are going to collect tolls from passengers over our possessed land," without knowing that it is not allowed by law. These comments show that organizing PUGs have brought more exclusionary attitudes toward the territory.

[11] We used "*tanai nutgiinhan*" ("members of your homeland" as literally translated) as "your community."

14.5.3 Inclusive Versus Exclusive Use of Pastureland and Roles of Local Governments

CBNRM projects often confine their concerns within their project sites. In To'v *aimag* we met herders who came from Dundgovi *aimag* with their livestock escaping from droughts. They said, "If a new pastureland law, which provides for exclusionary pastureland possession, does not allow outsiders to use possessed pastures any more, it will cause civil war." They meant by "civil war" not only that it would increase disputes and conflicts between herders, but also that they would revolt to protest against the government. This is a problem of survival for them because, if they cannot move to the north, they will have no other means but to leave their livestock starving to die. A new legislation may end or limit their negotiation power or a potential right to use pastures in To'v *aimag* as a second user (see Niamir-Fuller 2000).

Historically, in the collective era the government assisted herders to move, and even in the Qing dynasty when movements beyond banners were prohibited by law, local governments often urged herders to accept those of other banners escaping from droughts or *zud*. The governments intended to promote pastoral productivity as a whole or to mitigate a loss of livestock in case of disaster, by enhancing mobility and flexibility (Humphrey & Sneath 1999; Fernandez-Gimenez 2002). Although there may have been some compulsion by the government, herders also recognized the benefit for themselves: providing social safety nets would spread risk based on reciprocity as well as improve their livelihood.

However, since the dissolution of herding collectives in the early 1990s, the administration has withdrawn or "abdicated" (Mearns 2004) authority and responsibility for the pastoral sector because of the chaos of transition, lack of budget, and later the introduction of "the notion that the economic sphere could and should be emancipated from the political structure so as to allow the economy to assume its latent 'natural' form" (Sneath 2002: 192). In summer 2002 the then governor of Duut *sum* in Hovd *aimag* told me, "I would not regulate pastureland use or disputes because the administration should not intervene in such private economic activities. If I do so, I must be accused later." He never did use nor know his authority endowed by the revised 2002 Land Law; this was not an isolated case but a common case (Fernandez-Gimenez & Batbuyan 2004).

Local herders and officials think that the proposed pastureland law will legally devolve the authority, responsibility and duties of the government to PUGs. In autumn 2010 the governor of Bayan-O'njuul *sum* told me, "If pastures are possessed by PUGs, present agreements between *sums* on pastureland use will be unnecessary because then movements beyond PUG territories will disappear." Only a few *sums* in To'v *aimag* have pastures enough for every season in their own territories. Some *sums* are short of winter pastures while others are short of spring pastures. Those agreements allow herders to use the seasonal pastures in other sums that their own *sum* lacks. If pastures are possessed by PUGs as their property, such agreements should be made between them in case of need. PUGs will also bear all costs of making, monitoring, and enforcing rules regarding rangeland management, which at present local governments ought to bear. Nevertheless, it is very probable that the cost that the government should bear will rather rise as disputes and conflicts to be mediated decrease, and cases of punishment will be imposed by the government.

14.6 Conclusion

The privatization of pastureland in Mongolia was conceived as a part of the transition to a market economy in the early 1990s. The circumstances of the period appear to influence rationales for the legislation of pastureland law, such as stress on incentives to invest in pastures, orientation toward intensive livestock husbandry by formal business entities, preference for emancipation of the responsibility to protect pastureland and regulate the use from the government, and so on. This influence may be called "neoliberal bias."

Later, as the focus of pastureland issues shifted to environmental concerns, CBNRM approaches were introduced as common property-based resource management with new legislation on pastureland that provides for pastureland possession. However, the logic calling for PUG possession of pastureland is still based on a conventional model of human–land relationship: the result of human activities or investment to land is predictable, without reference to new research on arid and semiarid ecosystems. CBNRM projects also have organized small PUGs based on a "blueprint" approach with a conventional image of "community" or a wish to establish business entities. These entities have reduced mobility and flexibility and increased exclusionary attitudes toward outsiders, with reciprocity decreasing.

Even in relatively less-arid areas, pastoral mobility is still the most effective strategy to increase the number of livestock with spreading grazing pressure so as to improve herders' livelihoods as well as providing social safety nets in case of natural disasters. If the climate become more arid in Mongolia in the future, flexibility, mobility, and opportunistic management will be even more important to sustainable grassland management and herder livelihoods (Fernandez-Gimenez et al. 2008).

It is to be noted that local governments have played a significant role to enhance mobility and flexibility in herd management. Current problems regarding pastureland use, such as "open access" and abuse of winter and spring pastures in summer and autumn, are considerably attributable to the government's abdication of the authority and responsibility to the pastoral sector since the early 1990s. The present 2002 Land Law already endows local governments with the authority to allow a herder group to use pasture exclusively (Article 52.2). Therefore, the central government should provide *sum* governments with necessary financial and human resources to enforce the present Land Law, instead of introducing a new pastureland law. Moreover, in contrast to expectations, it is possible that pastureland possession by small PUGs may increase the cost borne by local governments with increasing disputes and conflicts to be mediated.

The problem of the concentration of human and livestock population into peri-urban areas should be resolved by other approaches than property rights approaches, such as revenue transfer to rural areas. Pastureland possession based on the model of a distinct community with "clear defined boundaries" as a legal entity, which most NBNRM approaches and the proposed Pastureland Law have adopted, does not appear to conform to the characteristics of the resource system of pastoralism in Mongolia. It is needed to intensify present institutions: the role of local governments and reciprocal ethics in particular.

References

ADB. 2002. *Program Performance Audit Report on the Agriculture Sector Program (Loan 1409-MON[SF]) in Mongolia*. PPA:MON 27536.

Agrawal, A. 2002. Common Resources and Institutional Sustainability. In: Ostrom, E., Dietz, T., Dolsak, N., Stern, Paul C., Stonich, S., and Weber, Elke U. (eds.) 2002, *The Drama of Commons*, National Academy Press, pp. 41–85.

Bedunah, Donald J. & Schmidt, Sabine M. 2004. Pastoralism and Protected Area Management in Mongolia's Gobi Gurvansaikhan National Park. *Development and Change*, 35(1): 167–191.

Behnke, R.H., & Scoones, I. 1993. Rethinking Range Ecology: Implications for Rangeland Management in Africa. In: Behnke, R.H., Scoones, I., and Kerven, C.(eds.) *Range Ecology at Disequilibrium: New Models of Natural Variability and Pastoral Adaption in African Savanas*. London: ODI, pp. 1–31.

Bruce, J. & Mearns, R. 2002. *Natural Resources Management and Land Policy in Developing Countries: Lessons Learned and New Challenges for the World Bank. Drylands Programme*. Issues Paper No. 115. London: International Institute for Environment and Development.

Ellis, J. E. 1995. Climate Variability and Complex Ecosystem Dynamics: Implications for Pastoral Development. In: Scoones, I. (ed.) *Living with Uncertainty, New Directions in Pastoral Development in Africa*. Intermediate Technology Publications, London, UK, pp. 37–46

Ellis, J.E. & Chuluun, T. 1993. Cross-Country Survey of Climate, Ecology and Land Use among Mongolian Pastoralists. *Conference on Grassland Ecosystems of the Mongolian Steppe, 4–7 November 1993*. Wingspread Center, Racine, WI.

Fernandez-Gimenez, M. E. 2002. Spatial and Social Boundaries and the Paradox of Pastoral Land Tenure: A Case Study from Postsocialist Mongolia. *Human Ecology*, 30: 49–78.

Fernandez-Gimenez, M. & Allen-Diaz, B. 1999. Testing a Non-Equilibrium Model of Rangeland Vegetation Dynamics in Mongolia. *Journal of Applied Ecology*, 36: 871–885.

Fernandez-Gimenez, M. E. & Batbuyan, B. 2004. Law and Disorder: Local Implementation of Mongolia's Land Law. *Development and Change*, 35(1): 141–166.

Fernandez-Gimenez, M. E., Kamimura, A., and Batbuyan, B. 2008. *Implementing Mongolia's Land Law: Progress and Issues: Final Report, a Research Project of the Central for Asian Legal Exchange (CALE)*, Nagoya University. Nagoya. (Web version).

Fernandez-Gimenez, M. E. & Swift, David M. 2003. Strategies for sustainable grazing management in the developing world. In: Allsopp, N, Palmer, A.R., Milton, S.J., Kirkman, K.P., Kerley, G.I.H., and Hurt, C.R. (eds.) *Rangelands in the New Millennium, Proceedings of the VII International Rangelands Congress, 26th July-1st August 2003*; Durban, South Africa: Document Transformation Technologies, pp. 821–831.

Fratkin, E. 1997. Pastoralism: Governance and Development Issues. *Annual Review of Anthropology*, 26: 235–261.

Hannam, I. 2008. *Report to United Nations Development Program Mongolia on Review of Draft Pastureland Law of Mongolia*. UNDP Sustainable Land Management for Combating Desertification in Mongolia, Ulaanbaatar.

Humphrey, C. & Sneath, D. 1999. *The End of Nomadism?: Society, State and the Environment in Inner Asia*. Duke University Press: Durham.

Jamsranjav, Chantsalkham. 2009. *Sustainable Rangeland Management in Mongolia: the Role of Herder Community Institutions*. Land Restoration Training Programme, Final project 2009. Keldnaholt, 112 Reykjavík, Iceland.

Leach M., Mearns, R., and Scoones, I. 1999. Environmental Entitlements: Dynamics and Institutions in Community-Based Natural Resource Management. *World Development*, 27(2): 225–247.

Marin, A. 2008. Between Cash Cows and Golden Calves: Adaptations of Mongolian Pastoralism in the 'Age of the Market.' *Nomadic Peoples*, 12(2): 75–101.

Mau, G. & Chantsalkham, J. 2006. *Herder Group Evaluation: A Study of Herder Groups, Their Present Status, and Future Potential. Mongolia*: UNDP Sustainable Grasslands Program, Ulaanbaatar.

Mearns, R. 1993. Territoriality and Land Tenure among Mongolian Pastoralists: Variation, Continuity and Change. *Nomadic Peoples*, 33: 73–103.

Mearns, R. 1994. Community-Based Management of Renewable Natural Resources: The Case of Pastoral Ecosystems Under Post-Socialist Transition in Mongolia. http://dlc.dlib.indiana.edu/dlc/handle/10535/4326

Mearns, R. 2004 Decentralisation, Rural Livelihoods and Pasture-Land Management in Post-Socialist Mongolia. *European Journal of Development Research*, 16(1): 133–152.

Niamir-Fuller, Maryam. 2000. Managing Mobility in African Rangelands, In: McCarthy, N., Swallow, B., Kirk, M. and Hazell, P. (eds.) *Property rights, risk, and livestock development in Africa*, International Food Policy Research Institute, pp.103–131.

Ostrom, E. 1990. *Governing the Commons: the Evolution of Institutions for Collective Action*. Cambridge University Press, Cambridge.

Ostrom, E. 2009. Social-Ecological Systems: a General Framework for Analyzing Sustainability of Social Ecological Systems. *Science*, 325:419.

Rossabi, M. 2005. *Modern Mongolia: From Khans to Commissars to Capitalists*. University of California Press, Berkeley.

Schmidt, S. 2006. Pastoral Community Organization: Livelihoods and Biodiversity Conservation in Mongolia's Southern Gobi Region. *USDA Forest Service Proceedings RMRS-P*, 39: 18–29.

Sneath, D. 2002. Mongolia in the 'Age of the Market: Pastoral Land-use and the Development Discourse. In: Mondel, R. & Humphrey, C. (eds.), *Markets & Moralities: Ethnographies of Postsocialism*. Berg: Oxford, N.Y., pp. 191–210.

Sneath, D. 2003. Land Use, the Environment and Development in Post-socialist Mongolia. *Oxford Development Studies*, 31(4): 441– 459.

Swift, J. 1995, Rural Development: The Livestock Sector. In: Grifin, K. (ed.) 1995, *Poverty and the Transition to a Market Economy in Mongolia*, St. Martin's Press, pp. 104–134.

Taylor, J.L. 2006. Negotiating the Grassland: The Policy of Pasture Enclosures and Contested Resource Use in Inner Mongolia. *Human Organization*, 65(4): 374–386.

Upton, C. 2005. Institutions in a Pastoral Society: Processes of Formation and Transformation in Post-Socialist Mongolia. *Comparative Studies of South Asia, Africa and the Middle East*, 25(3): 584–599.

Upton, C. 2008. Social Capital, Collective Action and Group Formation: Developmental Trajectories in Post-Socialist Mongolia. *Human Ecology*, 36: 175–188.

Wesche, K. & Retzer, V. 2005. Is Degradation a Major Problem in Semi-Desert Environments of the Gobi Region in Southern Mongolia? *Erforschung der Biologischen Ressourcen der Mongolei*, 9: 133–146.

World Bank. 2008. *MONGOLIA: Livestock Sector Study, Vol. 1 – Synthesis Report*. World Bank.

Chapter 15
A Mathematical Model of Population Shift Between Urban and Rural Areas

Norio Yamamura, Yutaka Kobayashi, Reiichiro Ishii, and Yasuhisa Saito

Keywords Economic and cultural values • Mathematical model • Migration costs • Population shift • Urban and rural areas

15.1 Introduction

Human migration problems are vivid and important for many countries throughout the world. The transition of populations between urban and rural areas has occurred every 10 years since 1960 in many large cities in the world (Ishikawa 2001). In the Tokyo, Osaka, and Nagoya areas in Japan, excess immigration (immigration–emigration) had minimums around 1955, 1975, and 1995, while it showed maximums around 1965 and 1985.

In Mongolia, a part of the people who lost governmental jobs in urban areas after the end of socialism moved to rural areas to be herders. In recent times in Mongolia, rapid introduction of foreign capital has increased jobs in urban areas, resulting immigration from rural areas. Nomadic people rearing livestock in grasslands also tend to shift to areas near Ulan Bator city. Thus, the population in Mongolia is

N. Yamamura (✉)
Research Institute for Humanity and Nature, Kyoto, Japan
e-mail: yamamura@chikyu.ac.jp

Y. Kobayashi
Center for Ecological Researches, Kyoto University, Kyoto, Japan

R. Ishii
Research Institute for Global Change, Japan Agency for Marine-Earth Science and Technology, Yokosuka, Japan

Y. Saitou
Department of Mathematics, Ajou University, Suwan, Republic of Korea

N. Yamamura et al. (eds.), *The Mongolian Ecosystem Network: Environmental Issues Under Climate and Social Changes*, Ecological Research Monographs,
DOI 10.1007/978-4-431-54052-6_15,

increasing rapidly only in the Ulan Bator area but that in other regions is almost constant (Mendsaikhan et al. 2009).

In population ecology, migration of animals has been extensively studied with mathematical modeling (Shigesada and Kawasaki 1997). However, human movement has special features, different from animal movement: (a) humans can obtain detailed information about the values of different areas, (b) rapid movement between distant areas is possible, and (c) movement requires severe costs such as establishment of life in the new area. To incorporate these features, various migration theories have been developed since the eighteenth century (see Akkoyunlu and Vickerman 2001 for history). The basic theoretical scheme is to assume the movement from a place with low utility to a place with high utility, where utility is determined by factors such as labor wage and cost of living (Harris and Todaro 1970). Following this scheme, but also taking the dynamics of the ecological state in the rural area into account, I constructed a model of population shift between urban and rural areas to analyze the characteristics of population shift in recent Mongolia. The rural area here is grassland where people raise livestock that graze on grasses. Later we argue that we can apply the model to more general cases where the rural area depends on any natural resource that is deteriorated by overuse.

15.2 Basic Model

We consider two areas, urban and rural areas, in the model. We denote total population and rural population by N and n, respectively, so that urban population is represented by $N–n$. The grassland in the rural area is degraded by overuse by livestock to change into unusable land, which is nearly bare or is overgrown by plants unpalatable to livestock. The proportion of usable grassland is denoted by x and the rate of change into unusable land is given by kxn, which is proportional to the number of rural population n. This assumption is reasonable if the rate of deterioration is proportional to the total livestock number and the average number of livestock per individual is constant. We assume that the unusable land $1–x$ recovers at a rate ax proportional to the area of usable land, taking into account the effect of seed dispersal of grasses from usable to unusable land. Thus, the change in the proportion of usable land is given by

$$\frac{\mathrm{d}x}{\mathrm{d}t} = a(1-x)x - kxn \tag{15.1}$$

The value of living in the rural area is represented by income obtained by selling livestock products such as living animals, meats, wool, fur, and milk, which is assumed to be proportional to the area of usable land and given by bx. On the other hand, the value of living in the urban area is represented by the average income from urban jobs, which is assumed to be a constant v. We assume that population shift occurs from the area with the lower value to the area with the higher value and that the rate of the shift is proportional to the difference of values between the two areas (Harris and Todaro 1970). For the first step, we also assume that costs of

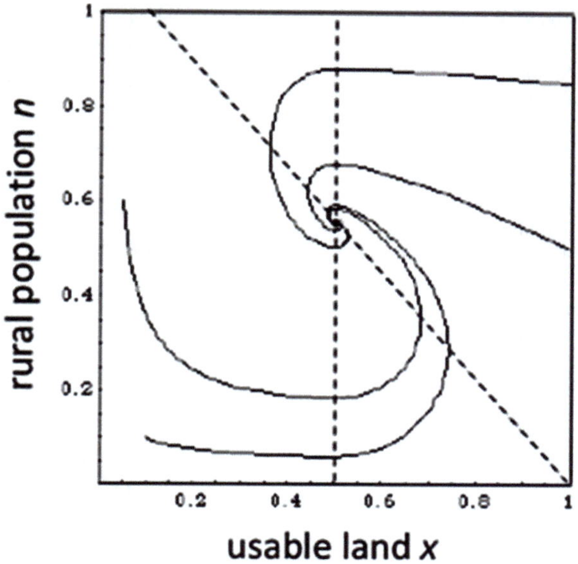

Fig. 15.1 In dynamic systems (15.1) and (15.2), the trajectories on (x, n) always approach to an equilibrium, determined by parameter values. Examples of numerical simulation are shown where $N=1$, $s=1$, $a=1$, $k=0.9$, $b=0.9$, and $v=0.45$

migration are absent or negligibly small. Thus, the change in rural population is represented as

$$\frac{dn}{dt} = -s(v - bx)n \qquad \text{for } v > bx$$

$$\frac{dn}{dt} = s(bx - v)(N - n) \quad \text{for } bx > v \tag{15.2}$$

The dynamic system of (15.1) and (15.2) simultaneously gives change in ecological conditions in the rural area and that in population distribution between the urban and rural areas. We can regard this system as an example of socioecological systems, whose analyses are most important in regional and global environmental problems (Ostrom 2009). The changes in the two variables (x, n) are represented by trajectories in the corresponding phase space (Fig. 15.1). Although the system obeys different equations in the two regions where $v<bx$ and $v>bx$, the trajectories are continuous and smooth even at the boundaries of the two regions.

The equilibrium state (x^*, n^*) is obtained by substituting 0 for the right-hand sides of (15.1) and (15.2), and it is $(1-kN/a, N)$ when $v<b(1-kN/a)$, $(v/b, a/k(1-v/b))$ when $b(1-kN/a)<v<b$, and $(1, 0)$ when $v>b$. In any case, trajectories $(x(t), n(t))$ in phase space (x, n) approach the equilibrium as shown in Fig. 15.1; that is, the equilibrium is always globally stable. We can prove this by using a Liapounov function (see Appendix).

The dependence of the equilibrium on the urban value v is shown in Fig. 15.2a. For low values of urban life people live only in the rural area, whereas for high values of urban life they live only in the urban area. For intermediate values they live in both areas, where rural population is lower for higher urban values.

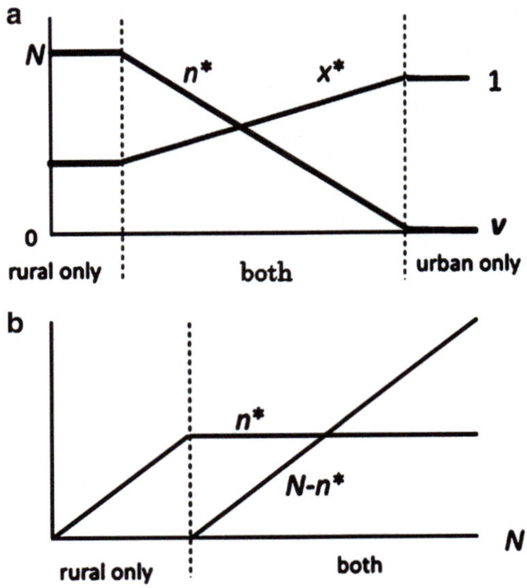

Fig. 15.2 (**a**) Parameter dependency of the equilibrium on urban value v. When $v<b(1-kN/a)$, people live only in the rural area, and when $v>b$, they live only in the urban area. In the intermediate value of v, they live in both areas. (**b**) Parameter dependency on total population N when $v<b$. When $N<(1-v/b)a/k$, people live only in the rural area. Otherwise, they live in both areas where rural population is a constant

The proportion of usable land is lower when rural population is higher. Imagine that the urban value increases gradually from 0. As the critical urban value for the urban population to occur is $b(1-kN/a)$ (as shown in Fig. 15.2a), the occurrence is earlier when the rural area is of low quality compared to the urban area: low income (small b), a high rate of degradation (large k), and low recovery rate (small a). High total population also contributes to early occurrence of the urban area. As the critical urban value for rural population to disappear is b, the disappearance is earlier for lower income in rural life.

The dependence of the equilibrium on total population N is shown in Fig. 15.2b: people live only in the rural area for small total population but live both in the rural and urban areas for large total population. Imagine that total population increases from 0 on a slower time scale than population–grassland dynamics, such that state variables (x, n) keep track of the gradually changing equilibrium. As the critical urban value for the urban population to occur is $a/k(1-v/b)$ (Fig. 15.2b), the occurrence is earlier when the rural area is of low quality: low income (small b), a high rate of degradation (large k), and a low recovery rate (small a). After the occurrence of the urban population, increase in total population is all absorbed in increase in urban population while the rural population is kept constant $a/k(1-v/b)$.

15.3 Model with Migration Costs

As stated in the Introduction, one of the most remarkable characteristics of human migration is to impose severe costs on migrants. Introducing the costs of migration, denoted by c, (15.2) is modified as

$$\frac{dn}{dt} = -s(v - bx - c)n \qquad \text{for } v > bx + c$$
$$\frac{dn}{dt} = s(bx - v - c)(N - n) \quad \text{for } bx > v + c \tag{15.3}$$
$$\frac{dn}{dt} = 0 \qquad \text{for } bx - c < v < bx + c$$

The first equation means that migration from the rural area to the urban area occurs only when the urban value exceeds the rural value by the shift costs. The second equation is interpreted analogously. The third equation means that nobody moves when the difference of the values is less than the shift costs. The dynamic system given by (15.1) and (15.3) is represented in phase space (x, n), and the equilibrium states constitute a linear line segment $n=(1-x)a/k$ for $(v-c)/b<x<(v+c)/b$. That is, the equilibrium is not unique and the trajectories approach different equilibrium states depending on the initial values.

When the urban value v gradually increases starting from 0, migration may occur only from the rural area to the urban area, and the net urban value for the migrants is $v-c$. Thus, the dependence of the equilibrium on v in this case is given by Fig. 15.2a, given that the entire graph is shifted by c to the positive direction. When total population N increases gradually from 0, migration occurs again from the rural to the urban area. As the net urban value is $v-c$, the dependence on N is given by substituting $v-c$ for v in Fig. 15.2b. Thus, the urban population occurs at a higher value of N and hence the maximum size of the rural population n^* is higher than in the case without migration costs.

As stated in the Introduction, population shift from rural to urban areas and that in the opposite direction are repeated roughly in a 20-year cycle in many large cities in the world, and this may occur with periodic business changes in the economy. To see clearly the effect of costs on the dynamics, we assume that the urban value fluctuates temporarily as

$$v(t) = v_0 + v_1 \sin(2\pi \frac{t}{T}) \tag{15.4}$$

where v_0, v_1, and T are the average, the amplitude, and the period of $v(t)$, respectively. In the dynamic system given by (15.1), (15.3), and (15.4), rural population n and the proportion of usable land x may fluctuate as shown in numerical examples in Fig. 15.3a, b: when the costs are high, the amplitudes of n and x are small and the timing of migration is delayed. We plotted $n(t)$ for different values of c (Fig. 15.3c). The amplitude of population fluctuation decreases monotonously with costs c, and for extremely high costs, there is no migration at all. In conclusion, costs contribute to reduce the scale of population shift and to retard the timing of the shift.

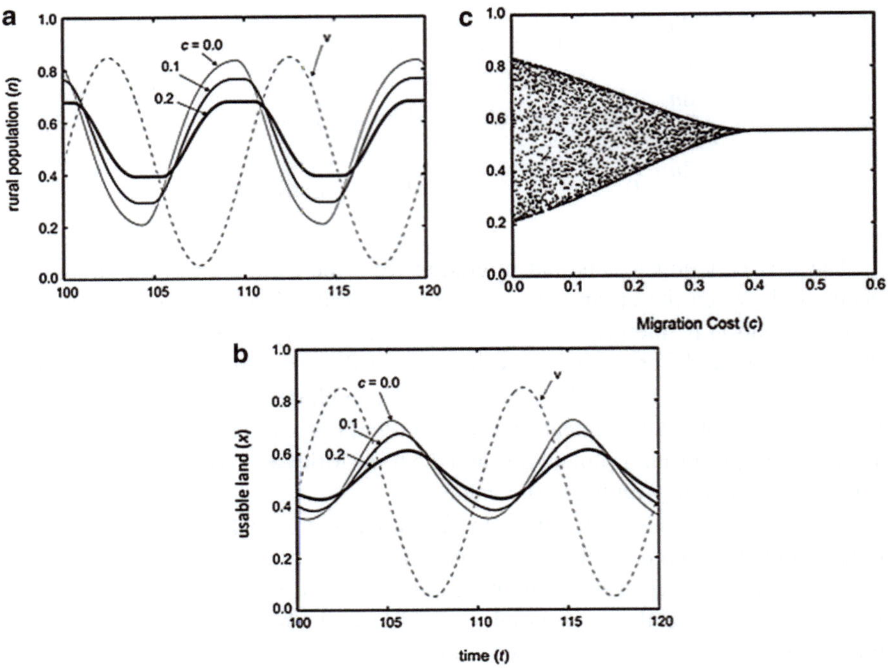

Fig. 15.3 Numerical simulations in dynamic systems of (15.1) and (15.3) with (15.4) where $N=1$, $s=1$, $a=1$, $k=0.9$, $b=0.9$, $v_0=0.45$, $v_1=0.4$, and $T=10$. (**a**) Rural population n and (**b**) proportion of usable land x where higher values of costs c make amplitudes of fluctuation smaller and retard the timing of population shift. (**c**) Dependency of amplitude in rural population n on shift costs c where there is no population for extremely high values of costs

15.4 Discussion

We explored a very simple model of population shift between rural and urban areas in which the dynamics of population distribution and that of the ecological condition of the rural area are both explicitly considered. This model may explain the recent population change in Mongolia where population has increased only in the Ulan Bator area while that in other areas is roughly constant. The mechanism is as follows. Obviously, population increases in both urban and rural areas. However, in the model, the rural value bx is dependent on the rural population numbers, that is, the value has negative density dependence on the population. On the other hand, we assume the urban value v is constant and independent of urban population. Therefore, the increase in total population is all absorbed in the urban population when people tend to move to the area with the higher value. In reality, the urban value may depend on its population, reflecting economy of scale (Fujita et al. 1999). That is, the value of an area increases with population when the population is small (economy of scale) but decreases with population when the population is large (diseconomy of scale).

An intermediate scale of population maximizes the value of the area, and the value will be roughly constant around the maximum. In other words, the urban value of Ulan Bator might be nearly constant in maximal value in these several years.

If we introduce economy of scale into the model, the equilibrium state may be unstable because economy of scale may introduce a positive feedback between population and the values of the areas. In this case, limit cycles, where population and the size of usable land fluctuate around the unstable equilibrium, or bistable fixed points, where urban population is small in one equilibrium but large in the other, may appear. Although we analyzed population fluctuation as the result of external fluctuation of the urban value, it may occur by internal forces, resulting in a limit cycle, as already mentioned. In the bistable case, there may generally be a jump from one stable equilibrium to the other, called a *regime shift* (Folke et al. 2004), when parameter values are changed. In the present model, changes in the urban area always slowly occur with slow parameter changes (as seen in Fig. 15.2). However, by introducing scale of economy, we could model the rapid appearance and disappearance of a city caused by regime shift.

In the model we assumed for simplicity that the urban and rural values are measured by income obtained in those areas. However, the urban value is given by not only income but also other economical and cultural factors. Especially in Mongolia, there are many cases where parents live in or near Ulan Bator city to remain in contact with their children studying in schools including universities. Also, they live there to contact their families working abroad or for the convenience of money transfer from them. These factors can be easily included in parameter v. The rural area may also have cultural values for a comfortable pastoral life. In this case, we can include the difference of cultural values in v.

In the model, we assumed for simplicity that migration occurs only in one direction from a low-value area to a high-value area. In reality, a part of the people may move in the opposite direction, although the net migration occurs toward high-value areas. To represent this fact concisely, we can modify (15.2) as

$$\frac{\mathrm{d}n}{\mathrm{d}t} = -M(v - bx)n + M(bx - v)(N - n) \tag{15.5}$$

where

$$M(z) = \frac{m}{1 + e^{-sz}} \tag{15.6}$$

Here, m is maximum migration rate when the value difference is infinity. This type of function is used for decisions on forest land use (Satake and Iwasa 2006). Function $M(z)$ is a monotonically increasing function of the value difference z, but positive even when z is negative. Using this function, analytical calculations will be more difficult, but we expect that the results are similar to those of the present model.

In the models in this Chapter, we considered rural living as raising livestock in grassland. The essential part in modeling is that the rural value has a strong negative dependency on rural population. This assumption would hold even though rural life is based on agriculture or fishery rather than pasturage. The agricultural production will depend strongly on the area of available land for agriculture, and per capita fish catch will sharply decrease with the number of fishermen. Thus, the present model may be applicable for population shift between an urban area and a rural area where people are living by agriculture or fishery.

We modeled here an example of socioecological systems, the analyses of which are considered to be very important for global and regional environmental problems in general (Folke et al. 2004). We consider that the formulation and analyses of simple mathematical models are very useful, as shown in several pioneer works (Satake and Iwasa 2006; Suzuki and Iwasa 2009). We expect this field of mathematical modeling will develop extensively in the near future.

15.5 Appendix

Here is given the proof of global asymptotic stability on dynamic systems (15.1 and 15.2) separately for three cases: (a) $v<b(1-kN/a)$, (b) $b(1-kN/a)<v<b$, and (c) $v>b$.

We easily show that there is a nonnegative solution for nonnegative initial conditions. Because the right-hand side of (15.1) and (15.2) satisfies the Lipshitz condition, the solution uniquely exists. Therefore, the model is well posed.

We first prove for case (b) $b(1-kN/a)<v<b$. To show global asymptotic stability for an interior equilibrium $(x^*, n^*)=(v/b, a/k(1-v/b))$, we construct two kinds of Lyapunov functions on two regions divided in $R^2_+=\{(x,n)\in R^2 | \ x>0, n>0\}$ with $n<N$ as follows:

$$V_1(x,n) = b(x - x^* - x^*\log\frac{x}{x^*}) + \frac{k}{s}(n - n^* - n^*\log\frac{n}{n^*}),\ v \geq bx$$

$$V_2(x,n) = b(x - x^* - x^*\log\frac{x}{x^*}) + \frac{k}{s}(-n + n^* - (N - n^*)\log\frac{N-n}{N-n^*}),\ v \leq bx$$

Calculating the derivative of V_i $(i=1, 2)$ along with solutions of (15.1) and (15.2), defined by $\dot{V}_i$, we have

$$\dot{V}_i(x,n) = -ab(x - x^*)^2 \tag{15.7}$$

Let $(x(t), n(t))$ be any solution of (15.1) and (15.2). In the case when $(x(t), n(t))$ crosses the line $x=v/b$ finite times, there exists $t_0\geq 0$ such that $x(t)<v/b$ or $x(t)>v/b$ for $t\geq t_0$. Suppose that $x(t)<v/b$, $t\geq t_0$ (the same argument can be made for the case $x(t)>v/b$, $t\geq t_0$.). Then, $G=\{(x, n)\in R_+^2 \mid x<v/b\}$ is invariant after t_0. By LaSalle's

invariance principle (see, for example, Smith and Waltman 1995), $(x(t), n(t))$ approaches M as $t \to \infty$, where M denotes the largest subset in

$$E = \{(x,n) \in \bar{G} \mid V_1(x,n) < \infty \text{ and } \dot{V}_1(x,n) = 0\}$$

that is invariant with respect to (15.1) and (15.2). From (15.7), $M = \{(x, n) \in R_+^2 \mid x = x^*\}$. Now we will show that $M = \{(x^*, n^*)\}$. Clearly, by definition, $M \supset \{(x^*, n^*)\}$. For $\varphi \in M$, the solution $(x(t, t_0, \varphi), n(t, t_0, \varphi))$ of (15.1) and (15.2) through (t_0, φ) remains in M for $t \geq t_0$ and satisfies, for $t \geq t_0$

$$\dot{V}_1(x(t,t_0,\varphi), n(t,t_0,\varphi)) = 0 .$$

Hence, $x(t, t_0, \varphi) = x^*$ for $t \geq t_0$ and then

$$\frac{\mathrm{d}x(t,t_0,\varphi)}{\mathrm{d}t} = 0$$

for $t \geq t_0$, which, by (15.1), implies $n(t, t_0, \varphi) = n^*$ for $t \geq t_0$. Therefore,

$$\varphi = (x(t,t_0,\varphi), n(t,t_0,\varphi)) = (x^*, n^*).$$

When $(x(t), n(t))$ crosses the line $x = v/b$ infinite times, we will show that there is no possibility of its converging any nontrivial periodic solution. Let $(\tilde{x}(t), \tilde{n}(t))$ be any nontrivial periodic solution of (15.1) and (15.2). Then there exists n_1, n_2 $(n_1 > n_2 > 0)$ such that, for some t_1, t_2, and t_3 $(t_1 < t_2 < t_3)$, $(\tilde{x}(t), \tilde{n}(t))$ starts the point $(v/b, n_1)$ at $t = t_1$, lies in the left-hand side of the line $x = v/b$ for $t_1 < t < t_2$ to cross the line $x = v/b$ at $t = t_2$ (i.e., $(\tilde{x}(t_2), \tilde{n}(t_2)) = (v / b, n_2)$), lies in the right-hand side of the line $x = v/b$ for $t_2 < t < t_3$ to go back to the point $(v/b, n_1)$ at $t = t_3$. Note that, by (15.5), $V_i(\tilde{x}(t), \tilde{n}(t))$ $(i = 1, 2)$ are strictly decreasing in the region of $x \neq x^*$. Hence, $V_1(v/b, n_1) > V_1(v/b, n_2)$ and $V_2(v/b, n_2) > V_2(v/b, n_1)$. Since $n_1 > n^* > n_2$, by the shape of $V_1(v/b, n)$, we obtain

$$n_1 - n^* > n^* - n_2 .$$

Similarly, we have $n^* - n_2 > n_1 - n^*$. Therefore, $n_1 - n^* > n^* - n_2 > n_1 - n^*$, which is a contradiction.

For other cases (a) $v < b(1 - kN/a)$ and (c) $v > b$, we have equilibria $(x^*, n^*) = (1 - kN/a, N)$ and $(x^*, n^*) = (1, 0)$, respectively. It is easy to see that there exists $t_1 > 0$ such that $x(t) > v/b$ for all $t \geq t_1$ in the case (a), or $x(t) < v/b$ for all $t \geq t_1$ in the case (c). Construct the following Lyapunov function on R_+^2 with $v \leq bx$ and $0 \leq n \leq N$:

$$V_3(x,n) = b(x - (1 - \frac{kN}{a}) - (1 - \frac{kN}{a}) \log \frac{x}{1 - kN / a}) + \frac{k}{s}(N - n)$$

for the case (a), while the following Lyapunov function on R_+^2 with $v \geq bx$ and $0 \leq n \leq N$:

$$V_4(x,n) = b(x - 1 - x\log x) + \frac{k}{s}n$$

for the case (c). Then, calculating the derivative of V_i ($i=3$, 4) along solutions of (15.1) and (15.2), we have that

$$\dot{V}_3(x,n) = -ab\big(x - (1 - kN/a)\big)^2 - k\big(b(1 - kN/a) - \nu\big)(N - n) \leq 0,$$
$$\dot{V}_4(x,n) = -ab\big(x - 1\big)^2 - k\big(\nu - b\big)n \leq 0$$

Clearly, $\dot{V}_3 = 0$ if and only if $(x,n)=(1-kN/a, N)$ for the case (a), while $\dot{V}_4 = 0$ if and only if $(x,n)=(1,0)$ for the case (c). Hence, by the Lyapunov stability theory (see, for example, Hale 1980), global asymptotic stability on these equilibria can be proven.

References

Akkoyunlu S, Vickerman R (2001) Migration and the efficiency of European labour markets. Working Paper, Development of Economics. The University of Kent at Canterbury.

Folke, C., Carpenter, S., Walker, B.. Scheffer, M. Elmqvist, T. Gunderson, L. and Holling, C. S. (2004) Regime Shift, Resilience, and Biodiversity in Ecosystem Management. Annu Rev Ecol Evol Syst 35: 557–581.

Fujita, M., Krugman, P. and Venables, A. J. (1999) The Spatial Economy: Cities, Regions, and International Trade. Massachusetts Institute of Technology, Cambridge.

Hale, J. K. (1980) Ordinary Differential Equations. Krieger, Malabar, FL.

Harris, J. R. and Todaro, M. P. (1970) Migration, Unemployment and Development: A two-Sector Analysis. Am Econ Rev 60: 126–142.

Ishikawa, Y. (2001) Migration turnarounds and schedule changes in Japan, Sweden and Canada. Rev Urban Region Dev Stud 13: 20–33.

Mendsaikhan, S., Gerelt-Od, G., Dagvadorj, Ch. And Bajiikhuu, Kh. (2009) Mongolian Statistic Yearbook 2008. National Statistical Office of Mongolia, Ulaanbaatar.

Ostrom, E. (2009) A general framework for analyzing sustainability of social-ecological systems. Science 325: 419–422.

Satake, A. and Iwasa, Y. (2006) Coupled ecological and social dynamics in a forested landscape: the deviation of individual decisions from the social optima. Ecol Res 21: 370–379.

Shigesada, N. and Kawasaki, K. (1997) Biological Invasions: Theory and Practice. Oxford University Press, New York.

Smith, H. L., and Waltman, P. (1995) The theory of the chemostat: dynamics of microbial competition. Cambridge University Press, Cambridge.

Suzuki, Y. and Iwasa, Y. (2009) The coupled dynamics of human socioeconomic choice and lake water system: the interaction of two sources of nonlinearity. Ecol Res 24: 479–489.

Chapter 16
Change in Livestock Species and Their Spatial Distribution

Izuru Saizen

Keywords Grassland management • Livestock distribution in Mongolia • Local indicator of spatial associations • Spatial clusters

16.1 Introduction

Uncontrolled changes in the domestic livestock population in Mongolia, a result of the country's shift from a planned economy to a market economy on switching to a capitalist economy in the early 1990s, is a very important topic of research (Yoshihara et al. 2008). Often unpredictable, this phenomenon raised some serious issues in terms of managing grasslands. Mongolia is located in the northeastern part of Asia where the climatic gradient from humid to arid conditions (forest–grassland–desert) causes the formation of ecotones. An ecotone is a transitional area between two adjacent ecological communities and is generally sensitive to any external disturbance of the environment, natural or human, such as climate change or human activities (Pogue and Schnell 2001; Peters 2002). Precipitation is concentrated in the summer months, whereas less than 10% of the annual total precipitation is observed during the cold season. Because the monthly temperature falls below 0°C from November to March, most of the precipitation during this season is in solid form. January is the coldest month, and the monthly mean temperature ranges from −30°C in the northern mountainous area to −15°C in the southern Gobi region (Morinaga et al. 2003). About 75% of the total area has been affected by the unrestrained grazing of animals, including cattle (cows, yaks, and cow-yak hybrids), camels, goats, horses, and sheep, following a nomadic pattern for thousands of years, and nomadism appears to be a hazardous enterprise (Goldstein et al. 1990). Accordingly, the

I. Saizen (✉)
Graduate School of Global Environmental Studies, Kyoto University, Kyoto, Japan
e-mail: saizen@kais.kyoto-u.ac.jp

N. Yamamura et al. (eds.), *The Mongolian Ecosystem Network: Environmental Issues Under Climate and Social Changes*, Ecological Research Monographs,
DOI 10.1007/978-4-431-54052-6_16,

severely cold climate, short and limited growing period, and especially grazing have led to sparse plant cover in the grasslands of Mongolia (Munkhtsetseg et al. 2007).

As an erstwhile socialist country, Mongolia transitioned to a free market economy in the beginning of the 1990s. As a result, all the livestock owned by herding cooperatives were transferred to herders as a result of privatization. No longer managed by the state, most herders were able to enlarge their nomadic activities, and hence, livestock population increased greatly (Chuluun and Ojima 2001). However, the increase in livestock population under the free market economy may have affected the growth, density, and distribution of grassland plants (Chen et al. 2007). Livestock distribution is considered the only predictable and controllable factor compared to other factors affecting grasslands conservation, such as climate change and economic conditions. An understanding of the spatial and time-series distributions of livestock plays a key role in the successful application of environmental management.

In this chapter, it is posited that particular spatial configurations of grazing animals in Mongolia are reflected in the presence of areas well suited for the different animals to live in and for herders to gain economic benefit from their nomadism. Then, the spatial clusters of each animal's distribution are explored using an exploratory spatial data analysis. Although spatial autocorrelation was defined decades ago, its application has been limited by computation capacity, software availability, and platform limitation of spatial datasets. Previous studies with spatial autocorrelation for spatial dependence in global and local scales had primarily focused on spatial econometrics (Anselin and Rey 1997; Pace et al. 1998). Thus, there have been few reports on the application of these statistics for understanding spatial livestock distribution in Mongolia (Saizen et al. 2010).

This chapter focuses on spatial livestock distribution to (1) identify and map the spatial patterns of livestock distribution and thus determine whether and where there are high (or low) spatial clusters of livestock distribution and (2) observe whether these clusters shift geographically and have specific characteristics of their spatial distribution by combining spatial analysis (geographic information system, GIS) with an empirical approach to clarify the characteristics of livestock distribution throughout Mongolia.

16.2 Background

16.2.1 Transition of the Mongolian Livestock Population

Until the early twentieth century, the majority of the livestock in Mongolia was owned by the upper class and Tibetan Buddhist monasteries in a few large herds. As a result of the revolution in 1921, Mongolia became a socialist state in 1924 (Walters et al. 1999), and all livestock were allocated more uniformly to small privately owned herds. In the late 1950s, the Mongolian government established livestock collectives throughout the country, and by 1960, all herders belonged to the

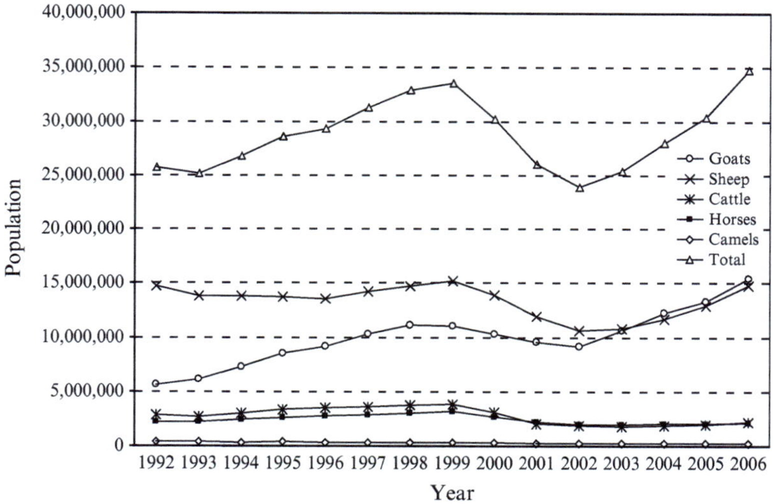

Fig. 16.1 Change in the livestock population of Mongolia (1992–2006)

collectives (Fernandez-Gimenez 2002). The government owned almost all the Mongolian livestock and paid herders a regular salary to feed their livestock. The collective administration allocated pastures and regulated pasture use by forming medium- and large-sized herds that were moved on a seasonal basis (Fernandez-Gimenez 2002). The Mongolian livestock population changed depending on the administrative structure, but not significantly, because the administrative structure had been designed and controlled by the state.

In the 1990s, livestock enterprise rapidly expanded, assisted by relatively good weather conditions and by many new entrants to the livestock economy (Lise et al. 2006). The transition of the Mongolian livestock population is shown in Fig. 16.1. After the change to a free market economy in 1992, the total livestock population increased until 1999, which indicated the first peak. This peak was followed by a rapid and steady decline from 1999 to 2002, when it reached its lowest point, and then another increase to its highest level in 2006. In this chapter, these four time points were focused to conduct a time-series analysis and to detect the geographic shift of spatial clusters: (1) the livestock population at the beginning of the free market economy in 1992, (2) the first peak in livestock population in 1999, (3) the lowest population in 2002, and (4) the second peak in 2006.

In 1992, there were 25.694 million domestic livestock in Mongolia (Table 16.1). By 1999, this number had grown to 33.569 million, indicating an increase of 30.6%. The increase did not result from a growing number of sheep, despite sheep accounting for the highest number of livestock: the number of sheep increased only slightly from 14.657 million to 15.191 million. In addition, the number of camels decreased slightly. Instead, the growth resulted from the increasing number of horses (from 2.2 million to 3.163 million), cattle (from 2.819 million to 3.825 million), and, in particular, the

Table 16.1 Livestock numbers and their changes in Mongolia (1992, 1999, 2002, and 2006)

	1992 (thousands)	1999 (thousands)	2002 (thousands)	2006 (thousands)	Change over 1992–1999 (%)	Change over 1999–2002 (%)	Change over 2002–2006 (%)
Goats	5,603	11,034	9,135	15,452	96.9	–17.2	69.2
Sheep	14,657	15,191	10,637	14,815	3.6	–30.0	39.3
Cattle	2,819	3,825	1,884	2,168	35.7	–50.7	15.1
Horses	2,200	3,163	1,989	2,115	43.8	–37.1	6.3
Camels	415	356	253	254	–14.3	–28.8	0.2
Households[a]	143	189	175	170	32.2	–7.4	–2.9
Total (livestock)	25,694	33,569	23,898	34,803	30.6	–28.8	45.6

Source: National Statistical Office of Mongolia

[a]"Households" refers to households actually leading a nomadic existence

number of goats almost doubling (from 5.603 million to 11.034 million). From 1992 to 2006, only the goat population increased dramatically and contributed to the increase in the total livestock population. The steep growth in the number of goats resulted from a strong demand for goat hair, or cashmere (Lise et al. 2006), which is the most important goat product. The yield, quantity, and quality of cashmere depend on the strain of goat, coat color, genetic potency, harvesting period, and technique of collection. Mongolian goat cashmere is considered to be among the finest in the world, owing to its exquisitely soft, light, and durable texture.

Between 1999 and 2002, the deterioration of winter conditions caused an extreme Zud, a Mongolian term describing livestock loss directly induced by harsh winter conditions but often influenced by drought in the previous summer (Retzer and Reudenbach 2005). Accompanied by deterioration in spring–summer rainfall, the results were disastrous for Mongolian livestock, with 12 million animals dying nationwide. Of an estimated 190 thousand herding households in 1998, 11 thousand lost all their livestock (Lise et al. 2006). In 2002, the total livestock population had declined to about 23.898 million, back to the level of the late 1980s (Lise et al. 2006). Compared to 1999, the losses were the most severe among cattle (–50.7%) and horses (–37.1%) and least severe among goats (–17.2%), indicating that goats are resilient to cold, and their survival rate was highest of all Mongolian domestic animals during this Zud.

16.2.2 Study Area

The study area covered all of Mongolia. The capital of Mongolia is its largest city, Ulaanbaatar (UB)—the political, economic, and scientific center of the country. More than one million inhabitants, that is, around one-third of the country's total population, live in UB, and the population has been increasing at a tremendous pace, primarily through migration of rural residents (Arguchintseva et al. 2008). Consequently, many herders looking to do business by selling livestock products tend to move and gather near UB.

16.3 Data and Methods

16.3.1 Spatial Variation of Livestock Distribution

Mongolia comprises 21 administrative divisions known as *aimags* (provinces), which are hierarchically divided into 317 *sums* (municipalities). The National Statistical Office of Mongolia provides statistical data of the livestock population for the units of Sums. The capital, UB, has provincial status but is administrated separately. In this chapter, the following two types of administrative areas were defined: (1) the 21 *aimags* and UB (AIMAG_U) and (2) the 317 *sums* and UB

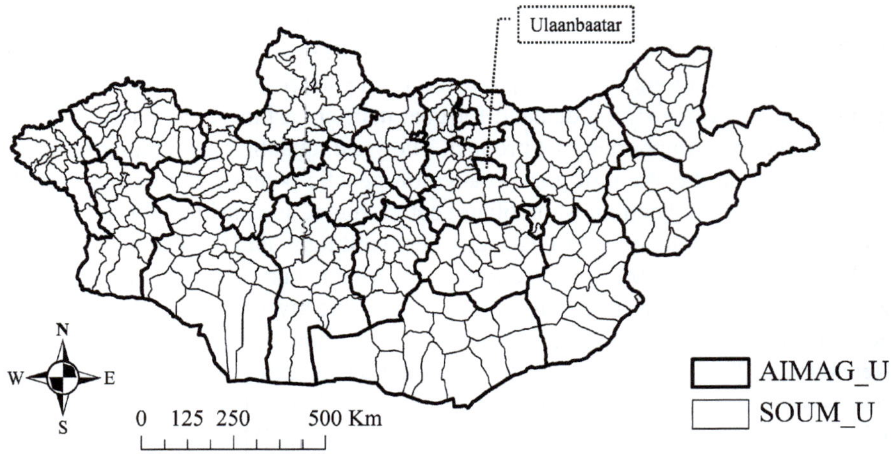

Fig. 16.2 Study area showing AIMAG_U and SUM_U boundaries and the location of Ulaanbaatar

(SUM_U), to cover all Mongolia. Figure 16.2 shows the map of the Mongolian administrative areas, AIMAG_U, and SUM_U.

Before conducting the spatial analysis, it was necessary to determine whether AIMAG_U and SUM_U were appropriate for the spatial analysis and which units reflected a more informative spatial variation for each domestic animal, because the smaller spatial unit (SUM_U) had high spatial accuracy but may have presented unreliable rates. The larger spatial unit (AIMAG_U) may have removed relevant geographic variation (Nakaya 2000).

Once the variance has been decomposed into each scale of possible administrative areas as in spectrum analysis, Moellering and Tobler (1972) argued that the scale with the highest variance should be applied for mapping and presented the concept of the scale variance component (SVC). Although Curry (1972), Batty (1976), and Batty and Sammons (1979) employed spatial entropy as a similar theory, the SVC is a highly valuable approach to derive a great deal of information on spatial variation from geographic datasets (Okamoto and Himiyama 1983). Further, the SVC requires hierarchical areal units, which are available in the present study. Although there is no formal criterion as to whether a low variance for the given scales should be discarded, the SVC method is useful to compare AIMAG_U and SUM_U in terms of choosing units that are more appropriate for the spatial analysis.

The SVC is defined by the following equations:

$$\mathrm{SVC}(1) = \frac{\sum_i^I (\bar{X}_{i.} - \bar{X}_{..})^2}{I-1} \tag{16.1}$$

$$\mathrm{SVC}(2) = \frac{\sum_i^I \sum_j^{Ji} (\bar{X}_{ij} - \bar{X}_{i.})^2}{\sum_i^I (J_i - 1)} = \frac{\sum_i^I \sum_j^{Jj} (\bar{X}_{ij} - \bar{X}_{i.})^2}{J-1} \tag{16.2}$$

Table 16.2 Scale variances of domestic animals (1992, 1996, 2002, and 2006)

Domestic animals	Year	SVC(1)	Percent sum of squares	SVC(2)	Percent sum of squares
Camels	1992	2,816	52.2	2,578	47.8
	1999	2,370	59.1	1,637	40.9
	2002	1,257	56.8	957	43.2
	2006	1,251	51.3	1,186	48.7
Horses	1992	7,854	39.7	11,933	60.3
	1999	19,333	43.6	24,981	56.4
	2002	13,057	54.1	11,090	45.9
	2006	11,759	46.5	13,511	53.5
Cattle	1992	13,511	30.6	30,703	69.4
	1999	33,570	41.0	48,320	59.0
	2002	13,784	40.8	20,039	59.2
	2006	13,392	33.3	26,784	66.7
Sheep	1992	229,965	33.8	450,037	66.2
	1999	264,260	34.7	497,182	65.3
	2002	174,853	36.5	303,627	63.5
	2006	310,323	34.9	577,584	65.1
Goats	1992	120,820	48.4	128,728	51.6
	1999	333,721	48.4	355,973	51.6
	2002	114,955	35.9	205,664	64.1
	2006	359,102	42.1	494,541	57.9

where I and J denote the numbers for AIMAG_U and SUM_U, respectively; $\overline{X}$, the average livestock population in Mongolia; $\overline{X}_{i.}$, the average livestock population for i AIMAG_U; and $\overline{X}_{ij}$, the livestock population for j SUM_U and i AIMAG_U.

The SVC was applied to each domestic animal over every observed year; the results are shown in Table 16.2. As may be seen, a larger scale variation took place on SVC(2) (SUM_U level) for cattle, sheep, and goats over the observed years. By contrast, the scale variation on SVC(2) was less than that on SVC(1) for camels over the observed years and for horses in 2002. The results may be interpreted as denoting the existence of a more informative variation of cattle, sheep, and goats on the SUM_U units as well as that of camels on AIMAG_U. Thus, an exploratory analysis was conducted for horse, cattle, sheep, and goats by SUM_U and for camels by AIMAG_U. Although the SVC(1) was larger than the SVC(2) for horses in 2002, the SUM_U was chosen with units to accomplish time-series analysis that would bring us fruitful findings about horse population changes over the observed periods.

16.3.2 Exploratory Spatial Data Analysis

Spatial autocorrelation can be defined as the coincidence of value similarity and locational similarity (Premo 2004). Therefore, positive spatial autocorrelation occurs when high or low values of a random variable tend to be spatially clustered, and negative spatial autocorrelation occurs when areal units tend to be surrounded by neighbors with very dissimilar values. Analyzing the geographic dimension of the spatial patterns of livestock should help us understand the characteristics of livestock distribution and transition and enable us to derive meaningful information for managing grasslands in Mongolia. Hence, the techniques of exploratory spatial data analysis were applied. These techniques serve to describe spatial distribution (clusters or dispersions) in terms of spatial association patterns such as global spatial association and local spatial association (Jacquez and Greiling 2003; Goovaerts and Jacquez 2004). Moran's I statistic is a representative measure of global spatial autocorrelation (Moran 1948; Anselin 1988). Moran's I is useful as a global test that may suggest randomness or non-randomness in the overall spatial pattern of livestock over the observed years in Mongolia but does not indicate where the clusters are located or what type of spatial autocorrelation is occurring spatially. Therefore, the local indicator of spatial autocorrelation (LISA) was applied as an indicator of local spatial association. The LISA, the so-called local Moran index, was developed based on Moran's I and is applied in multiple fields (Andresen 2009). These patterns are associated with a spatial weight matrix, where each unit is connected to a set of neighboring units. In other words, spatial connectivity is incorporated by means of a spatial weight matrix (Anselin 1995). In this study, a matrix of distances between the gravity points of each AIMAG_U/SUM_U was used to model the relationships between spatial units. The results presented are those obtained with a matrix of the ten nearest neighbor administrative units from which meaningful outputs were detected.

In this analysis, a Monte Carlo permutation approach is applied to verify the significance of the LISA. This permutation approach assumes that data are equally likely to be observed at any location. The observed values were randomly shuffled over all locations, and the LISA was recalculated for each permutation. Then, the significance of the LISA was determined by generating a reference distribution by using 999 random permutations. Finally, the LISA significance map was created by incorporating information about the significance of the local spatial patterns. Specifically, the LISA map shows the types of spatial relationships between a unit of place and its neighboring units, which allows us to visualize five types of local spatial associations between the observed units and their neighbors, each being located in a quadrant of the scatter plot. Therefore, in the present analysis, each neighborhood can be characterized by one of the following associations: (1) high–high (HH), indicating a clustering of a high livestock population density in an AIMAG_U/SUM_U (positive spatial autocorrelation); (2) high–low (HL), indicating that low values are adjacent to high values of livestock population density in an AIMAG_U/SUM_U (negative association); (3) low–low (LL), indicating clustering of low values of livestock population density in an AIMAG_U/SUM_U (positive association); (4) low–high (LH), indicating that high values are adjacent to low values of livestock population density in an AIMAG_U/SUM_U (negative association); and (5) nonsignificant (NS), indicating no spatial autocorrelation was detected by the LISA statistics.

The LISA was defined by Anselin (1995) and is given by

$$I_i = \frac{x_i - \overline{x}}{\sum_i^n (x_i - \overline{x})^2} \sum_j^n w_{ij}(x_j - \overline{x}) \tag{16.3}$$

where n equals the number of observed AIMAG_U/SUM_U units; w_{ij} denotes the weight between locations i and j; x_i and x_j, the values at locations i and j; and $\overline{x}$, the average over all locations of the variables.

The analyses were conducted using GeoDa (Anselin et al. 2006), a standalone software package that provides exploratory spatial data analysis techniques for areal data.

16.4 Results

16.4.1 Local Spatial Association of Livestock Distribution

Camels are ideal for the desert as a draft animal because they are extremely undemanding and well adapted to the extreme surroundings of the Gobi and other desert zones (Retzer et al. 2006; Moser 2007). As shown in Fig. 16.3, in 1992, 80.3% of camels were concentrated in eight *aimags*—Bayankhongor, Govi-Altai, Dornogovi, Dundgovi, Ovorkhangai, Omnogovi, Sukhbaatar, and Khovd—located in the south of Mongolia. By 2006, camels in the foregoing eight *aimags* still constituted 82.0% of the total camel population. Although the camel population had decreased since 1992 owing to the widespread use of trucks among herders (Oniki and Konagaya 2006), their distribution pattern had scarcely changed, and no spatial cluster in the

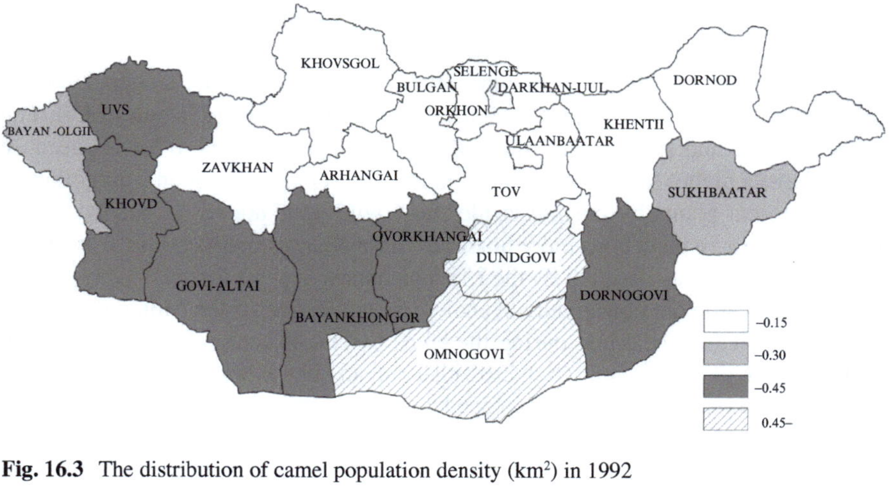

Fig. 16.3 The distribution of camel population density (km²) in 1992

observed years was detected with the LISA statistics based on the AIMAG_U units. Although camels display a voracious appetite for pasture plants, it seems reasonable to suppose that the camel's impact on pastures in Mongolia is decreasing in accordance with the decline in their population and that their distribution was affected not by the free market economy but by natural conditions.

Livestock population densities and the results of the LISA statistics for other animals are shown in Fig. 16.4. The distribution of their population density does not necessarily correspond with the LISA statistics. LISA statistics, as an exploratory spatial data analysis, should focus explicitly on the spatial aspects of the datasets in terms of spatial association and heterogeneity. In other words, this technique can describe spatial distributions, discover patterns of spatial association (spatial clustering), suggest different spatial regimes or other forms of spatial instability (non-stationarity), and identify atypical observations (Anselin 1996). In this case, LISA statistics can reveal local pockets of non-stationarity or hotspots that cannot be found on population density maps.

As shown in Fig. 16.4, LL clusters are mainly observed in the Gobi regions in the southeast of Mongolia. As already above, these regions are arid areas where annual precipitation is limited, making them an unsuitable habitat for these domestic animals. Because a detailed discussion of LL and LH clusters would carry us too far from the purpose of the present chapter, only the HH and HL clusters for each animal are examined closely.

Herders breed horses primarily for the purposes of riding, traveling, and racing and for horsemeat and *airag* (fermented mare's milk), a beverage commonly consumed during the summer (Degen 2007). The herders do not always consume the horsemeat and *airag* themselves but also sell them for profit. Figure 16.4a shows high values in the horse population density widely distributed in the central parts of Mongolia for every observed year, with a peak observed in 1999; however, the LISA cluster maps show that the distribution of HH clusters was locally concentrated south of UB during the period 1992–2006. There are possibly two main reasons for this distinctive distribution of HH types. First, horses used for horsemeat and *airag* would tend to be bred near UB because both goods spoil rapidly in summer. Most herders who live far from UB do not need cashable goods made from horses, using them mostly for traveling, riding, and sometimes racing. Second, livestock collectives specialized to breed horses used to be found in these areas, south of UB. The severe climate and sparse plant life in these areas is also suited to horses because horses have a higher tolerance for the cold in normal precipitation in Mongolia and are able to eat plants by pushing aside snow with their hooves (Kanaoka 2000). Historically, there have been more horses in these areas; moreover, this distribution still remains and was detected by the LISA statistics.

As far as cattle are concerned, an ecological adaptation of the Mongolian cattle, which helps reduce heat loss during the cold seasons, is coat color (Badarch et al. 2003). The coats are generally red, brownish red, and black, which absorb more heat than light colors. Another well-developed adaptation specific to Mongolian cattle is their ability to choose particular types of pasture hay and to wrap their tongues around the grass to tear it off (Badarch et al. 2003) and, compared to other cattle, Mongolian

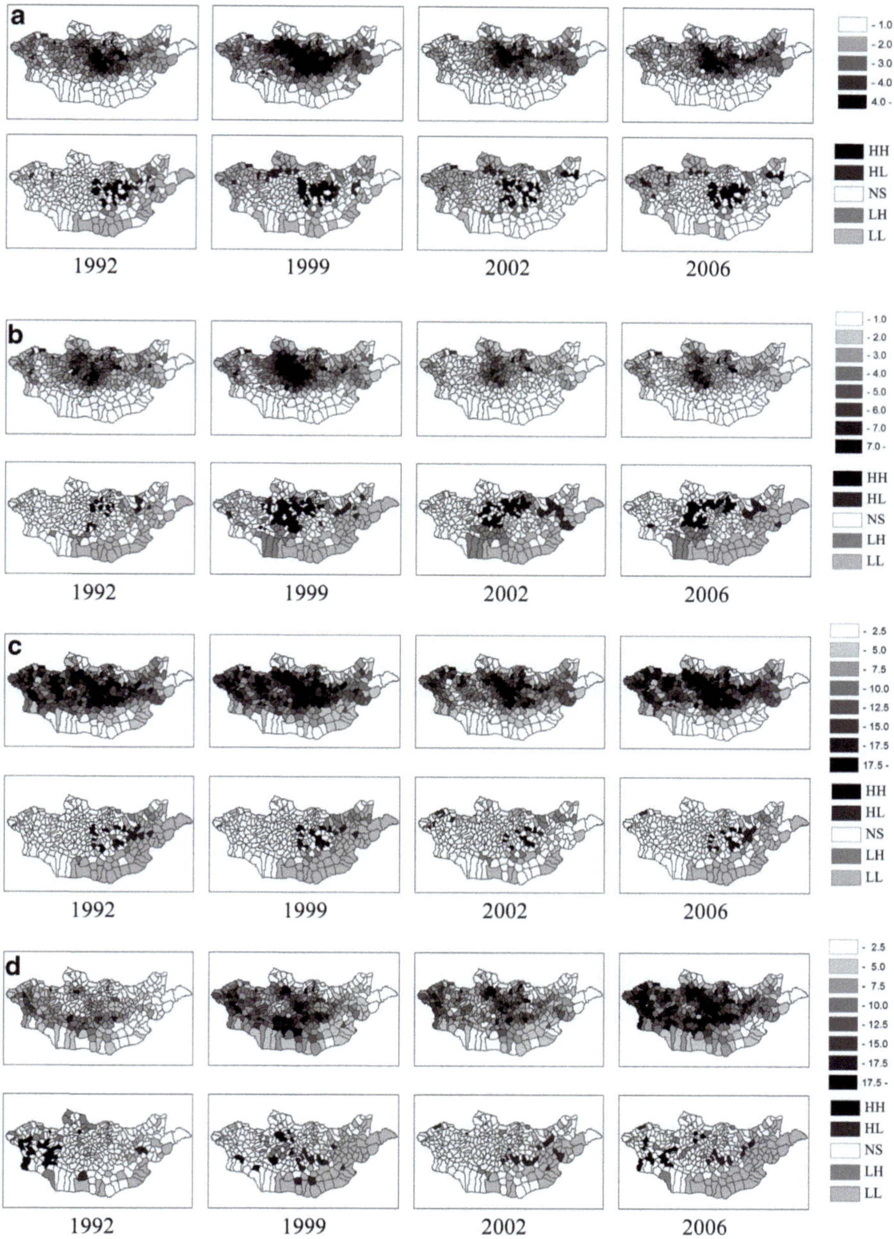

Fig. 16.4 Maps of livestock population density (km^2) and local indicators of spatial autocorrelation (LISA) cluster maps (1992, 1999, 2002, 2006). (**a**) *Top row*: Horse population density; *bottom row*: horse LISA statistics. (**b**) *Top row*: Cattle population density; *bottom row*: cattle LISA statistics. (**c**) *Top row*: Sheep population density; *bottom row*: sheep LISA statistics. (**d**) *Top row*: Goat population density; *bottom row*: goat LISA statistics

cattle are well adapted to high mountains. A map of cattle population density for all the observed years is presented in Fig. 16.4b. The high level of cattle population density is distributed from the west of UB to around the *Khangai* mountainous regions, which constitute the largest area of the so-called Mongolian Plateau to be covered with rich vegetation (Petit et al. 2002). LISA statistics also show similar characteristics in all the observed years, except in 1992. Meat is the principal product of Mongolian cattle (Badarch et al. 2003); therefore, HH clusters are observed around UB and in the Khangai mountain regions, which are physically accessible from the capital. For the same reason, the distribution of the HH clusters was unchanged, even after the rapid decrease in cattle population following its peak in 1999.

The distribution of sheep population density was largely spread from west to east in Mongolia (Fig. 16.4c). By contrast, HH and HL clusters were observed around or in the south of UB to a limited extent. The high values around UB reflect the city inhabitants' demand for mutton. Recently, the price of wool has been falling rapidly, but the demand for mutton, which is relatively difficult to transport over long distances because of its weight, still remains; therefore, HH clusters around UB would be observed and detected by LISA statistics as well.

As previously suggested, goats are the only animal species whose population increased during the observed period 1992–2006 owing to the strong demand for cashmere. Distribution maps of goat population density and spatial clusters are shown in Fig. 16.4d. In 1992, the goat population density was less than 10/km^2 for almost all of Mongolia. In 1999, the figure had increased considerably, especially in the central areas, by up to more than 17.5/km^2. Furthermore, with the impact of the Zud, population densities of more than 17.5/km^2 were widely observed in 2006, exceeding that of the first peak in 1999. Comparing the LISA cluster maps for the period 1992–2006, the HH cluster was observed in western Mongolia in 1992, and its distribution had shifted to the east in 1999. In 2002, the HH cluster had disappeared and was once again distributed in western Mongolia in 2006. These changes do not seem to be associated with the change in population density. Instead, it was shown that the HH goat cluster was shifting, and its distribution was different from those of other animals' HH clusters.

16.4.2 Relationship Between Goat Population Change and Sheep Units

To clarify the spatial characteristics of changes in the goat population, the LISA statistics of sheep units, which represent the degree of total domestic animal population, were calculated for a comparison with changes in the goat population. Table 16.3 shows the relationship between the changes in goat population during the period 1992–1999 and sheep units in 1992 as well as the relationship between these changes for the period 2002–2006 and sheep units in 2002. In 1992, the LISA statistics produced the following sheep units: 57 HH type, 1 HL type, 15 LH type, 73 LL type, and 172 NS type. Table 16.3 shows the characteristics of the different types

Table 16.3 Relationships between goat population change and type of sheep unit (SU) local spatial associations during 1992–1999 and 2002–2006

local indicators of spatial autocorrelation (LISA) statistical level types	High–high (HH)	High–low (HL)	Nonsignificant (NS)	Low–high (LH)	Low–low (LL)	Total
Number of SUM_U of sheep units in 1992	57	1	172	15	73	318
Rate of change (1992–1999) (%)	123.3	31.1	89.3	177.6	95.6	96.9
Goat population in 1992 (thousands)	752	21	3,192	99	1,539	5,603
Goat population in 1999 (thousands)	1,679	27	6,041	275	3,011	11,034
Gross area (km^2)	144,608	3,609	703,983	51,371	659,312	1,562,882
SU in 1992 (thousands)	11,317	190	30,164	1,820	10,600	54,090
SU in 1999 (thousands)	15,777	131	39,579	2,756	13,749	71,992
Number of SUM_U of sheep units in 2002	49	6	173	17	73	318
Rate of change (2002–2006) (%)	54.0	58.6	65.9	97.2	83.2	69.2
Goat population in 2002 (thousands)	1,343	209	5,060	260	2,263	9,135
Goat population in 2006 (thousands)	2,068	332	8,393	513	4,145	15,452
Gross area (km^2)	178,697	10,866	578,795	63,431	731,092	1,562,882
SU in 2002 (thousands)	8,509	872	26,387	1,872	7,712	45,351
SU in 2006 (thousands)	9,521	996	33,956	2,574	10,753	57,800

obtained from these calculations by each type of spatial grouping. On the whole, for both periods, areas with low livestock density showed a high increase rate of goat density as shown in LH and LL, whereas the second highest rate is observed in HH during the period 2002–2006. It should be noted that the highest rate of goat population change from 1992 to 1999 was observed in the LH type of sheep unit (178% increase), whereas the lowest rate was seen in the HL type of sheep unit (31% increase). In addition, in 2002, 49 HH, 6 HL, 17 LH, 73 LL, and 173 NS types of sheep units were obtained, and the highest rate of goat population change from 2002 to 2006 also occurred in the LH type of sheep unit (97.2% increase). Given that the LH type is the cluster with low values surrounded by high values of sheep units, this fact indicates that the goat population tended to increase in areas with relatively low sheep units and especially surrounded by high sheep units; in other words, goats were moving into these areas. Also, the relationship between the LISA statistics of sheep units and other domestic animals was observed; however, any significant aspect observed in case of goats was not detected. Compared to other domestic animals, goats are not required to be bred in specific conditions such as physical nearness to UB or near rich vegetation owing to the characteristics of their main product—cashmere—and their survival ability, and accordingly, the LISA statistics reflect this aspect in the case of the sheep units.

16.5 Discussion and Conclusion

A herder's mobility is a key factor in ensuring the sustainability of a livestock production system in semiarid and arid landscapes (Ellis and Swift 1988; Sneath 1998). This system was suitable in Mongolia until the 1990s, when the goat population and its spatial distribution were most influenced by the shift to a free market economy. After 1992 and excluding the period between 1999 and 2002, livestock density maps of horses, cattle, and sheep show geographic shifts depending on the changes in their population throughout Mongolia, and in contrast, the LISA statistics corresponding to these animals do not represent any significant change of their spatial distribution whereas only goats have the characteristic of moving to the areas with LH clusters. As already mentioned, cashmere can be easily transported from any location, near or far, from UB and maintained in good condition even in summer, as compared to the perishable products of other animals, such as meat and milk. Therefore, herders are not limited by geographic conditions when choosing locations for breeding goats. In addition, the highly attractive prospect of cash income from cashmere accelerates this tendency. Meanwhile, herders do not tend to expand the number of livestock other than goats in areas far from UB because of poor profits and limitations of their suitability to natural condition; accordingly, the HH and HL clusters of these animals did not change significantly during the observed years, as per the results of the LISA statistics.

During the period 1999–2002, with the occurrence of several record-breaking disastrous Zuds, livestock mortality was associated with the low normalized difference vegetation index (NDVI) in summer, high snow water equivalent (an index calculated by multiplying snow depth by its density) in winter, a high previous year's mortality, and high previous year's livestock population (Tachiiri et al. 2008). Templer et al. (1993) identified the Zud risk and clarified that the eastern region had the lowest risk by summarizing the historical frequency of the Zud, whereas *aimags* in the other regions had medium and high levels of risk based on *aimag* units. During this study, a rapid decrease of livestock was also observed mainly in these areas. By using more detailed units, namely, *sums* where geographic variance is higher, it was possible to clarify that the characteristics of spatial distribution for camels, horses, cattle, and sheep had not changed before and after the Zud; however, their populations had changed in accordance with the impact of the Zud. The high spatial clusters of these animals depended mostly on their suitability for the severe natural environment of Mongolia and their historical backgrounds.

The forage preferences of these domestic animals are well known to be substantially different. These preferences result from differences in the body size; reticulorumen capacity; anatomy of teeth, lips, and mouth structures; grazing ability; agility; and digestive systems of the different species (Vallentine 2001). The grazing animals in this study can be divided into three groups based on their forage preferences: (1) grazers, (2) intermediate feeders, and (3) browsers. Grazers primarily consume graminoids and strongly avoid woody species. Occasionally, they consume locally available forbs and woody species when graminoids are not available. Horses and cattle are grazers (Vallentine 2001). Horses mostly consume graminoids and occasionally eat forbs and woody species. Cattle prefer grasses, eat forbs and woody species seasonally, and can switch to woody species or forbs if grasses are not available. Intermediate feeders eat large amounts of graminoids, forbs, and woody species and are known for their ability to adjust their feeding habits to whatever is locally available. Sheep are intermediate feeders and are known to be highly selective foragers. Sheep consume large amounts of forbs and grasses and smaller amounts of woody species. Their diets also consist of a large proportion of grasses compared to that of goats (Gordon and Illius 1992). Browsers primarily consume forbs and woody species. They commonly consume green grass during growing seasons but avoid dry mature grass (Vallentine 2001). Most studies agree that goats are browsers. Goats prefer the leaves and tender twigs of woody species and consume the young growth of many woody species that are less palatable to other species (Vallentine 2001). They have the ability to consume forage of lower quality and can consume needle leaves of conifer trees up to 220 cm tall by bending the trees to reach the leaves on higher branches (Child et al. 1985). Goats are not recommended to be used to maintain tree and shrub in rangelands, and goat grazing is often restricted in tree plantations where tree seedlings are short or thin enough for goats to bend over (Child et al. 1985). In addition, goats consume grasses and forbs (Gordon and Illius 1992) and are less sensible to the Zud effect because of their browsing grazing habit, which is woody plants having usually deep roots that make them less sensible to the effect of summer drought. Therefore, of all Mongolian livestock, goats can be considered to have the greatest effect on vegetation even after severe natural disasters, such as the Zud.

The spatial analysis was tried by administrative units, that is, *sums*. Nomadic people in Mongolia are eligible to various public services such as education, medical treatment, suffrage, etc., from the Sum in which they are registered as residents. They normally do not move far away without changing their resident registration. In case they decide to move and change the pasture area for their livestock for whatever reason, be it to make money by selling livestock products or to seek a better place for nomadic life far away from the Sum they have lived in so far, they will register their residency in the Sum to which they intend to move. Further, they are required to pay a tax based on their livestock population so that the livestock population datasets provided by the Sum, which was used in this study, reflect the almost exact number of livestock and can partly explain social aspects of Mongolia (Yoshino and Ganbat 2006).

Although livestock distribution is changeable depending on natural conditions such as precipitation, vegetation, temperature, altitude, and water availability, this study revealed that the characteristics of livestock distribution could be partly explained by the demand for products made from each type of animal. Generally speaking, it is difficult for us to obtain time-series statistical datasets of natural conditions spatially throughout Mongolia, which affect livestock distribution. In contrast, datasets of livestock population are easily available for this kind of spatial analysis. Our procedure designated in this study would contribute a macroscopic spatial analysis in Mongolia.

These results should be considered in maintaining a sustainable nomadic enterprise. The local spatial associations of cattle, horses, sheep, and goats were detected by LISA statistics, clarifying that the LH type of local spatial associations are areas where the goat population will potentially increase in the near future and severe land degradation will occur because of overgrazing. LISA statistics may prove to be a useful method for predicting areas of grazing impact and will be increasingly helpful in managing sustainable nomadic enterprises in Mongolia.

References

Andresen MA (2009) Testing for similarity in area-based spatial patterns: a nonparametric Monte Carlo approach. Appl Geogr 29: 333–345.

Anselin L (1988) Spatial econometrics: Methods and models. Dordrecht, The Netherlands: Kluwer Academic Publishers.

Anselin L (1995) Local indicators of spatial association-LISA. Geogr Anal 27: 93–115.

Anselin L (1996) The Moran scatterplot as an ESDA tool to assess local instability in spatial association. In: Fischer M, Scholten H, Unwin D (eds) Spatial analytical perspectives on GIS. Taylor & Francis, London, pp 111–125.

Anselin L, Rey SJ (1997) Introduction to the special issue on spatial econometrics. International Regional Science Review 20: 1–7.

Anselin L, Syabri I, Kho Y (2006) GeoDa: Introduction to spatial data analysis. Geographical Analysis 38: 5–22.

Arguchintseva AV, Arguchintsev VK, Ubonova LV (2008) Modeling of the distribution of anthropogenic pollutants in the atmospheric boundary layer of the city of Ulan Bator. Geography and Natural Resources 29: 146–149.

Badarch D, Zilinskas RA, Balint PJ (2003) Mongolia today. Routledge Curzon, London.

Batty M (1976) Entropy in spatial aggregation. Geographical Analysis 8: 1–21.

Batty M, Sammons R (1979) A conjecture on the use of Shannon's formula for measuring spatial information. Geographical Analysis 11: 304–310.

Chen Y, Lee G, Lee P, Oikawa T (2007) Model analysis of grazing effect on above-ground biomass and above-ground net primary production of a Mongolian grassland ecosystem. J Hydrol 333: 155–164.

Child DR, Byington EK, Hansen HH (1985) Goats in the mixed hardwoods in the southeastern United States. In: Baker FH, Jones RK (eds) Multispecies grazing. Winrock International Institute for Agricultural Development. Morrilton, AK, pp 149–158.

Chuluun T, Ojima D (2001) Sustainability of pastoral systems in Mongolia. In: Abstracts of open symposium on change and sustainability of pastoral land use systems in temperate and central Asia. Ulaanbaatar, Mongolia, p. 75.

Curry L. (1972) A spatial analysis of gravity flows. Regional Studies 6: 131–147.

Degen AA (2007) Sheep and goat milk in pastoral societies. Small Ruminant Research 68: 7–19.

Ellis JE, Swift DM (1988) Stability of African pastoral ecosystems: alternate paradigms and implications for development. J Range Manag 41: 450–459.

Fernandez-Gimenez ME (2002) Spatial and social boundaries and the paradox of pastoral land tenure: a case study from post-socialist Mongolia. Hum Ecol 30 (1): 49–78.

Goldstein MC, Beall CM, Cincotta RP (1990) Traditional conservation on Tibet's northern plateau. National Geographic Research 6 (2): 139–156.

Goovaerts P, Jacquez GM (2004) Accounting for regional background and population size in the detection of spatial clusters and outliers using geostatistical filtering and spatial neutral models: The case of lung cancer in Long Island, New York. International Journal of Health Geographics 3:14.

Gordon IJ, Illius AW (1992) Foraging strategy: From monoculture to mosaic. In: Speedy AW (ed), Progress in sheep and goat research. CAB International, Wallingford.

Jacquez GM, Greiling DA (2003) Local clustering in breast, lung and colorectal cancer in Long Island, New York. International Journal of Health Geographics 2:3.

Kanaoka H (2000) The Sixty Chapters to Understanding Mongolia. Akashi-Shoten, Tokyo (in Japanese).

Lise W, Hess S, Purev B (2006) Pastureland degradation and poverty among herders in Mongolia: Data analysis and game estimation. Ecol Econ 58: 350–364.

Moellering H, Tobler W (1972) Geographical variances. Geographical Analysis 4: 34–50.

Moran P (1948) The interpretation of statistical maps. J R Stat Soc B 10: 243–251.

Morinaga Y, Tian S, Shinoda M (2003) Winter snow anomaly and atmospheric circulation in Mongolia. Int J Climatol 23: 1627–1636.

Moser A (2007) Mongolia: Land of Genghis Khan. Bucher.

Munkhtsetseg E, Kimura R, Wang J, Shinoda M (2007) Pasture yield response to precipitation and high temperature in Mongolia. J Arid Environ 70: 94–110.

Nakaya T (2000) An information statistical approach to the modifiable areal unit problem in incidence rate maps. Environment and Planning A 32: 91–109.

Okamoto G, Himiyama Y (1983) Application of Scale-Variance to Agricultural Geography. The Tohoku Geography 35 (3): 116–117 (in Japanese).

Oniki T, Konagaya Y (2006) The policies for sustainable development of animal husbandry in Mongolia. In: Konagaya Y (ed) Mongolian environment conservation handbook. Kenbunsya (in Japanese, with English abstract).

Pace RK, Barry R, Sirmans CF (1998) Spatial statistics and real estate. Journal of Real Estate Finance and Economics 17 (1): 5–13.

Peters DPC (2002) Plant species dominance at a grassland-shrubland ecotone: An individual-based gap dynamics model of herbaceous and woody species. Ecological Modelling 152: 5–32.

Petit C, Déverchère J, Calais E, San'kov V, Fairhead D (2002) Deep structure and mechanical behavior of the lithosphere in the Hangai-Hövsgöl region, Mongolia: New constraints from gravity modeling. Earth and Planetary Science 197: 133–149.

Pogue DW, Schnell GD (2001) Effects of agriculture on habitat complexity in a prairie-forest ecotone in the Southern Great Plains of North America. Agric Ecosyst Environ 87: 287–298.

Premo LS (2004) Local spatial autocorrelation statistics quantify multi-scale patterns in distributional data: an example from the Maya Lowlands. Journal of Archaeological Science 31: 855–866.

Retzer V, Nadrowski K, Miehe G (2006) Variation of precipitation and its effect on phytomass production and consumption by livestock and large wild herbivores along an altitudinal gradient during a drought, South Gobi, Mongolia. Journal of Arid Environments 66: 135–150.

Retzer V, Reudenbach C (2005) Modelling the carrying capacity and coexistence of pika and livestock in the mountain steppe of the South Gobi, Mongolia. Ecological Modelling 189: 89–104.

Saizen I, Maekawa A, Yamamura N (2010) Spatial analysis of time-series changes in livestock distribution by detection of local spatial associations in Mongolia, Applied Geography 30: 639–649.

Sneath D (1998) State policy and pasture degradation in Inner Asia. Science 281: 1147–1148.

Tachiiri K, Shinoda M, Klinkenberg B, Morinaga Y (2008) Assessing Mongolian snow disaster risk using livestock and satellite data. Journal of Arid Environments 72: 2251–2263.

Templer G, Swift J, Payne P (1993) The changing significance of risk in the Mongolian pastoral economy. Nomadic Peoples 33: 105–122.

Vallentine JF (2001) Grazing management. Academic Press, San Diego.

Walters B, Hall D, Nixson F, Stubbs P (1999) Institutional change in a transitional economy: the reform of economics higher education in Mongolia. International Journal of Educational Development 19: 423–439.

Yoshihara Y, Chimeddorj B, Buuveibaatar B, Lhagvasuren B, Takatsuki S (2008) Effects of livestock grazing on pollination on a steppe in eastern Mongolia. Biological Conservation 141 (9): 2376–2386.

Yoshino E, Ganbat J (2006) Mongolian income tax system and estimates of its gross domestic product. Keizaigaku Kenkyu 55 (4): 31–42 (in Japanese).

Chapter 17
The Cash in Cashmere: Herders' Incentives and Strategies to Increase the Goat Population in Post-Socialist Mongolia

Ai Maekawa

Keywords Cash income • Cashmere • Goat • Livestock ownership • Mongolia • Price of livestock products

17.1 Introduction

There is a direct correlation between the transition to a market economy and the steep increase in the number of goats compared to the other four main kinds of livestock in Mongolia (Fig. 17.1). The environmental impact of this development has been identified as one of the major causes of the degradation of grassland, which has fueled discussion among herders, officials, local and international media, and aid agencies and academics (Badarch and Ochirbat 2002; Lise et al. 2006; World Bank 2003). Concurrently, the benefit of mixed herds for the sustainable use of grasslands is emphasized as a traditional form of knowledge utilized by Mongolian nomads that has been ratified by scientific research (World Bank 2002). All parties concerned acknowledge the unhealthy environmental consequences of the increased number of goats.

Although considered a potentially hazardous phenomenon, this trend for the increased husbandry of goats has continued unabated for decades. To understand why this might be the case, it is necessary to examine the meaning of goat possession for herders. This chapter discusses the growth of the goat population by examining the following points: (1) the historical changes to the labor style of Mongolian herders in modern times; (2) a comparison of livestock products to verify the significance and advantages of breeding goats; and (3) the relationship between

A. Maekawa (✉)
Utrecht, The Netherlands
e-mail: ai@inmongolia.com

N. Yamamura et al. (eds.), *The Mongolian Ecosystem Network: Environmental Issues Under Climate and Social Changes*, Ecological Research Monographs,
DOI 10.1007/978-4-431-54052-6_17,

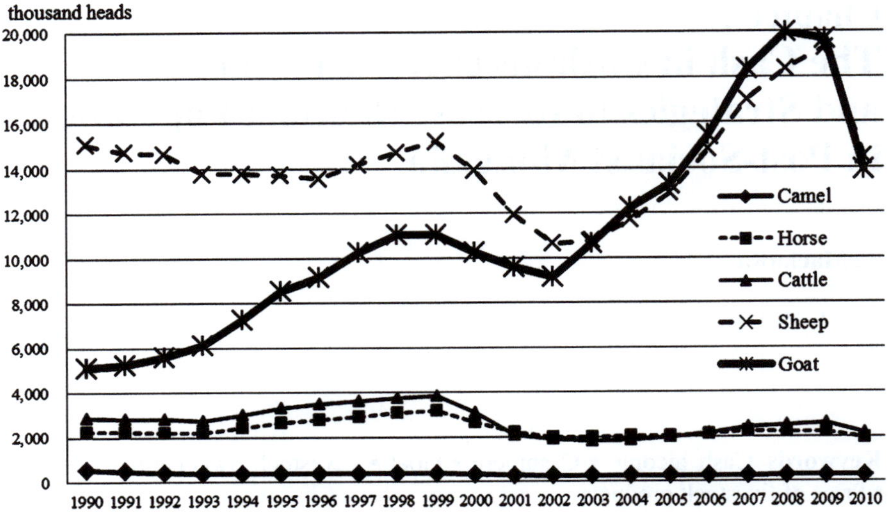

Fig. 17.1 Number of livestock after the transition in Mongolia (1990–2010)

herd size and the proportion of goats in a herd, in response to economic fluctuations observed by fieldwork and questionnaires. Chapter 16 (this volume) provides further understanding of the related issues.

17.2 Historical Background of the Herders' Labor Style in Mongolia

The ownership of livestock is often considered the main cause of the social changes undergone by Mongolian pastoralism. "Livestock privatization" has featured prominently in accounts of the transition to a market economy in the 1990s (Fernandez-Gimenez 2001). Contradicting this approach, this research claims that changes to the style of labor were more decisive.

17.2.1 The Shift of Livestock Ownership and Labor in Modern Mongolia

General Mongolian pastoralists first experienced collectivization in 1930 under the direction of the Comintern, which was repealed after 2 years (Futaki 1993). The second wave of collectivization was completed at the end of the 1950s. It lasted for 30 years and led to the formation of herders' cooperatives, known as *Negdel*.

Livestock ownership has shifted three times, from the ruling, governing, or wealthier classes to ordinary herders. The first shift resulted from the socialist revolution in the 1920s, whereas the second came about because of the policy reversal in 1932. The final shift was caused by the abandonment of the socialist system.

Whether it was deprivation or granting, each shift of ownership was probably considered a significant social achievement at that time. Both devolutions of livestock ownership are indicative of the deprivation of ordinary people's right to choose.

Labor changes were imposed by the power of the time and had a fundamental impact on sustenance. The biggest transformation brought by collectivization in the 1950s was the establishment of a monetary economy among pastoralists based on equity. Consequently, subsistence herding vanished, and nomads became cash earners by belonging to the herding cooperatives, *Negdel*. Although herders always had a right to possess a certain amount of private livestock (Futaki 1993), each of them was obligated to become salaried members of the cooperative, thereby depriving them of individual choice.

The labor quota was heavy for *Negdel* members, and their salary was based on an achievement-oriented system (Kazato 2009). Furthermore, the socialist government kept herders' salaries low compared to workers in the newly built cities. From the 1970s until 1990, *Negdel* members were earning (on average) only 50–60% as much as city workers were earning each month (Futaki 1993). Herders had stable incomes; nevertheless, their work was treated as less profitable because the prices of livestock products were rigid and controlled by the government. As a result of the collectivization completed by the end of the 1950s, all Mongolian herders in effect became salaried workers. In a sense, then, conditions were ripe for the introduction of a monetary economy.

17.2.2 The Collapse of the Cash Salary Supply System: Dismantlement of the Herder's Cooperatives

By the end of 1991, most *Negdels* were dismantled by the transition to the market economy, which is commonly regarded as the achievement of the citizens' spontaneous political movement that rose to prominence in 1989. However, if we look only at the practical financial consequences for ordinary herders, it soon becomes apparent that they simply lost their monthly cash income because of the decisions of authority. They suddenly became self-employed freelance individual proprietors.

Herding was thereby turned into a subsistence level of economic activity, even though Mongolian society had completed realization of a monetary-based economy by the collectivization process. Needless to say, simple herding does not bring a monthly cash income to the owner of the livestock without economic activity in the market, yet livestock owned by the cooperatives was privatized along with the transition process. All herders had to undertake this unfamiliar business activity in which they became responsible for the generation of a cash income. In short, they became "free" herders in the market economy.

Discrepancies between the rich and the poor became prominent because of the various conditions attached to commercial activity. Kazato (2009) noted that many herders deemed the dividends of cooperatives' livestock unfair during the first period of privatization. Differing degrees of adaptive capability to the new economic and trading behavior among individual herders was another significant variable, which manifested as income disparities among households. Additionally, the relative distances from the market contributed to regional income disparities that were mostly alleviated by state control during the socialist era (Konagaya 2007).

Furthermore, Mongolian herders needed a greater cash income after losing their stable cooperative salary as a result of the democratization process, because the socialist state had provided many fundamental social services free of cost, such as medical treatment and education. However, all citizens had to pay in cash for these social services after the transition.

This research has argued that the most significant change to ordinary herders' lives caused by the transition was the loss of a regular, guaranteed cash salary. This economic transformation explains the behavior of herders, such as their internal migration patterns, their fluctuating proportions of livestock, and their acute dependence on cashmere. Statistical data and concrete examples of herders' behavior offer further support to this argument.

17.3 Characteristics and Price of Livestock Products

Following the collapse of the cooperatives in 1991, herders converted their actual livestock or livestock products into commodities. This phase meant they had to choose the most profitable option available to generate their cash income.

17.3.1 Differences of Livestock Products in Commercial Value

Table 17.1 shows the general characteristics of the main livestock products. Raw milk is unsuitable for transportation. Herders located very close to cities can sell this product; however, its price drops easily because the milk supply season is concentrated in summer. Second, sheep wool has almost no international and little domestic demand because of its character, which limits its use, while supply is abundant. Therefore, the market price is very low. Cashmere has plural benefits as a commercial product. Although demand is affected by the global economic climate (Fig. 17.2), the final products have attained a high status in the international market for luxury goods. Additionally, this product does not require temperature control on its way to the markets. This is a particularly beneficial side benefit for those herders who inhabit remote areas because it makes it possible for them to compete with their counterparts who have ready geographic access to the market.

Table 17.1 General characteristics of livestock products in the Mongolian context

	Reproducible products (milk and animal hair)			Nonreproducible products (meat and skin)	
	Raw milk	Sheep wool	Cashmere (goat wool)	Mutton/beef	Sheep (live/head)
Transportation difficulty to the market	Difficult	Possible	Possible	Possible but perishable in summer	Possible but high cost labor
Weight	Heavy	Light	Light	Heavy	Heavy
Destination for consumption	Domestic	Domestic	Domestic and international	Mainly domestic	Domestic
Business opportunity	Mainly in summer	Once a year	Once a year	Year-round	Year-round
Scarcity as resource	Low, but higher in cities and in winter	Very low	High	Higher in cities	Higher in cities

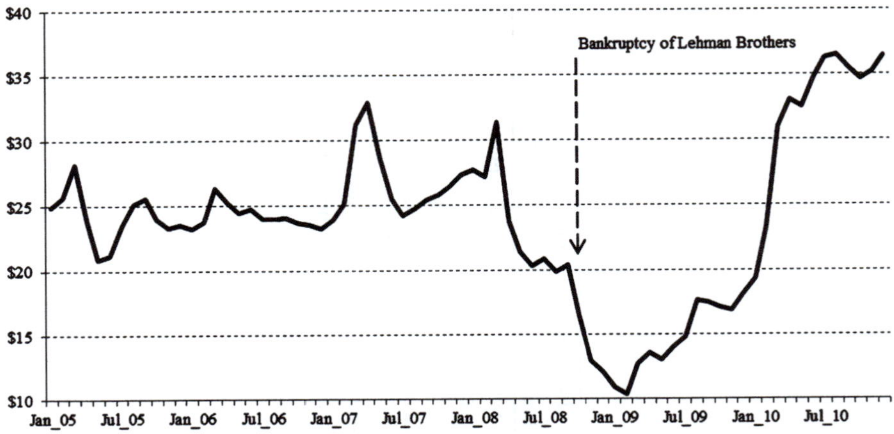

Fig. 17.2 Average price of cashmere in the domestic market, January 2005 to December 2010, monthly, per kilogram, US$ (price is converted into US$ by the average exchange rate of the month). *Source*: National Statistical Office of Mongolia (2005–2011), Mongol Ulsin niigem, ediin zasgiin baidal. 2005 January to 2011 January, in total 73 reports (in Mongolian)

The demand and prices for meats are relatively stable in the cities. Given the weight of this product, it is more ideally suited to herders living closer to the sedentary areas. Livestock are often brought to the cities even from the remote areas because of the discrepancies in meat prices between the cities and rural areas. Once livestock are sold as meat, herders have lost some of their property. Therefore, selling meat or live animals is not the most favored option to earn cash for households who only breed a relatively small number of livestock.

17.3.2 Price Difference of Livestock Products

Table 17.2 shows the average price of livestock products from 2005 to 2010. It is obvious that the price of cashmere surpasses its competitors, which offers a great incentive for herders to breed goats. Although a goat produces a very limited yearly quantity of cashmere (approximately 200–500 grams, g), the cash potential is extremely high. If the yield of one goat is estimated as 250 g, herders can receive about 8 US$ per animal, according to the 2010 price level (National Statistical Office of Mongolia 2011).

Table 17.2 does not include the price of sheep wool, but it was reported as trading at 0.21 US$ at the livestock market of Ulaanbaatar in March 2010 (Mongoru Tsuushin 2010a,b). One sheep yields approximately 1,500 g wool each year. If herders can sell it at the average market price, they can earn 0.3 US$ per sheep annually.

Thus, herders can anticipate a 25 times larger cash income per year (approximately) by breeding one goat, compared to having one sheep, unless they slaughter the animal. The value of goat meat per kilogram was 80% of sheep meat at the end

Table 17.2 Annual national average price of livestock products in US$[a]

	2005	2006	2007	2008	2009	2010
Raw milk/liter	0.46	0.48	0.54	0.80	0.94	1.14
Cashmere/kg	24.14	24.20	26.80	21.21	14.71	32.37
Mutton/kg	1.55	1.74	1.97	2.45	1.73	2.42
Beef/kg	1.65	1.80	2.04	2.53	2.31	3.01

Source: National Statistical Office of Mongolia (2005–2011) Mongol Ulsin niigem, ediin zasgiin baidal. 2005 January to 2011 January, in total 73 reports (in Mongolian)
[a]Price is converted into US$ by the average exchange rate of the month

of 2010 (National Statistical Office of Mongolia 2011). Therefore, the price difference between sheep and goat meat cannot be a major incentive to have sheep if the herders focus on increasing their cash income.

The prices of other livestock products (Table 17.2) reinforce the fact that herders seem to prefer goats because they enable them to sell cashmere. As already mentioned, other products have multiple disadvantages for their sellers that were not caused by the price difference. It is clear that efforts to increase the proportion of goats originated from herders' rational evaluations of their standing as self-employed members of the capitalist economy.

The problem associated with cashmere is the price fluctuation (Fig. 17.2), which is influenced by (1) the regulation of exports by the Mongolian government for the protection of the domestic cashmere industry, (2) the favorability shown by the global economic market toward luxury goods, and (3) the massive depletion of the goat population caused by harsh winter conditions. More recently, it has been reported that cashmere has become a target for speculative purchasing by Chinese traders (Mongoru Tsuushin 2010a,b). Indeed, the steep price rises in 2010 were attributed to the combination of a severe winter and these speculative dealings.

In any case, herders have no means available to minimize the price fluctuations, partly because their chance to sell cashmere is often limited by their distance from the market. In this sense, herders can be regarded as passive players in the cashmere trade, even though they are producers. To maximize their profit, they are always very aware of price fluctuations and act flexibly by changing their herding strategy, if possible. The next section describes the concrete strategies employed by herders in response to price fluctuations.

17.4 Differentiation of Herders' Strategies and Behaviors Reflecting the Size of Herds

17.4.1 The Relationship Between Herd Size and Goat Ratio

Figure 17.3 is based on surveys conducted by the author in 2008 and 2009. A logarithmic relationship is shown between the total number of small livestock (sheep and goat combined) and the number of goats. The slope of the regression line was 0.825.

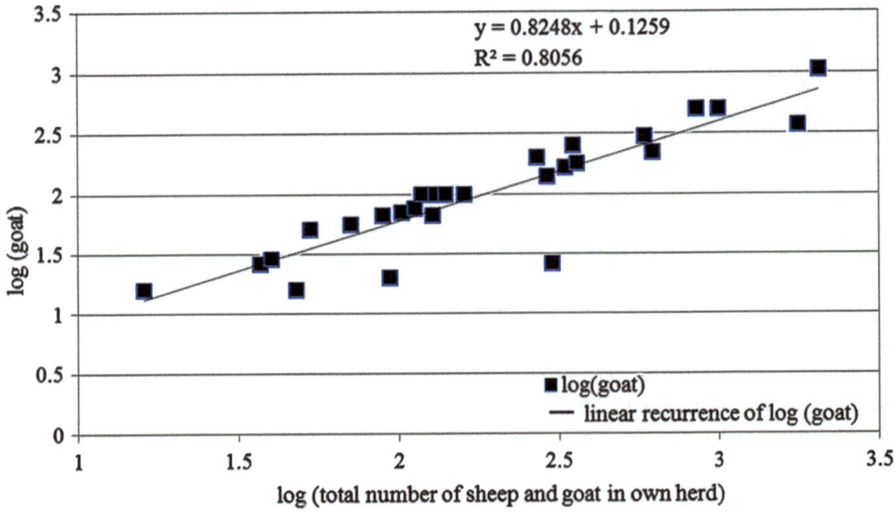

Fig. 17.3 Relationship of herd size to number of goats

If the slope were 1, the number of goats would increase at the same rate as the small livestock herd's total number. As the slope is less than 1, we can say that when the total number of sheep and goats was higher, the proportion of goats was lower. Hence, Fig. 17.3 indicates that herders with larger herds had a smaller ratio of goats.

It is natural to assume that the owners of larger herds are wealthier. We have to find a reason then why wealthier herders have a smaller proportion of goats, and why the owners of herds predominantly composed of goats can be poorer, despite goats being a resource for a cash income.

There are two possible explanations. First, if herders have a larger number of livestock, it is likely they have additional resources at their disposal, which they are able to convert into cash as needed. Selling livestock is a very flexible way to make money in contemporary Mongolia, because the domestic demand for sheep meat is stable as an increasing proportion of the general population do not own livestock, especially in the cities. Although meat prices change on a seasonal basis, these fluctuations are not to the same degree as cashmere. If the herd is sufficiently large, a certain amount of livestock reduction is not detrimental to the maintenance of the size of their herd or the resources at their disposal, because they can expect a natural increase in livestock through births. Thus, there is ordinarily an opportunity for households who have more than a certain number of livestock to make cash. The size of the herd that makes this possible correlates to the size of the family and the respective region. Obtaining cash by selling cashmere is a choice for people who wish to avoid reducing the number of livestock they possess. Cashmere sales do not decrease the size of herds until herders decide to slaughter or sell, or the goats die of natural causes.

The second reason is that people are aware of the environmental risk of having too many goats because it is a principle of their traditional knowledge as well as being current and common discourse in society more generally. Although selling cashmere is a rational economic decision to maximize income, many herders consider the breeding of too many goats harmful to the maintenance of good pastures. Hence, they prefer to have more sheep than goats.

The national number of goats surpassed sheep in 2003 (Fig. 17.1) for the first time in the history of the Mongolian plateau. Traditionally, the goat population had been stable at around 20%, even after modernization introduced by the socialist regime. Many of the goat-breeding herders who responded to the questionnaires expressed their concerns about this abnormal situation (Table 17.3). Their answers showed that a dependency on a cashmere income is a morally challenging choice in spite of its exceptional value as a Mongolian livestock product. It appears that this choice contributes to their wealth, but not to their moral well-being, or happiness.

17.4.2 Differing Strategies in Response to Economic Fluctuations Among Herders

Table 17.3 shows the results of a survey of herders conducted in 2008 and 2009 on their preference for proportions of sheep and goats in their herds. It was conducted in Dundgovi Aimag, Tsagaan-Sain Sum, and in the western outskirts[1] of the capital city, Ulaanbaatar. The herders interviewed in the capital arrived from all over Mongolia.

The clearest result was that none of them wanted to increase the number of goats, a notable change affected by the global financial crisis that began in September 2008. As a consequence, cashmere prices rapidly decreased during the time of the research (Fig. 17.2). The herders were obviously anxious about the future price of cashmere and were forced to consider a shift to other livestock products to lessen their dependency on cashmere. Irrespective of how many livestock they had, there was a consensus that the outlook had worsened considerably.

The herders could be divided into four distinct groups. The first group consisted of several herders who mentioned the damage caused to pastures by goats. They expressed their intention to increase their number of sheep. Although it appears as though they were motivated by a sense of ecological awareness, it must also be

[1] This place is called the "22nd check point" or "22 *Tovchoo*" in Mongolian. There is a regulation that restricts the entrance of live animals for commercial purposes beyond this border. Therefore, people selling live animals personally to city consumers directly gather outside of the border annually. Those herders are relatively wealthy because coming to the capital city is expensive. Often it takes more than 1 month to reach Ulaanbaatar, and furthermore, it takes many weeks to sell all their livestock. Therefore, only when they can anticipate additional benefits by going to Ulaanbaatar will they start traveling with their sheep and goats. Cars are usually used to transport these animals so as to avoid the extra cost of hiring trucks.

Table 17.3 Herders' preferences and reasons for increases to their species of livestock

	A	B	C	D	E	F (%)	G	H[a]
1	Nov-08	Dundgovi	Sain Tsagaan	100	250	71.4	2	Goats are very bad for the pasture.
2	Nov-08	Dundgovi	Sain Tsagaan	32	16	33.3	2	I will increase my number of sheep to sell its meat.
3	Nov-08	Dundgovi	Sain Tsagaan	30	100	76.9	2	It depends on the pasture's capacity. But sheep is better for pasture, so I prefer sheep to goat.
4	Nov-08	Dundgovi	Sain Tsagaan	2	51	96.2	2	Price of meat does not go down much, but the price of cashmere is unstable.
5	Nov-08	Dundgovi	Sain Tsagaan	285	300	51.3	2	Goats are bad for the pasture. The price of goat meat and cashmere went down, so I will increase sheep.
6	Nov-08	Dundgovi	Sain Tsagaan	180	180	50.0	2	N/A
7	Nov-08	Dundgovi	Sain Tsagaan	40	100	71.4	2	N/A
8	Nov-08	Dundgovi	Sain Tsagaan	36	76	67.9	3	I will keep the size of my herd of sheep and goat within 200 in total. Plus, I breed two productive milk cows.
9	Nov-08	Dundgovi	Sain Tsagaan	19	99	83.9	2	The price of cashmere is declining.
10	Nov-08	Dundgovi	Sain Tsagaan	163	167	50.6	3	I want to produce something which I can sell for a high market price.
11	May-09	Dundgovi	Sain Tsagaan	0	16	100.0	2	Goats are bad for the pasture.
12	May-09	Dundgovi	Sain Tsagaan	11	29	72.5	3	The number of livestock number is small.
13	May-09	Dundgovi	Sain Tsagaan	60	100	62.5	2	Mutton is good for your health and easy to sell.
14	May-09	Dundgovi	Sain Tsagaan	11	26	70.3	3	The number of livestock is small
15	May-09	Dundgovi	Sain Tsagaan	23	66	74.2	2	I will increase my holdings of livestock which have a less harmful effect on the pasture.
16	May-09	Dundgovi	Sain Tsagaan	44	0	0.0		The number of livestock is small.
17	May-09	Dundgovi	Sain Tsagaan	275	26	8.6	3	N/A
18	May-09	Dundgovi	Sain Tsagaan	62	66	51.6	3	N/A
19	May-09	Dundgovi	Sain Tsagaan	74	20	21.3	2	Sheep is profitable if it is sold as meat. The price of cashmere went down.
20	May-09	Dundgovi	Sain Tsagaan	31	71	69.6	3	The number of livestock is small.
21	May-09	Dundgovi	Sain Tsagaan	15	56	78.9	2	N/A
22	Aug-09	Govi-Altai	Delger	70	200	74.1	2	Goat meat is cheap and the price of cashmere recently went down.
23	Aug-09	Ovor khangai	Burd	500	500	50.0	2	Sheep survives more when Zud comes.

24	Aug-09	Arkhangai	Tsetserleg	800	0	0.0	2	I don't have goats at all, because I dislike them. I am not happy that goats are increasing in my Sum.
25	Aug-09	Zavkhan	Bayan-Khairkhan	1400	370	20.9	other	If the price of cashmere goes up, I will increase the number of goats; otherwise I will increase the sheep.
26	Aug-09	Selenge	Orkhontuul	400	220	35.5	2	I will increase sheep, because goats require much more care than sheep.
27	Aug-09	Govi-Altai	Togrog	150	140	48.3	3	Now I want to extend the herd size, so I will increase both goats and sheep.
28	Aug-09	Zavkhan	Tsetsen-Uul	350	500	58.8	2	Right now, mutton is more profitable, that's why I increase sheep. When the price of cashmere recovers, I will increase goats.
29	Aug-09	Uvs	Ondor-Khangai	1000	1050	51.2	2	Mutton is more expensive.

A: Research month and year
B: Present residence Aimag
C: Present residence Sum
D: Possessing sheep number
E: Possessing goat number
F: Goat proportion in total of sheep and goat
G: Question "Which small livestock would you like to increase now?"
1. Goat
2. Sheep
3. I won't change the proportion of livestock
H: Reasons for answer to "G"

[a]Original descriptions are written in Mongolian. They are translated by the author

emphasized that goats accounted for more than 50% of their herds at the time of the questionnaire. That is to say, they were conscious of the heavier impact goats had on the pasture, but continued to breed goats for economic reasons. It was not expressed as such, but it can be assumed it was the drop in cashmere price, rather than these ecological arguments, that motivated them to change their strategy.

The second group explicitly indicated the lower—and falling—price of cashmere as a justification for increasing the proportion of sheep in their herds. Their answer was clearly and overtly a reaction to the changed economic situation. They argued that selling mutton would be more profitable and stable than dependency on the highly fluctuating economic benefits of cashmere.

The third group indicated they would increase their proportion of sheep as well. However, the telling difference was that they would increase their number of goats if the price of cashmere recovered in the future. Their views were backed by a long-term herd management strategy. Additionally, their herd sizes were relatively larger, making them successful competitors in the market economy.

The fourth group was adamant that they would not change the proportion of animals in their respective herds. Most of them regarded their herds as small, and thus their primary focus was on increasing their number of livestock in general, regardless of species. Expanding the size of their property remained a long-term goal.

In essence, the survey results were consistent with broader market trends, which is why generally a shift from goats to sheep emerged very strongly as a management strategy. This finding supports the notion that herders are continuously making an effort to adjust their herding style to the dynamics of the market economy.

17.5 Discussion and Conclusion

It has frequently been remarked that herders increased their numbers of livestock after the transition because they retrieved the private ownership of livestock and were motivated to herd freely without cooperative control (Komiyama 2005; Badarch and Ochirbat 2002). Although certainly an explanatory factor, it must be considered alongside herders' critical needs for cash income. The latter is decisive in the modern Mongolian society, where a money economy had already been well established during the socialist collectivization. It is not appropriate to presume that it mainly derived from a sense of liberty from the negative collectivization by the socialist regime.

Each herder's pragmatic necessity for cash has been deciding the overall outcomes of animal husbandry in Mongolia subsequent to the transition. Herders evaluate the most rational way to maximize cash income, as described in this chapter. They are sensitive to economic trends in their efforts to avoid any risks to their income. Therefore, they are attentive to the price fluctuations of livestock products. The ecological sustainability of pasturelands is accorded lower priority, despite their expressed awareness of the impact of goat browsing.

It can be concluded that only when an alternative to cashmere is found, as occurred during the period of price instability during 2008–2009, is there any likelihood of a decline in the proportion and absolute number of goats in Mongolia. However, as Fig. 17.2 has shown, the price of cashmere boomed once again after the period of price instability. We can easily predict that this acts as an incentive for herders to breed goats, at least in the short term.

The development of an alternative cash resource for the herders, specifically for those with small herds, should be the subject of future discussion. Such initiatives would solve the problem of overgrazing by goats, which presents a real danger to pasturelands (as so convincingly described by Saizen in Chap. 16, this volume). Failure to act will mean that goat numbers will continue to fluctuate in accordance with global and domestic economic trends. It is beyond dispute then that herders will face an unhappy future so long as they remain dependent on goats.

References

Badarch D, Ochirbat B (2002) On sustainable development of cashmere production and goat population in Mongolia. In: Oka H (ed) Mongolian Studies at CNEAS. CNEAS Monograph Series 6:187–200.

Futaki H (1993) Basic structure and reform of agriculture. In: Aoki N (ed) Mongolian economy under the transition. Institute of Developing Economies Kenkyu Sosho 438:103–133 (in Japanese).

Kazato M (2009) Ethnography of contemporary Mongolian nomads – living in the post-socialist era. Sekai shiso sha (in Japanese).

Komiyama H (2005) Actual damage to Mongolian animal husbandry due to 2000–2002 Dzud disaster. Bull JAMS 35: 73–85.

Konagaya Y (2007) Characteristics and transformation of pastoral system in Mongolia. E-journal GEO 2(1): 34–42 (in Japanese).

Lise W, Hess S, Purev B (2006) Pastureland degradation and poverty among herders in Mongolia: data analysis and game estimation. Ecol Econ 58: 350–364.

Fernandez-Gimenez, M.E (2001) The effects of livestock privatization on pastoral land use and land tenure in post-socialist Mongolia. Nomadic Peoples 5(2):49–66.

Mongoru Tsuushin (2010a) Montsame 2010. March 5. Montsame (in Japanese).

Mongoru Tsuushin (2010b) Montsame 2010. April 23. Montsame (in Japanese).

National Statistical Office of Mongolia (2005–2011) Mongol Ulsin niigem, ediin zasgiin baidal (Monthly report of Mongolian society and economic status). 2005 January to 2011 January, in total 73 reports (in Mongolian).

National Statistical Office of Mongolia (2011) UNE110119. http://www.nso.mn/v3/files/UNE110119.xls. Accessed 23 Feb 2011.

World Bank (2002) Mongolia Environment Monitor. http://siteresources.worldbank.org/INTEAPREGTOPENVIRONMENT/Resources/MongEnvMonitor2002eng.pdf. Accessed 23 Feb 2011.

World Bank (2003) From goats to coats: institutional reform in Mongolia's cashmere sector. Report no. 26240-MOG. http://www-wds.worldbank.org/external/default/WDSContentServer/WDSP/IB/2006/05/02/000012009_20060502111130/Rendered/PDF/262400ENGLISH0rev0MG0Cashmere01PUBLIC1.pdf. Accessed 23 Feb 2011.

Chapter 18
Marketability of Dairy Products in Relation to Distance in Northern Mongolia

Shinichi Mori

Keywords Dairy • Farm • Market • Milk • Mongolia • Semi-intensive

18.1 Introduction

The land in northern Mongolia, which includes among others the provinces of Selenge, Bulgan, and Tov, is suitable for dairy farming because of the predominance of relatively grassy pasturelands resulting from higher precipitation than in other areas. Before a market economy system was introduced in the 1990s, there were 45 dairy farms maintaining a total of 16,000 cows that provided raw milk to dairy factories in urban areas (Komiyama 2006). These large-scale state dairy farms were dismantled in the early 1990s on the introduction of the market economy system. Many cowsheds were dismantled, and the milk and dairy products distribution system collapsed. Under the state and collective farming systems, herders were divided into several functional units in which they concentrated on specific tasks, most herders therefore only having partial knowledge about dairy farming. These herders, who after the dismantling of the state farms were given a couple of cows, knew neither how to breed good-quality cows, as they lacked proper knowledge on artificial insemination, nor how to improve a dairy farm's productivity. As a result, the quality of cows throughout the country has degraded and milk production has rapidly decreased. Because of insufficient supplies of raw milk in addition to the collapse of its collective distribution channels, many of the dairy factories in urban areas ceased functioning, causing Mongolia, a country previously with animal husbandry at its core that was once self-sufficient for dairy products, to become an importer of powdered milk and long-life milk.

S. Mori (✉)
International Management Group, Inc, Yokohama, Japan
e-mail: smori@xp.wind.jp; smori@ga2.so-net.ne.jp

N. Yamamura et al. (eds.), *The Mongolian Ecosystem Network: Environmental Issues Under Climate and Social Changes*, Ecological Research Monographs,
DOI 10.1007/978-4-431-54052-6_18,

There are some successful cases that have recently been created by the combination of private initiatives and international donor community support, where dairy farms with improved breeds, created by artificial insemination, have become capable of supplying milk to the market throughout the year, although most of these farms are located close to urban areas. Herders who reside far from urban areas have difficulty selling their dairy products because of high transportation costs, which results in an increase in their dependency on cashmere for their income. This chapter analyzes the advantages and disadvantages brought by the distances[1] from urban markets in the production of dairy products in northern Mongolia, based on which income-generating opportunities from dairy products for the herders who reside far from the urban markets are examined.

The author of this chapter has, since 2003, been engaged in the establishment and management of a model semi-intensive small-scale dairy farm with a cheese factory in Sant Sum,[2] Selenge Province. Most of the discussion in this chapter is based on empirical evidence gained though what has been learned, witnessed, and experienced through running of this dairy farm as well as though his own research about dairy production and marketing methods conducted in different Sums and provinces in northern Mongolia.

18.2 Price and Cost Factors of Raw Milk

In Mongolia, herders usually control the reproductive cycle of their livestock to avoid the risk of losing newborn animals during the harsh winters, meaning that deliveries of most newborn animals take place in spring. As a result, the peak milking period is inevitably concentrated during summertime, causing an oversupply of milk that leads to a fall in the milk price in the summer months. In contrast, milk prices sharply rise during the winter months because of the shortage of raw milk throughout the country. For example, the prices of raw milk brought by herders to be sold at Ulaanbaatar markets in 2009 were about 600 Togrog (0.5 USD) per liter in the summer and about 900 Togrog (0.75 USD) in the winter.

Herders residing close to main paved roads or railway stations are often observed delivering their raw milk directly to the market or consumers. There are also milk collection points on main roads where dairy companies' trucks circulate collecting raw milk from herders. If herders do not reside close to main roads or railway stations, transport costs increase sharply because a vehicle's fuel and depreciation expenses increase.

---

[1] The concept of "distance" in this chapter applies not only to physical distance but also to the level of transport costs and the amount of extra time incurred by poor road conditions.

[2] Sant Sum is located at approximately 350 km to the north of Ulaanbaatar, and 100 km from both Darkhan and Erdenet, the second and third largest cities in Mongolia. Because the Sum is 15 km on a rough road distant from the main road, transport costs are high.

Shifting the peak milking period to winter is one way to increase a herder's income, although it is not easy to achieve because other elements, such as artificial insemination techniques, feed costs, and improved dairy farm management, have to be factored in.

18.2.1 Artificial Insemination Techniques[3]

There are several challenges in the application of artificial insemination (A.I.) in Mongolia. First, to achieve A.I.'s cost efficiency, a certain scale of the service area needs to be ensured, because semen needs to be preserved and transported in a tank filled with liquid nitrogen that must be refilled every 3 months. Second, for A.I. to be applied immediately when heat is occurring,[4] (1) A.I. equipment and semen need to be maintained within the Sum, (2) several herders or veterinarians within the Sum must acquire A.I skills, and (3) individual herders must continue monitoring the heat cycle of their cows to avoid the heat being overlooked; failure to apply A.I. at the time of heat means the cow produces no milk for another month. For the Sum to overcome these challenges, interventions by the government or international donors will usually be needed as A.I. skills and delivery mechanisms need to be established.

18.2.2 Feed Costs

As there are relatively abundant rangelands in northern Mongolia, herders living in these areas are able to collect hay with minimal transport costs; conversely, dairy farms located close to urban areas must pay a significant amount of money to buy and transport hay from northern Mongolia.[5] During the summertime, herders usually graze their cattle on pasturelands at almost no cost, whereas they must feed hay to their penned-up cows during the wintertime. Because it is imperative to provide a large amount of feed to cows producing milk to maintain high production, shifting the milking period to winter means that herders must provide a large amount of hay to feed their cows over the winter period.

18.2.3 Improved Dairy Farm Management

Newborn calves need more intensive care in winter than in summer. Moreover, a warm, comfortable cowshed is needed to maintain high milk yields. Therefore, good dairy farm management is a prerequisite for shifting the milking period to winter.

---

[3] Artificial insemination is applied at 99% of dairy farms in Japan.

[4] Alternatively, it is also possible to artificially induce heat by an ovulation inducer drug.

[5] As of summer 2009, the price of a pack of hay (25 kg) was 1,300 Togrog (0.9 USD) in Sant Sum and 2,000 Togrog (1.4 USD) in Ulaanbaatar.

Many herders in northern Mongolia use a simple pen for their animals to get through the winter, which is not appropriate for newborn calves or cows producing milk. Herders who wish to shift the milking period to winter need to build cowsheds with the correct insulation.

Taking the foregoing factors into consideration, and as it requires more money, more techniques, and more intensive care, dairy farms located close to urban areas have fewer incentives to shift the milking period to winter, compared to those located far from urban areas. One main factor to be considered when making the shift to winter is that a considerable portion of the cash income generated from milk sales during the wintertime will be consumed by feed costs.

Another way to increase a dairy farm's income is to process milk into value-added products, such as butter and cheese. However, in Mongolia there is an abundant supply of homemade butter and cheese produced in traditional ways, and thus the profit margins on these products is quite small compared to the labor required. Mongolian homemade butter also faces severe competition from cheap Russia-made butter. An increase in the middle-income population in urban areas, mainly brought about by booming mining industries and from the remittances sent from migrant Mongolian workers in foreign countries such as Korea, is creating market opportunities for non-traditional, high-valued dairy products, including European-type cheese.[6] Because cheese production is generally a labor-intensive industry, in those areas where manual labor is relatively cheap because of the long distance from urban areas, they have a comparative advantage in production cost terms, which could offset part of their additional costs for transport to markets and help redress the disadvantages of their locality. The difference observed in profitability of raw milk and cheese in relationship to distance from markets is further discussed in the next section.

18.3 Difference in Profitability of Dairy Products in Relationship to Distance to Urban Markets

In this section, the relationship between the cost factors and a herder's distance to markets, which combined determine the profitability of raw milk and high-valued cheese, is explained in charts. In this chapter, to carry out a comparative analysis of Mongolian dairy farms, the farms have been categorized into three groups: farms at a "short" distance from markets that are able to sell raw milk in summer; farms at a "medium" distance from markets that are only able to sell raw milk in winter; and farms at a "long" distance from markets that cannot sell raw milk in any season.

Feed, labor, and transport costs are the main variables that are affected by distance. As Fig. 18.1 shows, costs of feed and labor decrease in proportion to greater distance to markets. Conversely, the transport costs of raw milk increase relatively and proportionally to greater distance to markets.

[6] Although there is a variety of traditional cheese in Mongolia, their fermentation periods are mostly short and their water content is low compared to European-type cheese.

Fig. 18.1 Distance and cost factors

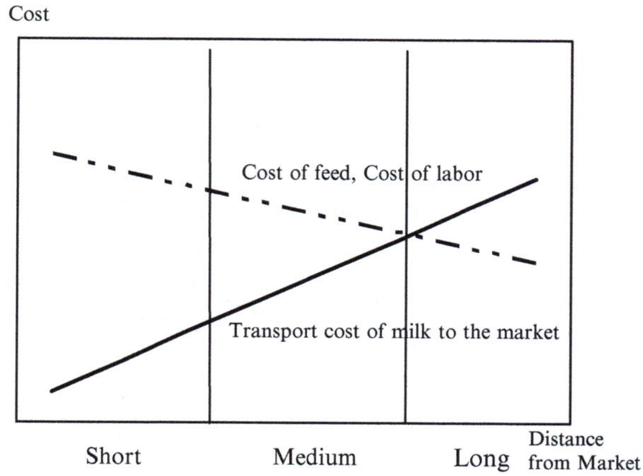

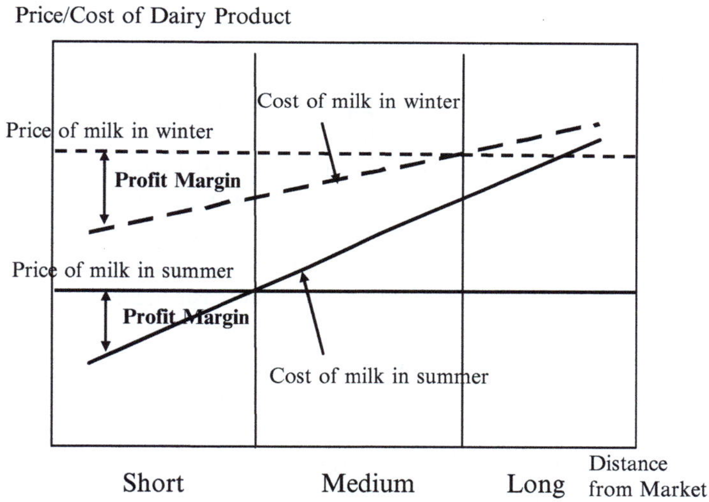

Fig. 18.2 Distance and profit margin from milk sales

In the wintertime, both the price and cost of raw milk are higher than in the summertime because of the shortage of milk and the increased feed cost in winter months, respectively. As dairy farms at locations remote from urban areas need to spend lesser amounts of money on feed because abundant pasturelands are close by, compared to those farms located close to urban areas, their disadvantage caused by additional transport costs is partly offset. As a result, the profit margins on raw milk sales, which are negative in summer for the farms in the "medium" distance category, become positive in winter. Farms in the "long" distance category are not able to sell raw milk even in winter, because the transport costs still outweigh the benefits of the increased wintertime profit margins (Fig. 18.2).

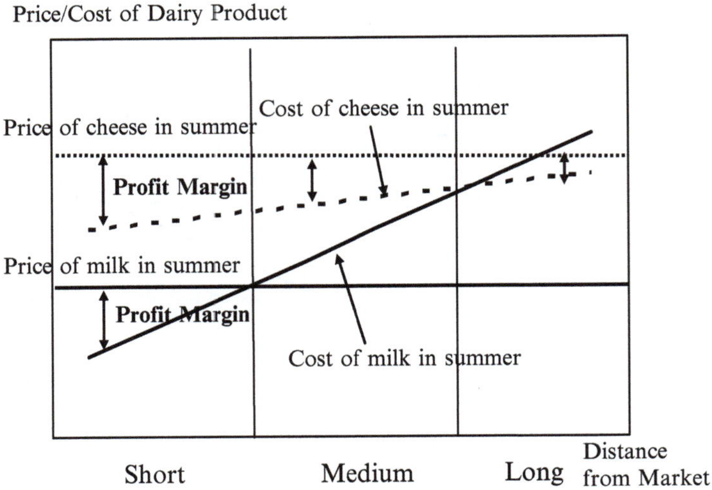

Fig. 18.3 Distance and profit margin from cheese sales

When milk is turned into high-valued cheese[7] (such as European-type cheese) with the use of additional labor, new opportunities arise for herders residing far from urban areas. As labor costs are lower in rural areas because employment opportunities are fewer, the disadvantages caused by high transport costs are further mitigated. Even those herders who reside at a long distance from urban areas have an opportunity to gain profit from this type of cheese production. Herders residing at medium and long distances from markets, who cannot make any profit from raw milk during summertime, are now able to generate cash income if they produce high-valued cheese (Fig. 18.3).

Those herders who are already able to sell their raw milk in urban areas have less incentive to shift their resources into high-valued cheese production because this requires not only an investment in equipment and storage facilities, but also proper cheese-making skills, intensive labor, and quality control measures.

In conclusion, for herders in disadvantageous locations in relationship to markets to generate a cash income from dairy production, they have to either shift their milking period to winter or proceed with the production of high-valued products. Table 18.1 summarizes the product most suited to each category of herders in terms of generating a cash income and the input needed to realize it.

As Table 18.1 shows, additional facilities, equipment, skills, and efforts are required for those herders who are located at a medium or long distance from urban areas to produce a dairy product suitable for generating a cash income. Without support from the government or international donor agencies in mobilizing these resources, the daily farming carried out by these herders will remain at a subsistence level, and thus they will continue depending on cashmere as their only cash income source.

[7] As fermentation does not take place at low temperature, cheese can only be produced in the summertime in Mongolia.

Table 18.1 Dairy product and necessary input by distance from market

Distance	Short	Medium	Long
Most suited product	Raw milk in summer	Raw milk in winter	High-valued cheese
Necessary input	Nothing particular	Hay making Artificial insemination Additional feed in winter Improved dairy farm management including a warm cowshed	Investment in equipment and facilities Production skills Intensive labor Quality control measures

Reference

Komiyama H (2006) Economic analysis of settled and semi-settled animal husbandry in Mongolia: a focus on dairy farming. Dissertation, Tokyo International University

Chapter 19
The Impact of Agricultural Development on Nomadic Pastoralism in Mongolia

Yuki Konagaya

Keywords Agricultural development • *Atar* • Mongolia • Nomadic pastoralism • *Sharilj*

19.1 Agriculture in Mongolia During the Pre-Socialist Period

According to the findings of archeology, cultivation in Mongolia began in the Neolithic period and developed during the Bronze Age (Dorj 1971). Much later, in the ninth century C.E., the Kirgiz people, who had been living and cultivating in the upper basin of the Yenisey River, moved into Mongolia. They probably maintained their own agricultural knowledge and technology on the Mongolian plateau. After that, in the fourteenth century, Mongolian leaders directed Chinese soldiers to make farms (Perlee 1959). In this way, there is much historical evidence of agriculture on the Mongolian plateau. We must take climate change into account when we are concerned with agriculture as human intervention in the natural environment. Therefore, here I cover only the period from the latter half of the nineteenth century to the end of the twentieth century, when the climate seems to have been about the same as it is today.

During this period, well-known Russian explorers noticed that there were agricultural fields near the temples and especially in the western part of Mongolia. For example, Potanin, who traveled through western Mongolia from 1876 to 1877, wrote about the Muslim people called Khoton (Potanin 1881). He described them as living around Lake Uvs and growing barley around the temple of Ulaangom, using irrigation water drawn from the Terin River. He wrote also about the *Tarachin* (tariachin in Mongolian means men of crops, or cultivators), who lived around Lake Khar Us,

Y. Konagaya (✉)
National Museum of Ethnology, Suita, Japan
e-mail: yuki@idc.minpaku.ac.jp

N. Yamamura et al. (eds.), *The Mongolian Ecosystem Network: Environmental Issues Under Climate and Social Changes*, Ecological Research Monographs,
DOI 10.1007/978-4-431-54052-6_19,

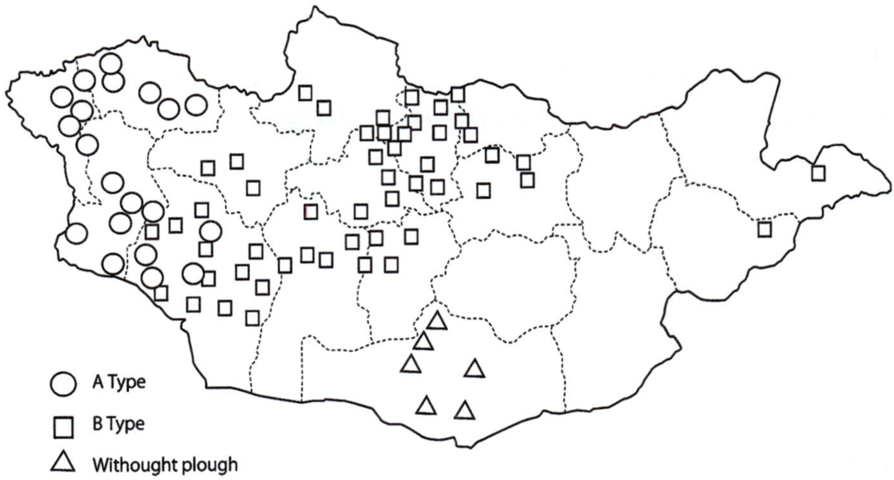

Fig. 19.1 Distribution of types of plough. *Source*: Academy of Science in Mongolia (1979). *Atlas of Ethnography and Linguistics in the Mongolian People's Republic*

and that they were poor employees on the official farms supervised by a Chinese director. In a further example, Pozdneev wrote in his expedition report in 1847 about a cultivating system with a fallow period (Pozdneev 1896–1898).

On the basis of such information in these travel reports, Mongolian ethnographers have done surveys in the field, and the data were compiled into an atlas of ethnography (Academy of Science in Mongolia 1979). Here I would like to recompile some of the maps relating to cultivation into two maps (Figs. 19.1 and 19.2). From these maps we can grasp the outline of traditional agriculture in Mongolia. From the point of view of the types of ploughs, we can distinguish three types of traditional cultivation in Mongolia: one is the western type, another is the central type, and yet another is the southern type that was practiced without ploughs. From the aspect of crops, we can also identify three types of traditional cultivation: one is foxtail millet, which was widespread in central Mongolia, another is wheat or barley, in western Mongolia, and a third is vegetable farming by the Chinese.[1]

The official ethnography of the socialist period, which was published in 1996 (Badamkhatan 1996), reported that some western Mongolian ethnic groups, for example, the Dorbod, Oold, Torguud, Bayad, Uriankhai, and Zakhachin, but excluding the Myangad, maintained agriculture in the traditional manner until after the socialist revolution. From the descriptions, we can understand the following: (1) they had irrigation

[1] In the Orkhon-Selenge River Basin, foxtail millet or wheat was cultivated using a large plough, and in the Gobi Region foxtail millet was cultivated without a plough. In the West Altai Region, barley was cultivated using a small plough, and also in the tributary region of the Zavkhan River wild millet was gathered.

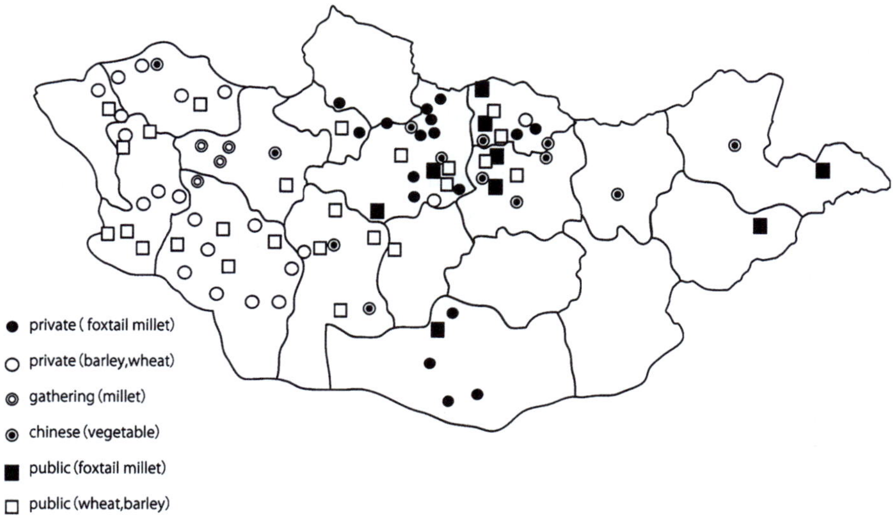

Fig. 19.2 Distribution of types of management and crops. *Source*: Basic Atlas of Mongolian Peoples' Republic (1990)

systems; (2) they used riverside areas for cultivated fields, far away from the summer rangeland; (3) they cultivated mostly barley and naked barley; and (4) they utilized fallow periods.

This traditional agriculture must have been sustainable because it utilized land with water resources and coexisted with the herding of animals (Erdenebaatar 1996: 101–104).[2]

19.2 The Policy of Agricultural Development as Socialist Modernization

In the former Soviet Union, after the death of Stalin in 1953, Khrushchev presented the famous project of virgin lands development of fourteen million hectares. This enormous project was initiated beginning in Kazakhstan. Modeling itself after the Kazakhstan project, in Mongolia agricultural development became one of the most important official targets. Dry farming which is quite different from the traditional way had been introduced. In 1959, at the Third General Assembly of the Central Committee of the Mongolian People's Revolutionary Party, the Virgin Lands Campaign was begun.

[2] It is said to originate from the seventeenth century under the Jungar Khan dynasty. Also, it may have been begun by Turkish Muslims who were moved for agricultural production purposes.

Thereafter, in 1976 at the Eleventh General Assembly of the Central Committee of the Mongolian People's Revolutionary Party, the policy entitled "Cultivating the uncultivated land in order to increase production of crops, especially cereals" was affirmed and begun, which was the start of the second movement of agricultural development.

In the Mongolian language uncultivated land is called *Atar*, and cultivation of it as a movement of agricultural development is called *Atar ezemshik*, which means to hold uncultivated land, usually called just *Atar* for short. Hence, we can refer to the policy of agricultural development as *Atar*.

The chief executive in charge of *Atar* during its first stage was Mr. Lookhuuz, who was purged for 25 years beginning in 1964. From his narrative of his life history, we can understand the following: (Lkhagvasuren and Konagaya 2007).

1. In the beginning, the Mongolian delegation went to the Soviet Ministry of Agriculture and often visited the state farms in Kazakhstan as a model.
2. The Mongolian government invited Russian specialists in the management of state farms from the Soviet Ministry of Agriculture as foreign advisers who had a great influence on political decisions.
3. The Russian pedologist who was a member of the Soviet Council for Foreign Economic Cooperation led the soil investigation and indicated clearly suitable places for cultivation.
4. The Mongolian Ministry of Foreign Affairs and Trade also committed to introducing machines for agriculture from the Soviet Union.
5. The Mongolian Revolutionary Association for Youth organized the *Atar*, sending young people to the fields. At that time, young herders who were losing animals because of enforced integration of cooperatives were pouring into the capital city, and the government had to take measures to create employment opportunities for these unemployed youths.

In Mongolia, in contrast to the Soviet Union, agriculture developed as a quite new modern sector, mostly taking the form of state farms. Therefore, we can understand where agriculture developed through the allocation of state farms (Fig. 19.3). Figure 19.4 shows that area and production had increased considerably, and after 1980 the state farms had the potential to export wheat, up until the democratization movement in 1990.[3]

[3] More than 80% of the cultivated area is planted with wheat, so only the production of wheat is shown here. Until 1979, the production of wheat was not stable, but from 1980 to 1989 wheat flour was exported.

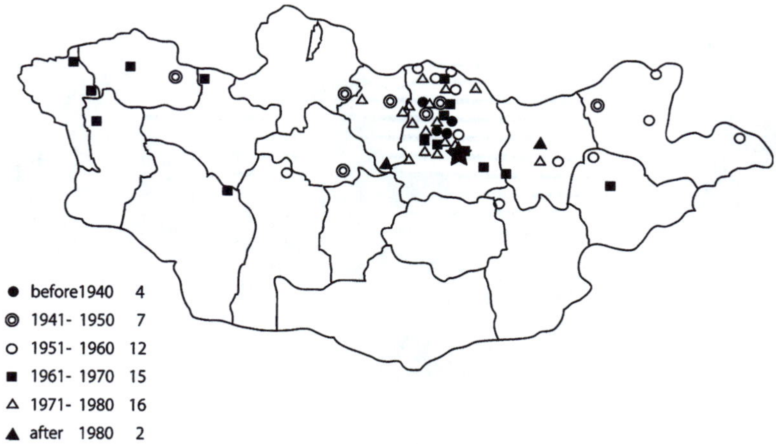

Fig. 19.3 Sites of Mongolian state farms (up to 1990). *Source*: Basic Atlas of Mongolian Peoples' Republic (1990)

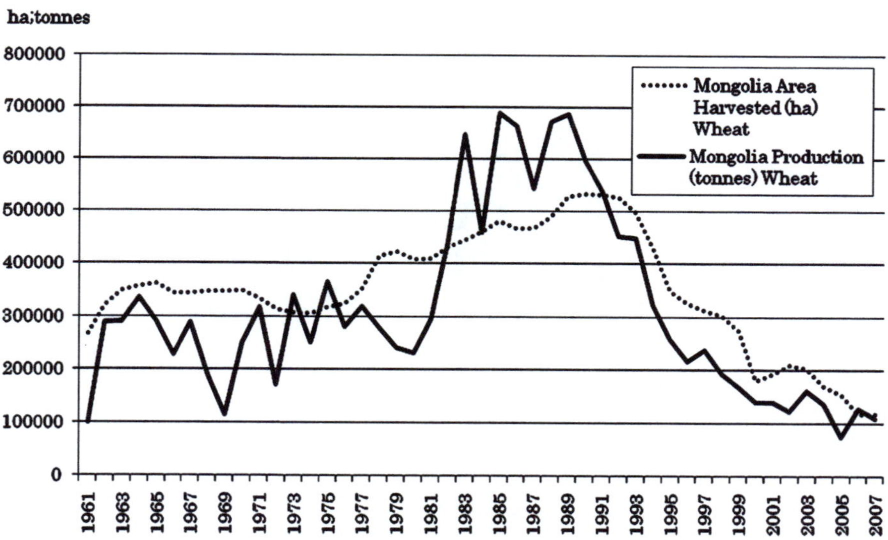

Fig. 19.4 Area harvested and production of wheat (tonnes per hectare, ha) in Mongolia by year. *Source*: FAOSTAT

19.3 The First *Atar* and the Second *Atar* During the Socialist Period

In the first *Atar* (starting in 1959), the area newly cultivated was suitable for cultivation on the basis of soil investigation. According to the work of Rolomjav, who was a specialist in Mongolian agricultural history, Soviet scholars started field surveys

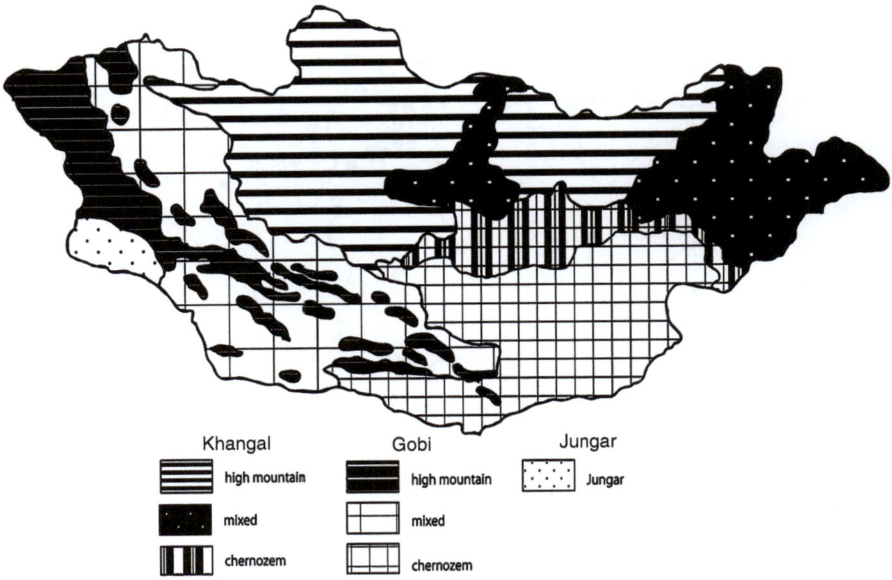

Fig. 19.5 Soil types in Mongolia. *Source*: Basic Atlas of Mongolian Peoples' Republic (1990)

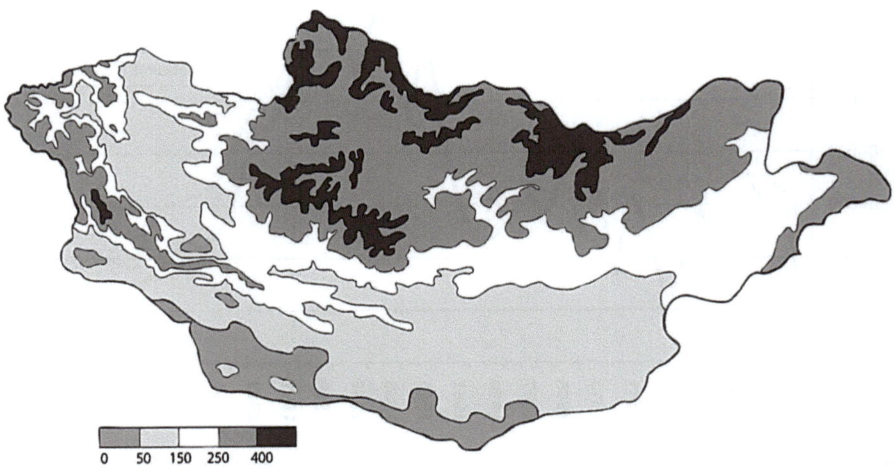

Fig. 19.6 Climate zones in Mongolia (precipitation in mm/year). *Source*: Basic Atlas of Mongolian Peoples' Republic (1990)

in Mongolia early in the 1920s (Rolomjav 1984: 76), and Bespalov wrote a book entitled *Mongolian Soil* in Russian (Bespalov 1951). Bespalov concluded that the area suitable for cultivation in Mongolia is the basin of the Orkhon-Selenge Rivers (Rolomjav 1984: 78). We can now understand that the first phase of modern cultivation in Mongolia was rational from the point of view of ecology, as is evident by looking at the maps of soils and precipitation (Figs. 19.5 and 19.6).

Syubin's book entitled *The History of Mongolian Agriculture* was also important in supplying academic support for the first phase of *Atar*. This book was published in 1953, just when Khrushchev declared the virgin lands development. In this book, Syubin summarized that there were four traditional cultivating areas: Orkhon-Selenge, western Mongolia, the Gobi-Altai lake basin, and eastern Mongolia (Shubin 1953: 81). What is interesting is that on the map of these traditional cultivation areas he never distinguished Chinese-cultivated areas from Mongolian, and just classified them according to the size of the cultivated area. We can see that the author's concern with their size reflected the spirit of modern times. Based on the results of the first 5 years of *Atar*, Mr. and Mrs. Konokov published a book entitled *Mongolian Agriculture* in 1964 in Russian. They concluded that 1.5 to 1.7 million ha can potentially be cultivated in Mongolia. In reality, only 0.32 million ha were being cultivated at most. The researchers may have overestimated the area.

However, the area suitable for cultivation parallels that with forest and is always good for herding livestock as well, which means that agriculture was competing for land with livestock herding. On the other hand, the newly developed agriculture provides forage for domestic animals. Thus, agriculture could eliminate the insecurity of nomadic pastoralism, opening the way for the shift from nomadic pastoralism to sedentary livestock farming.

In the second *Atar* (starting in 1976), to provide food for the increasing numbers of citizens in the capital city, the area around the capital was developed as state farms. Thus, cultivation began of ecologically unsustainable areas (Konagaya 2010: 56).

19.4 Case Study of the State Farm Named *Atar*: From Its Establishment to Its Collapse

For the case of the second *Atar*, I introduce here the state farm named *Atar*, which may be a good representative. According to the narrative of the leading manager or the farm foreman of this state farm (Konagaya 2010: 107–133),[4] the national purpose of the second *Atar* was to provide food for the growing urban population, which was realized in the following ways:

1. The Central Committee of the Mongolian People's Revolutionary Party, as the executive decision-making body, nominated the farm foreman who had much experience of management in the Mongolian Revolutionary Association for Youth.
2. The farm foreman bought the machines, for example, trucks and combines, etc., from the mechanical support center attached to the Mongolian Ministry of Agriculture.

[4] The interview was conducted on 12 August 2008, assisted by Dr. Lkhagvasuren and financially supported by the RHIN project.

3. At the first stage, a Russian specialist in the Mongolian Ministry of Agriculture advised on the area and the seed for cultivating.
4. Mongolian specialists in agriculture and animal husbandry were also recruited.
5. The drivers of machines were employed from among the graduates of the technical college attached to the state farm named *Bayanchandamani*, located north of the capital city.
6. In the area planned for construction of the state farm, herders had been living, so that the uncultivated land was not virgin but had been used for herding.
7. The Soviet Trust for construction made efforts to build some facilities such as a hospital, schools, dormitories, a telephone office, and a cultural club.
8. Some prefectures gave herds of domestic animals to the state farm as social capital.
9. The variety of wheat grown, called "Saratov-2," was not irrigated but was grown relying on rainfall alone.
10. Annually, the specialist of agriculture, called in Mongolian *Agronom*, decided the start of furrowing and sowing, which were extremely critical for unirrigated farming.
11. Trucks coming from the transportation station in the capital city carried wheat to the milling factory by contract, and if the yield was more than planned, the cooperative got a special bonus.
12. On the farm, potatoes, turnips, carrots, onions, garlic, and cabbages were cultivated in addition to wheat.
13. They had a dairy farm with 400 cattle, and managed a carpet factory and a construction cooperative.

Thus, building a state farm was making a new community where people came together from many different regions, and agricultural development has rearranged or disturbed the society as well while cultivation has disturbed the vegetation.

Because the state farm *Atar* performed well, people chose that as a model that worked for cooperatives, even after the end of socialism. However, they failed in making a transition to a market economy. Now the uncultivated fields have been overrun with the weed *Artemisia*, which is an allergen that adversely affects the health of the citizens (Fig. 19.7).

19.5 The Deterioration of Grasslands in the Former Cultivated Area

The "shock therapy" in the economy destroyed the agricultural cooperatives during the transition period. The government, receiving the advice of international organizations, has stopped any financial support, and the banks followed belatedly, and with a high rate of interest for short-period loans. Therefore, the socialist modern cooperatives could not buy seeds in time and had poor yields. Also, the maintenance of machines required was costly. Without financial support, the cooperatives definitely could not maintain themselves.

Fig. 19.7 Abandoned fields with *Artemisia*

Fig. 19.8 Recovering through plant succession

In the two decades of transition to a market economy, many cultivated fields that were abandoned have now become weedy plains, always with *Artemisia*. Natural disturbances such as drought lead to a successional recovery with meadow grasses, although *Artemisia* is prevalent (Fig. 19.8). In contrast, artificial disturbances such as cultivation remove the subterranean roots, which does not easily lead to succession, so the fields stay weedy with *Artemisia* in the long term. Technical methods are required to create succession and recover the grasslands.

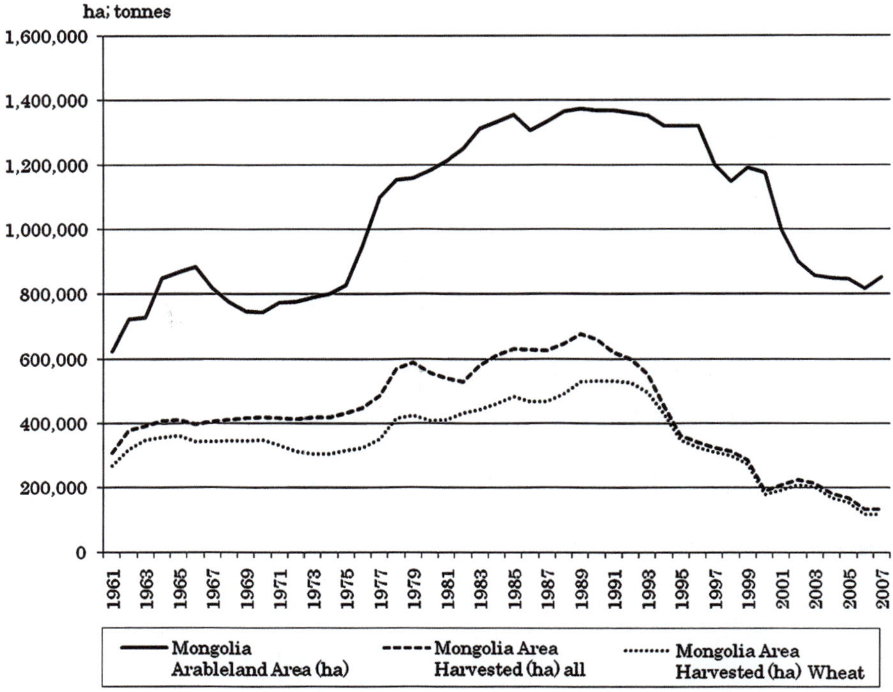

Fig. 19.9 Arable and harvested areas (tonnes per hectare, ha) in Mongolia by year. *Source*: FAOSTAT

Figure 19.9 shows the increasing amount of arable land area, and we can clearly identify the two older periods. The first wave was from 1959 to 1965, and the second from 1976 to 1988. Almost 600,000 ha that were developed during the socialist period are no longer being used for cultivation. Many hectares must be more or less infested with *Artemisia*.

19.6 The Third *Atar* and Its Hidden Problems

The Mongolian government issued, in January 2008, a challenge to investors to reconstruct the agricultural sector. This new deal was called *The Third Atar* and it meant agricultural redevelopment. A new program to give financial support for buying new agricultural machines came from examining the mistakes made during the transition period. People who have worked in cultivation were delighted with this policy, although some people were opposed. In general, agricultural redevelopment in the area suitable for cultivation was approved. However, almost all the agricultural redevelopment was being conducted near the capital city because of accessibility to the market, and that approach means that ecologically no one can guarantee the sustainability of new cultivation in the urban area.

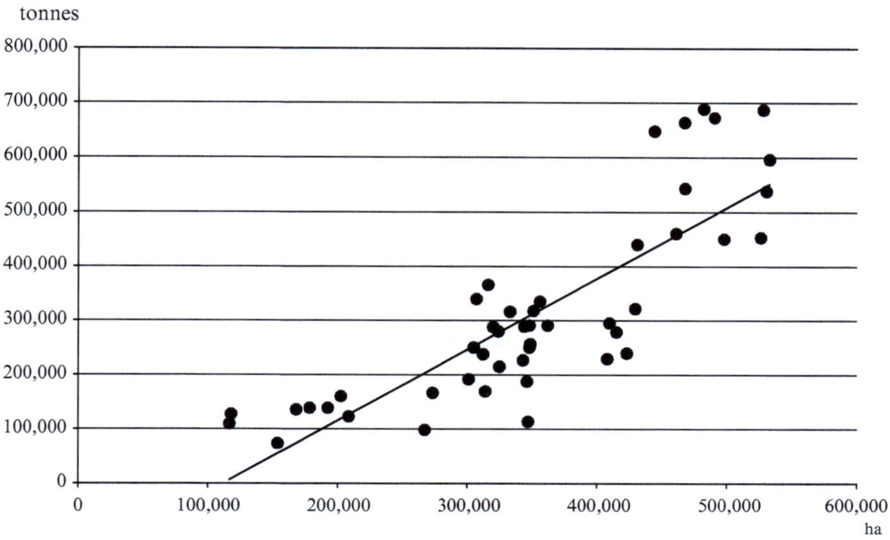

Fig. 19.10 Correlation between production (tonnes) and harvested area (ha) of wheat in Mongolia. *Source*: FAOSTAT

On 11 December 2009, the Mongolian prime minister made a speech about the results of the third *Atar*. The cultivated area now totaled more than 350,000 ha, and 389,100 tons of wheat and 156,000 tons of potatoes were harvested. It is amazing that in such a short period they were able to produce enough to meet almost all the demand for crops. In this current situation, I would like to propose a limitation to cultivation by applying the concept of carrying capacity.

The yield of wheat can be fairly well correlated with the harvested area: the coefficient of correlation is 0.83 (Fig. 19.10). However, for more than 400,000 ha, the correlation is weaker, at 0.67, whereas for an area less than 400,000 ha it is 0.72. In other words, 400,000 ha may be the limit of sustainable cultivation in Mongolia. The concept of carrying capacity is usually used for domestic animals in pastoralism. In this arid area, however, this concept needs to be applied much more to cultivation, because extreme cultivation brings about a negative impact on the grasslands.

Modern agriculture was introduced in Mongolia as a state project during the socialist period. This concept was a huge discontinuity from the local agricultural practices of the past. In a way, the introduction bore resemblance to the introduction of the mining industry. Both projects started by buying machines, then excavating the soil with those machines, and getting yields from the earth. Because modern agricultural practice was formulated by the state as a new industry with an economic quota, environmental awareness for sustainable use of the natural resources did not develop among the people. Thus, modern agriculture in Mongolia has never been a subsistence economy.

Additionally, the land law has made agricultural redevelopment much more dangerous from the aspect of environmental protection because the land law allows the people engaged in agriculture only the rights of utilization but no rights of possession. In other words, businessmen engaged in agriculture hold the land without any responsibility for keeping the land productive as their own property. Therefore, they use the land without any environmental consideration. To force mining companies to take responsibility for environmental protection, the mining law has been revised. Similarly, certain regulations are needed for agricultural redevelopment to sustain productive and responsible use of the land.

19.7 Conclusion: Concerning Environmental Consciousness

In Mongolia, the policy of agricultural development has been one of the means of socialist modernization. In the first phase (1940s and 1950s), state farms were constructed and performed well. During the second phase (after 1976) new state farms constructed closer to the urban markets did perform well economically but at a considerable ecological cost. After the end of the socialist regime, because of the lack of financial support, the state farms began to collapse. Therefore, in 2008 the government again began to give financial support to agricultural business. This new policy has split public opinion in two. The negative response comes from concern about ecological conditions. Agricultural business that relies on rainfall is too insecure, and to make it reliable independently of rainfall, it must rely on underground water, which is a scarce resource, limited, and can never be forecast exactly. Thus, the policy of agricultural redevelopment, as well as mining-related environmental destruction, brought environmental consciousness to Mongolian society.

Generally speaking, nomads move to access fresh resources for the sake of their own animals, which is concern for the environment from a utilization perspective, but not a consciousness aimed at protection. The nomads can move their herds even if they are not aware of the modern global idea of environmental consciousness, and through these moves they can preserve their environment. Therefore, we can understand that their means of environmental conservation comes not from environmental consciousness but from their own empirical way of living. However, recently the mining problems have brought on environmental consciousness, and mining policy has added regulations for the protection of the environment. In contrast, although the agricultural policy also brought on environmental consciousness, agricultural business is allowed to proceed freely without any concern for environmental consequences.

What is the impact of agricultural development on nomadic pastoralism in Mongolia? In some regions, such as Selenge Prefecture, where the former state farms are dominant, competition between agriculture and nomadic pastoralism is a serious problem. However, such impacts may not be equal all over the nation. It is important to note that modern agriculture caused Mongolian pastoralism to become less mobile, with an unchanged lack of environmental consciousness. Nowadays nomads with less mobility have to be conscious of the environment as never before, as do agricultural businessmen. We must doubt that the nomads now understand environmental conservation simply because nomadic pastoralism has changed to a less mobile form under the socialist modernization accompanying agricultural development.

References

Academy of Science in Mongolia (1979) *Atlas of Ethnography and Linguistics in the Mongolian People's Republic* (in Mongolian and French). Ulaanbaatar, Mongolia

Badamkhatan S (1996) *Ethnography of Oirad* (in Mongolian). Ulaanbaatar, Mongolia

Bespalov ND (1951) *Mongolian Soil* (in Russian). Moscow, Russia

Dorj D (1971) *East Mongolia in the Neolithic Age* (in Russian). Ulaanbaatar, Mongolia

Erdenebaatar B (1996) Socio-Economic Aspects of the Pastoral Movement Patterns of Mongolian Herders. In: Humphrey C, Sneath D (eds) *Culture and Environment in Inner Asia*, vol. 1.White Horse Publication, Cambridge

Konagaya Y (2010) The History of Agricultural Development in Mongolia: seeking a tradeoff between development and conservation. (in Japanese). Bull Natl Mus Ethnol 35(1):9–138

Lkhagvasuren I, Konagaya Y. (2007) *The Twentieth Century in Mongolia* (2): Political Life in Socialist Mongolia. Senri Ethnographical Report 72 (in Mongolian). National Museum of Ethnology, Osaka

Perlee K (1959) *Mongolian Old Cities in the Ancient and Middle Ages*. (in Mongolian). Ulaanbaatar, Mongolia

Potanin GI (1881) *Reportage on North-west Mongolia*, vol 2 (in Russian). St. Petersburg, Russia

Pozdneev AM (1896–1898) *Mongolia and Mongolians*, vol 1 (in Russian). St. Petersburg, Russia

Rolomjav B (1976) Form of the Mongolian tradition of cultivation. Study of History 12(8):100–105 (in Mongolian)

Rolomjav B (1983) Transition and Challenges of Development in Mongolian Traditional Cultivation. Study of History 16(6):54–61 (in Mongolian)

Rolomjav B (1984) Soviet Studies on Agricultural Development of Mongolian Peoples' Republic. Study of History 18(10):76–82 (in Mongolian)

Rolomjav B (1987) *Brief history of agriculture in Mongolia* (in Mongolian). Ulaanbaatar, Mongolia

Syubin VF (1953) *History of cultivation in the Mongolian Peoples' Republic* (in Russian). Moscow, Russia

Chapter 20
Conflict Between Mining Development and Nomadism in Mongolia

Yukio Suzuki

Keywords Desertification • Mining • Pasture degradation • Permafrost • Placer gold • River flow

20.1 Introduction

The United Nations, various international organizations, and institutions of foreign countries have issued research reports on the impacts of mining development on Mongolia's pastureland and water. The World Bank reported the destruction and pollution of the natural environment caused by the discharge of tailing by mining companies and Ninjas[1] (informal miners) in various areas, which has typically involved permafrost destruction, decreased river water, and mercury pollution (World Bank 2006). The joint pilot studies conducted by Korea and Mongolia surveyed placer gold mining and indicated why it is essential to establish a comprehensive environmental management system. They alluded to the serious devastation of the river banks and contamination of river water by large-scale placer gold mining and illegal mining activities (Korea Environment Institute and Mongolian Nature and Environment Consortium 2004, 2005).

The United Nations Human Rights Team Group have carried out a fact-finding survey on herders, Ninjas, local government staff, and mining company executives (United Nations Human Rights Team Group 2006). Their findings reveal that local residents are not able to participate in the decision-making processes concerning the commencement of mining in their territories. Moreover, local governments and

[1] Informal micro-scale miners are called "Ninjas" in Mongolia because of the basins they wear on their backs, thought to make them look like characters in the American cartoon, "Ninja Turtles." They use the basins for gravity separation to extract placer gold from sand and gravel in water.

Y. Suzuki (✉)
Ministry of Agriculture, Forestry and Fisheries, Tokyo, Japan
e-mail: yukio.suzuki2010@gmail.com

N. Yamamura et al. (eds.), *The Mongolian Ecosystem Network: Environmental Issues Under Climate and Social Changes*, Ecological Research Monographs,
DOI 10.1007/978-4-431-54052-6_20, © Springer Japan 2013

residents usually cannot oppose the central government and mining companies' power to initiate mining. In addition, countermeasures were found to be necessary. Provision should be made for the participation of local residents throughout all stages of the decision-making process, including exploration. Furthermore, transparent procedures must be put in place before the commencement of any mining operations in their territory.

Tumenbayar has argued that the rights of herders under the current Mongolian legal system are so obscure that they are vulnerable in comparison to the mining rights prescribed in Mongolia's minerals law. Consequently, herders have difficulties maintaining effective pastureland management (Tumenbayar 2002).

Placer gold mining in rivers is said to have an especially serious impact on water flows. A research team that included members of the Mongolian University of Science and Technology issued a report focusing on the decreased water flow of the Ongi River. This is a known dry river on the way from Ovorkhangai Prefecture ("Aimag" in Mongol, hereinafter referred to as "Aimag") to its destination, the Ulaan Lake in Omnogovi Aimag (Mijiddorji and Bayasgaralan 2006).

In light of the conflicts between the mining and nomadic animal husbandry sectors, which have been compounded to some extent by the global economic crises in 2008 and their recourse to the legal systems, a more specific and comprehensive analysis is carried out in this chapter, with additional reference to the natural environment.

20.2 Two Main Sectors in Mongolia

20.2.1 Trend of the Two Main Sectors

In the mid-1990s, the animal husbandry and crop farming sector accounted for 40% of Mongolia's gross domestic product (GDP). Until that time nomadic animal husbandry had traditionally been the dominant sector, making use of the vast pasturelands that account for 80% of Mongolia's total land area. Recently, however, it has declined to 20% of total GDP. In contrast, the mining sector has grown rapidly since 2003, comprising more than 30% of total GDP in recent years (Fig. 20.1).

Mongolia produces a wide range of livestock products, many of which are exported, including raw and processed cashmere, live animals, meats, wool, and hides. The mining sector, meanwhile, has extracted underground resources such as copper, gold, and coal, mostly for export. An examination of Mongolia's exports for 2000 shows that livestock products and mining products, respectively, accounted for approximately half of total export value. Since then, however, mining sector products have grown rapidly to more than 80% of total exports, whereas livestock products had shrunk to about 10% in 2008[2] (Fig. 20.2).

---

[2] Looking at the number of major five livestock species in Mongolia, the number of goat, sheep, cattle, horses, and camels has increased from about 24 million in 2002 to 43 million in 2008. However, the weight accounting for the Mongolian economy has been shrinking.

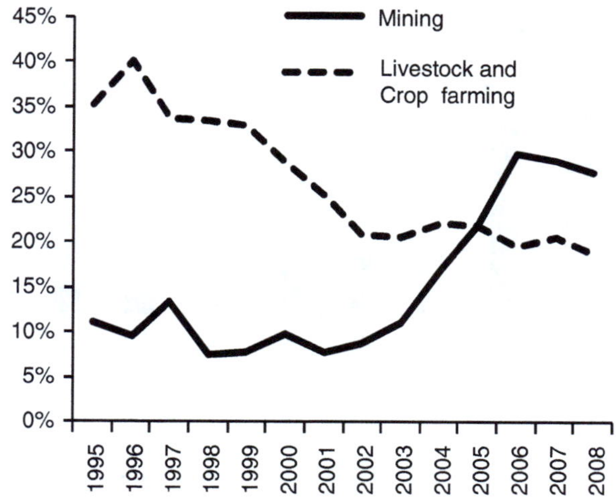

Fig. 20.1 Ratio of each sector of total gross domestic product (GDP). *Source*: National Statistical Office of Mongolia

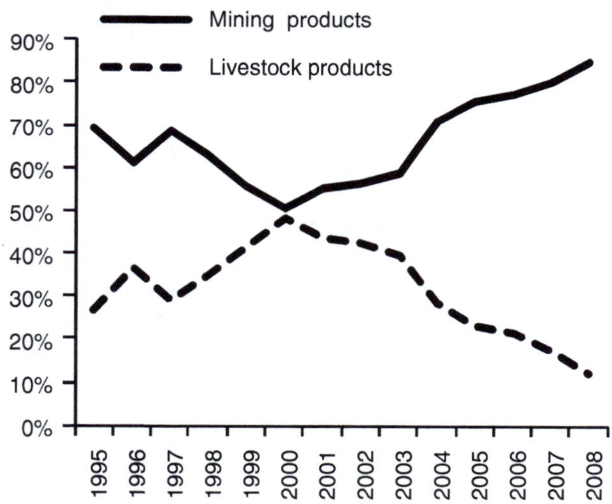

Fig. 20.2 Ratio of each sector of total export value. *Source*: National Statistical Office of Mongolia

Most of the products of Mongolia's mines have been exported. Figure 20.3 shows the value of exports for 2008. Copper and gold are predominant. Coal, zinc, molybdenum, petroleum, iron ore, and fluorite come next. Cashmere is the fourth largest export item, having about the same value as coal and zinc, and the largest value among livestock products.

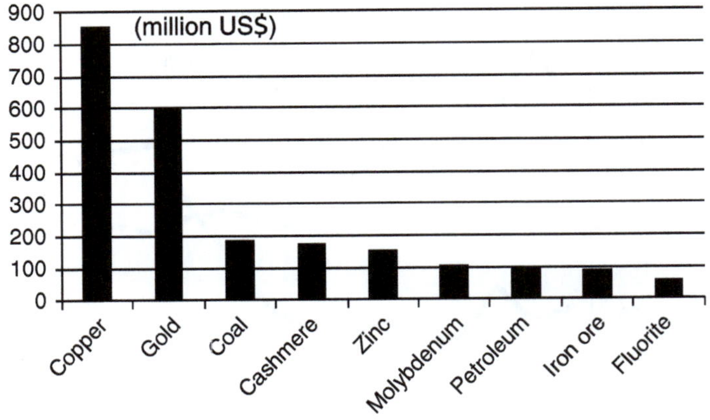

Fig. 20.3 Export value of each item in 2008. *Source*: National Statistical Office of Mongolia

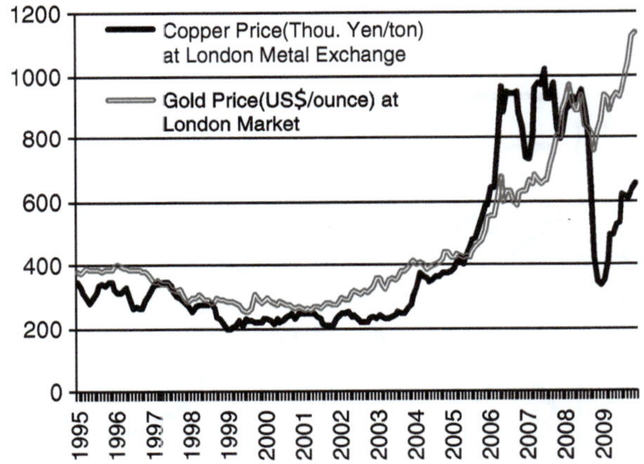

Fig. 20.4 Trend of monthly price of copper and gold. *Source*: Copper prices at London Metal Exchange; gold prices at London Market. Copper prices: http://www.jcma2.jp/dou.htm. Accessed 16 Jan. 2010. Gold prices: http://gold.tanaka.co.jp/commodity/souba/m-gold.php. Accessed 16 Jan. 2010

20.2.2 Impact of the Mining Industry in the Mongolian Economy

Copper and gold are Mongolia's predominant exports, as just mentioned. Figure 20.4 shows the price of copper and gold on the international market in London, with prices increasing sharply between 2003 and summer 2008. Prices dropped drastically, however, after the global financial crisis triggered by the Lehman Brothers collapse in mid-September 2008. The price of copper plunged to one-third of its peak price from September to December 2008.

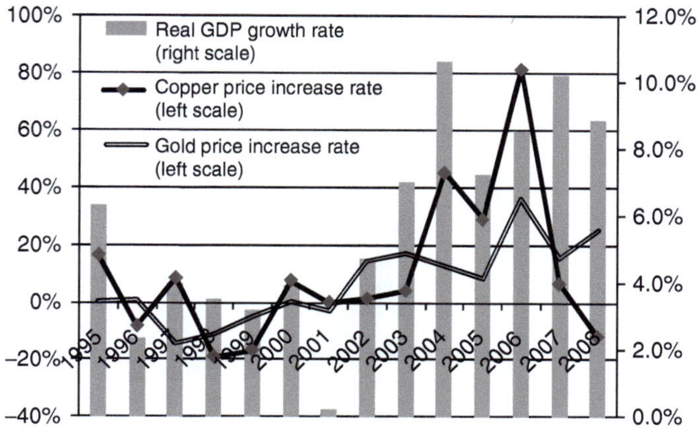

Fig. 20.5 Correlation between prices and GDP. *Source*: GDP: http://www.imf.org/external/pubs/ft/weo/2010/02/weodata/index.aspx. Accessed 16 Jan. 2010. Copper prices at London Metal Exchange; gold prices at London Market. Copper prices: http://www.jcma2.jp/dou.htm. Accessed 16 Jan. 2010. Gold prices: http://gold.tanaka.co.jp/commodity/souba/m-gold.php. Accessed 16 Jan. 2010

The price of the country's major mining products on the international market seems to have a significant impact on Mongolia's economy, which enjoyed high growth rates of more than 7% after 2003.

Figure 20.5 shows the correlation between the increase in copper and gold prices and the growth of GDP, using annual data from 1995 to 2008. Mongolia's GDP is greatly affected by fluctuations in the international prices of these commodities. The outcome of a multi-regression analysis is as follows:

Multi-regression analysis between copper and gold price and GDP:

- Multiple correlation coefficient: $R=0.73$
- Multiple determination coefficient: $R^2=0.53$
- Multi-regression formula:

$$\text{Growth rate of real GDP} = 0.045 + 0.02 \times \text{increase rate of copper price} + 0.137 \times \text{increase rate of gold price}$$

Since September 2008 the rapid fall in export revenue, which is largely dependent on copper and gold, has caused financial crises in Mongolia including (1) a shortage of foreign currency, (2) a sudden drop in the foreign exchange rate of the national currency, the togrog, and (3) a lack of financial resources. The lack of financial resources in banks has brought an end to loans to private companies.

Banks stopped lending to construction projects that had not reached 80% completion. As a result, construction work on many buildings has stalled halfway because of the lack of financing triggered by the global economic crisis.

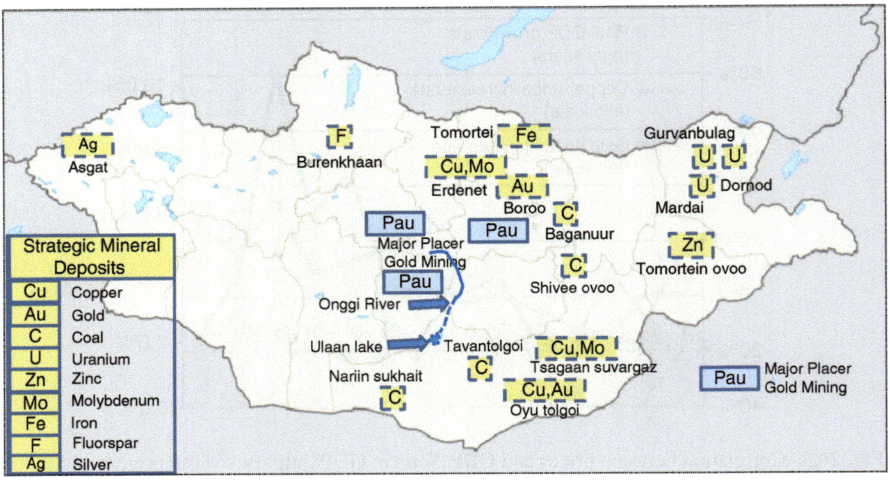

Fig. 20.6 Large-scale mining sites and future mining projects. *Note*: *Dotted squares* indicate the Strategic Mineral Deposits designated by the Government

The International Monetary Fund (IMF) decided in March 2009 to give Mongolia a standby credit to alleviate the country's foreign currency problems. The international price of copper started to rise again in spring 2009. At the end of 2009, the international price of copper had recovered from one-third to two-thirds of the price just before the global economic crisis.

20.2.3 Mining Deposits

Mongolia is assumed to have several world-class underground deposits of resources such as copper, coal, uranium, gold, and silver. Figure 20.6 shows large-scale mining sites and predicted future mining reserves in Mongolia. Major reserves include the Oyu Tolgoi Project in Omnogovi Aimag, which finally reached an investment agreement with the government in October 2009. This project involves 30 million tons of copper deposit, about 6% of the world's total copper reserves, as well as a large deposit of gold, said to be as large as 1,000 tons, about 2% of the world's gold reserves (U.S. Department of the Interior 2010). This project is the largest new copper mining development in the world.

Mongolia also has major uranium deposits. Identified reserves include about 60,000 tons in Dornod Aimag, the easternmost Aimag of Mongolia. Addition to that Mongolia seems to have further reserves of uranium remain to be discovered in unexplored areas throughout Mongolia (World Energy Council 2007). It is possible that these reserves will amount to as much as 1.4 million tons, even larger than Australia's 1.24 million tons, the largest reserves in the world so far.

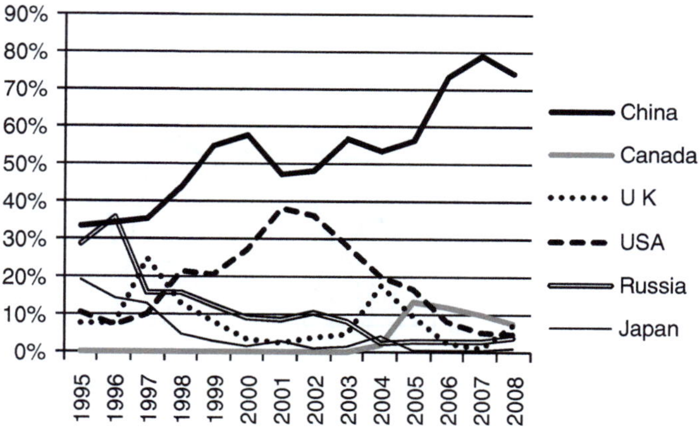

Fig. 20.7 Export value ratio by destination. *Source*: National Statistical Office of Mongolia

There are also major reserves of coal in the Gobi region, including the Tavan-Tolgoi and Nariin-Sukhait Projects in Omnogovi Aimag. The Tavan-Tolgoi Project is the largest in Mongolia, having about 6 billion tons of coal, about a quarter of which is high-quality coal suitable for coking.

In the westernmost part of Mongolia, there is a silver deposit at Asgat in Bayan-Olgii Aimag. This silver deposit contains about 70,000 tons, which accounts for about 18% of the world's silver reserves (U.S. Department of the Interior 2010).

Mongolia therefore offers great potential for further mining development. According to the provision of the Minerals Law, the Government designated 15 deposits as Strategic Mineral Deposits for Mongolia in 2007. The strategically important mining projects are shown by dotted squares in Fig. 20.6.

20.2.4 Trading Partners

In the socialist era that lasted until 1990, Mongolia's main exports were products of the livestock and crop farming sector, which traded with the countries belonging to COMECON (Council for Mutual Economic Assistance), including the Soviet Union and Eastern European countries.

After the transition to a market economy, the destination countries for Mongolian exports have changed dramatically. Instead of the former COMECON counties, China and the United States now make up a much larger share. As a neighboring country, China is a particularly important trading partner, and exports to China have comprised more than 70% of total exports in recent years (Fig. 20.7).

As already mentioned, the biggest export items are mining products. World-class mining projects such as Oyu-Tolgoi and Tavan-Tolgoi located in Omnogovi Aimag in southern Mongolia have an advantage in terms of transportation to China.

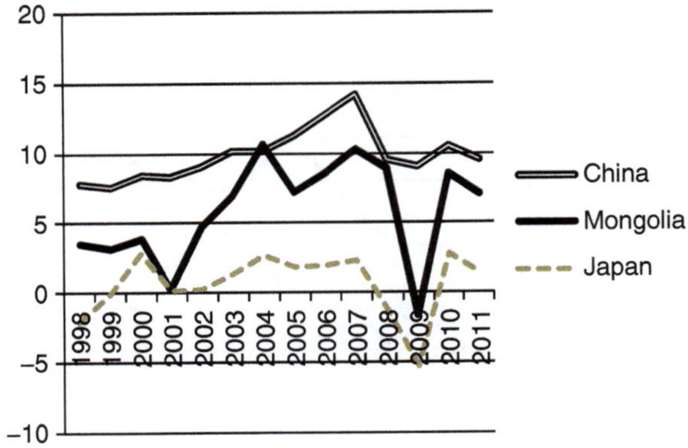

Fig. 20.8 Comparison of real GDP growth rate (percent change). *Source*: IMF, http://www.imf.org/external/pubs/ft/weo/2010/02/weodata/index.aspx. Accessed 5 March 2011

The growth rate of China's economy is higher than that of Mongolia (Fig. 20.8). China is expected to have the world's highest economic growth in the years to come. One of the reasons for this is that China has recently emphasized economic policies aimed at expanding domestic demand instead of export-oriented policies that are facing difficulties after the global financial crisis. It is considered that China's domestic demand will be stimulated by its large population and growing income levels in rural areas. Chinese economic policies have so far been successful in minimizing the shock caused by the global economic crisis.

The Mongolian economy will also be stimulated by strong demand in China, especially in the mining sector. It therefore seems likely that Mongolia will depend increasingly on the Chinese economy and its mining sector in the future.

20.3 Conflicts Between Mining and Nomadic Grazing Over Pastureland

20.3.1 Issue of Mining Licenses in Nomadic Grazing Areas

About 80% of Mongolia's territory is pastureland, as a result of natural conditions including an arid climate and low temperatures. Nomadic grazing is suitable to the natural environment, and livestock raising on the vast grasslands has traditionally been the predominant industry in Mongolia. Grazing in Mongolia requires wide areas of pastureland because herders have to move seasonally to spread out the grazing pressure on thin pastures.

Most mining in Mongolia is surface mining that excavates pasturelands. Mining and nomadic grazing therefore compete directly against each other. To locate mineral deposits, experimental digging licenses are issued for explorations throughout the huge pasturelands that cover almost half the country. If the explorations

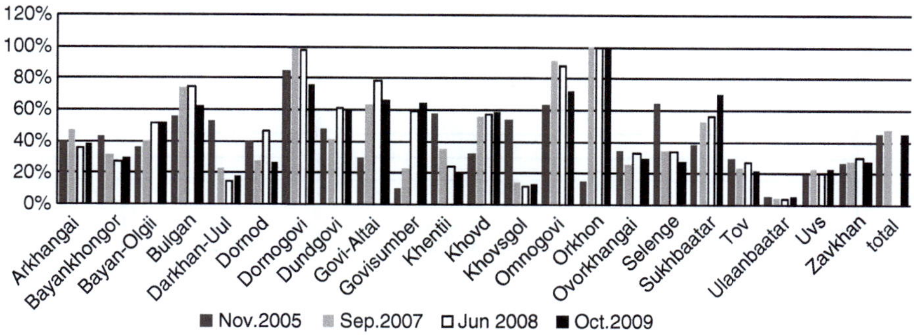

Fig. 20.9 Ratio of exploration license area in each Aimag. *Source*: Mineral Resources & Petroleum Authority of Mongolia

identify areas likely to hold mineral resources, mining licenses are issued to mining companies. A mining license can be issued without the consent[3] of local herders and allows mining companies to exploit even good pasturelands and water resource areas. Mining areas and their surroundings are then no longer available for nomadic grazing.

Figure 20.9 shows the percentage of exploration license areas in each Aimag. In Dornogovi, Omnogovi, Orkhon, and Sukhbaatar Aimag, exploration licenses cover more than 70% of the entire territory of each Aimag. On a nationwide level, exploration licenses cover almost half of Mongolia's total land area. From 2005 to 2008, the percentage increased from 45% to 51%. Since then, however, this figure has decreased to 45% (as of October 2009). This decline seems to be caused by several factors, including the invalidation of exploration licenses acquired for resale purposes.

It is difficult for nomadic herders to object to the granting of exploration licenses on the grasslands they are using. In many cases, exploration licenses are allocated without the herders knowing anything about it.

The areas where mining licenses are issued cover less than 1% of the entire Mongolian territory, about 420,000 hectares (ha). Omnogovi Aimag has the largest area of licensed mining at 180,000 ha, followed by Tov and Selenge Aimag. Between 2005 and 2007 mining license areas increased rapidly in Omnogovi Aimag, following the discovery of large deposits of coal in that region (Fig. 20.10).

[3] According to the provisions of Article 19 of the Mineral Law of Mongolia, the mining authority is required to obtain the approval of the Sum and Aimag assemblies before issuing an exploration license to a company. The Aimag governor should provide a response within 30 days of receiving the license application. According to interviews at several Sums, however, circumstances often make it difficult to provide this confirmation, with documentation sometimes not arriving at the Sum until just a day or two before the deadline. Mining licenses, meanwhile, are not governed by any legal procedures similar to those mentioned above for exploration licenses.

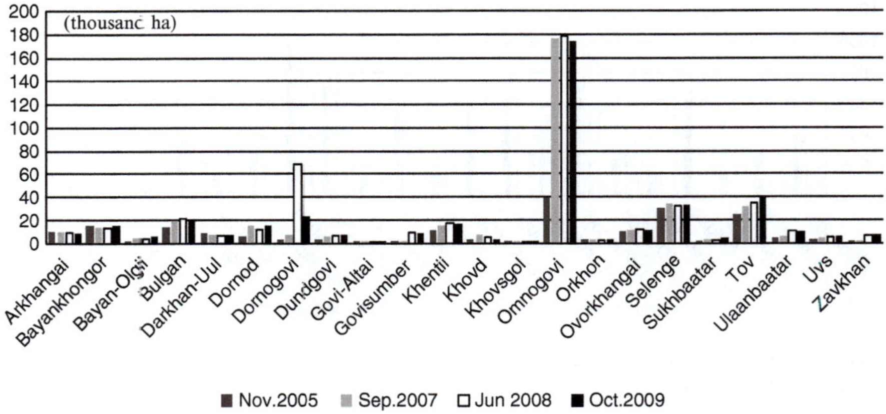

Fig. 20.10 Mining license areas by Aimag. *Source*: Mineral Resources & Petroleum Authority of Mongolia

If initial explorations seem promising, a mining license will be issued to a company without the consent of the nomadic herders, who then have to give up the use of the grasslands. If they have winter or spring camps with facilities for livestock in the licensed area, they will be obliged to move them out of the mining license area, even they have a contract with the County ("Sum" in Mongol, hereinafter referred to as "Sum") government granting them winter or spring camping rights to a site. For pasture usage, Mongolian herders graze on their pasturelands according to the traditional custom rules in their Sum, not a written legislation. So it is significantly difficult for herders to counter the mining right (see Sect. 20.4.2 for details).

20.3.2 Potential Pastureland Degradation Induced by the Mining Industry

Mongolia has undertaken large-scale mining of its mineral resources, including copper, gold, coal, zinc, and molybdenum. Large- and small-scale placer gold mining also takes place, along with micro-scale mining, much of it informal.

The upper pictures in Fig. 20.11 show gold mining locations. The one on the left shows placer gold mining at Zaamar Sum in Tov Aimag, where large-scale companies including Russian investment have mined placer gold. The one on the right shows hard rock gold (gold ore) mining undertaken by Boroo Gold, in Bayangol Sum in Selenge Aimag. This is the largest gold mining company in Mongolia, producing about half of all Mongolia's gold. The picture on the lower left shows a coal mine at Tavan-Tolgoi in Omnogovi Aimag, where there are reserves of about 6 billion tons of high-quality coal such as coke coal. Most of this is exported to China on large trucks.

Fig. 20.11 Examples of major large-scale open-pit mining. (Photographs by author). (**a**) Placer gold, Zaamar Sum, Tov Aimag. (**b**) Gold ore, Bayangol Soum, Selenge Aimag. (**c**) Coal, Tavan tolgoi, Omnogovi Aimag. (**d**) Copper, Erdenet, Orhon Aimag

The picture on the lower right is a copper mine in Erdenet, the largest copper project in Asia. As can be seen from the pictures, most of the mining in Mongolia is surface mining that involves digging up grazing grounds, therefore competing directly with nomadic pasture use.

Most roads in Mongolia are wheel-rutted roads across pastureland. The wheel-rutted roads tend to expand across wide areas of pastureland to avoid mud after rain or to avoid cars coming from the opposite direction. The expansion of roads across pastureland causes the deterioration and desertification of grazing grounds.

The Tavan-Tolgoi Project excavated about 10,000 tons of coal a day in 2007, most of which is exported to China. Large trucks transport the coal from Tavan-Tolgoi in Omnogovi Aimag to the Chinese border at Gashuun-Suhayt in Khanbogd Sum in the same Aimag, covering a distance of 270 km. One hundred trucks, each carrying a load of about 100 tons of coal, make the trip both ways every day along a pastureland road (Fig. 20.12). The road has spread widely and it is possible to say that the frequent transportation of large trucks have caused deterioration to the

Fig. 20.12 Coal transportation, Tavan Totologoi, Omnogovi Aimag

Fig. 20.13 Spread of roads in pasturelands, Tavan Totologoi, Omnogovi Aimag

pasturelands (Fig. 20.13). It has been one of the causes of desertification. The construction of a railway along this route is currently being considered as part of the Tavan-Tolgoi Project.

According to the Minerals Law and other relevant Mongolian regulations, mining companies are required to restore their mined sites. However, many companies, especially placer gold mining companies, fail to fulfill this obligation. At most, companies simply fill in the mined holes and do not carry out vegetative restoration (Fig. 20.14). Figure 20.15 shows restored mining sites with recovered vegetation (note the patchwork patterning). Although the law and relevant regulations set down required methods of restoration in some detail, the number of companies doing vegetative restoration is very limited. According to the Ministry of Nature and Environment of Mongolia (interview in May 2012), 26% of the mined holes area of placer gold has been filled in and only 8% of the area has been restored with vegetation. It is therefore difficult for herders to use the pastureland again after mining.

Fig. 20.14 Restoration (no vegetation), Zaamar Sum, Tov Aimag

Fig. 20.15 Vegetative restoration, Bayangol Sum, Selenge Aimag

20.3.3 Mining in River Basins

The impact of placer gold mining in rivers seems notably disruptive to the natural environment, water resources, and pasturelands compared to the impact of other environmental damage of mining already described such as (1) large-scale surface mining digging into grazing grounds, (2) pasture degradation by the expansion of pasture roads where large trucks transport mining products, and (3) breach of regulations requiring restoration after mining.

According the following case studies, it is possible to say that three aspects of mining are responsible for decreasing river flow: first, digging up riverbeds causes water to flow underground; and second, the river water used for gravity selection in placer gold extraction is then scattered around the mining site. Most of the scattered water evaporates, or vanishes into the air, because of Mongolia's very dry climate, and third, destroys or melts the permafrost layer to mine placer gold by mining

Fig. 20.16 Midstream mining of Tuul River, Zaamar Sum, Tov Aimag

companies and Ninjas. Because most placer gold is contained in alluvial soil, major placer gold mining in Mongolia is done in river basins. Therefore it is considered that the adverse impact of placer gold mining in rivers is larger than other negative impacts by the mining industry.

20.3.3.1 Case Study on the Tuul River

The Tuul River is the fifth longest river (704 km) in Mongolia, running through Ulaanbaatar to the north. Figure 20.16 shows the midstream section of the river. This mining site is at Zaamar Sum in Tov Aimag. The mining facility, which looks like a building or a ship, works by taking in gravel from the riverbed and by selecting placer gold from the gravel by an automatic gravity selection system called "Drag," a big machine made in Russia. Five "Drags" have been used for placer mining at Zaamar Sum.

These machines extract gravel, including placer gold, by digging up the riverbed. It should be recognized as a serious environmental problem that this digging up of riverbeds causes water decreases in rivers, as the riverbeds are broken down, and surface water flows sink deep underground.

Figure 20.17 is a map showing the areas of exploration and mining licenses around Zaamar Sum in Tov Aimag and neighboring Sums (the exploration and mining licensed areas are shown by the dark gray color and diagonal lines respectively in the photograph). Zaamar Sum is a large and well-known placer gold mining site in Mongolia. The exploration licenses areas are allocated across a wider area.

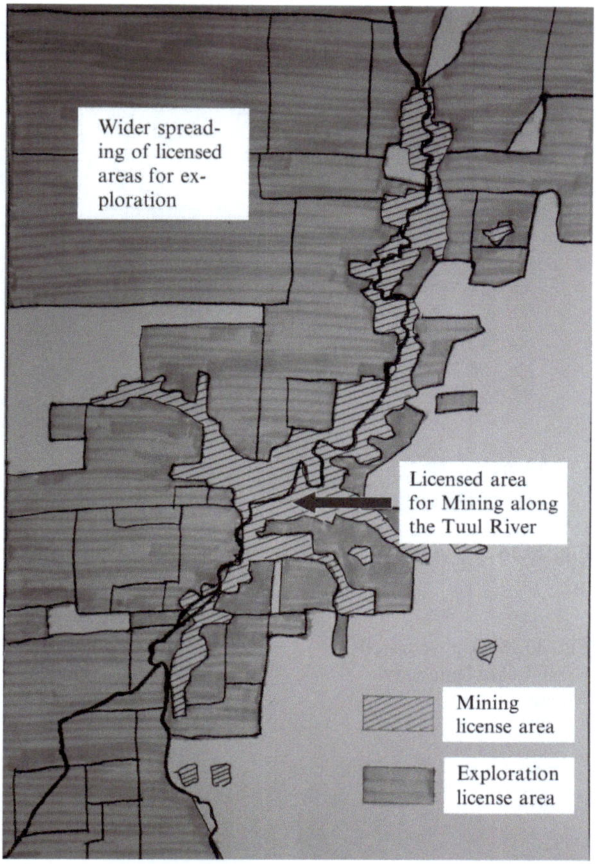

Fig. 20.17 Spread of exploration and mining licenses, Zaamar Sum, Tov Aimag and neighboring Sums. *Source*: Illustrated by author from the photograph at the administration office of Zaarmar Sum

Mining licenses are allocated along the Tuul River. We can see from the map that placer gold is extracted mainly from the river basin. In Zaamar Sum, about one-third (90,000 of a total of 280,000 ha) of the entire territory is licensed for exploration or mining.

20.3.3.2 Case Study on the Ongi River

The Ongi River was previously the twelfth longest river (435 km) in Mongolia, flowing through eight Sums in three Aimags, from Oyanga Sum in Ovorkhangai Aimag, through Dundgovi Aimag, to the Ulaan-nor (lake) in Omnogovi Aimag.

Figure 20.18 shows drilling in an upstream section (the Oolt Stream) of the Ongi River, at Uyanga Sum in Ovorkhangai Aimag. Many miners, including a large company, several small-scale companies, and numerous Ninjas, have been excavating the bottom of the stream.

Fig. 20.18 Upstream of Ongi River, Uyanga Sum, Ovorkhangai Aimag

Fig. 20.19 The vanished Ulaan Lake, Omnogovi Aimag

The river's ultimate destination was not the sea but a large lake. The Ulaan Lake was the largest lake in Omnogovi Aimag, about 200 km^2. However, the Ongi River dried up halfway to the lake. The lake disappeared completely about 15 years ago (Fig. 20.19). This fact is well known in Mongolia. It is claimed that the drying up was caused first by climate change and second by placer gold mining in the river. When I surveyed the site in September 2008, the river disappeared near Arvayheer, the Aimag center (prefectural capital) of Ovorkhangai Aimag, making it roughly one-third of the length of the former Ongi River (Fig. 20.20).

Fig. 20.20 Partially dried up Ongi River, suburb of Arvayheer, Ovorkhangai Aimag

Mongolia is located in a dry climatic zone, having less than about 300 mm rainfall per year. Rivers are significant water resources for nomadic herders and their livestock, providing drinking water and supplying moisture to good pastureland. The drying up of the river and the lake has adversely affected people and animals in the eight Sums directly, and it is possible to say that it has accelerated pasture degradation and desertification.

As already mentioned, the depletion of the river is caused by climate change and placer gold mining. According to the researcher involved in the mining impact survey and related officials in the Sums where placer gold mining operations are carried out, it seems likely that two aspects of mining are responsible: first, digging up riverbeds causes water to flow underground; and second, the river water used for gravity selection in placer gold extraction is then scattered around the mining site. Most of the scattered water evaporates, or vanishes into the air, because of Mongolia's very dry climate.

The research team including the eco-development center of the Mongolian University of Science and Technology carried out research into "the drying process of the Ongi River Basin"[4] and issued a report in 2006 (Mijiddorji and Bayasgaralan 2006). The research suggests that climate change and other factors are responsible for 80% of the decrease of the river's flow. Placer gold mining, meanwhile, contributes 16%. Mining appears not to be as big a factor as expected. However, the report also reveals that in the two upper streams (the Buuruljuut Stream and the Oolt

[4] The research is widely thought to be subject to the influence of mining companies, including the first President of Mongolia, Dr. Ochirbat, who is honorary President of the National Mining Association of Mongolia and one of the authors of the research report. Even in this report, however, there is clear evidence of the adverse impact that mining activities in the Ongi River are having on the flow of the river.

Fig. 20.21 Upstream of Orkhon River, Bat-Olzii Sum, Ovorkhangai Aimag . Photograph by the Ongi River Movement

Stream), where the largest amount of mining has been carried out in the Ongi River Basin, the two streams made up 21% of the total flow of the Ongi River before mining. From these data it can be estimated that about three-quarters (16% of a total of 21%) of the decrease in the flow of the two upper streams was caused by placer gold mining. So, it is possible say that this ratio shows mining to be a significant contributory factor.

20.3.3.3 Case Study on the Orkhon River

The Orkhon River is the longest river (1,124 km) in Mongolia, flowing from the central part of Mongolia (the Khangai Mountains) to Russia. Figure 20.21 shows large-scale placer gold mining in one of the upper streams of the river. This site is at Bat-Olzii Sum in Ovorkhangai Aimag. Because this area is far upstream in the forest-steppe zone, the forests around the stream and the riverbed have also been destroyed by mining activities.

In northern and central regions of Mongolia, permafrost (a layer of frozen soil) is widely distributed. Permafrost is a kind of underground reservoir of water. As Mongolia is situated in the arid zone, permafrost is invaluable in terms of maintaining the country's ecological environment. In recent years, however, the permafrost has begun to melt in response to climate change and is gradually shrinking in thickness. Human destruction of permafrost is also a problem. Figure 20.22 show placer gold being extracted from permafrost after destruction of this layer. The site is at Tsenkher Sum in Arkhangai Aimag. Ninjas, who conduct informal micro-scale individual mining after the mining companies leave, are melting permafrost by

Fig. 20.22 Placer gold mining of permafrost by Ninjas, Tsenkher Sum, Arkhangai Aimag. (**a**) Permafrost. (**b**) Ninjas are burning cattle dung to melt the permafrost for mining placer gold. Photograph by B. Tumenbayar

Fig. 20.23 Placer gold mining, including permafrost, by a mining company, Bodont, Tsenkher Sum, Arkhangai Aimag. Permafrost was found at *a* in a digging site (at this time no mining operation was occurring as it was the winter season)

burning cattle dung. Figures 20.23 and 20.24 show a mining site of the company extracting placer gold after digging up the permafrost in Tsenkher Sum. This method is quite different from the melting method used by the Ninjas; here the company is destroying the permafrost using heavy machines. One kilometer downstream from the permafrost mining, the Bodont Stream has no water (Fig. 20.25). Because permafrost is a huge storage warehouse of water in Mongolia, it is assumed that the destruction of permafrost causes deterioration and desertification in the grassland vegetation. Mining developments destroying permafrost are carried out not only in Arkhangai Aimag but also in Selenge, Ovorkhangai, and Bulgan Aimags (Mongolian Business Development Agency 2003).

Fig. 20.24 Close-up picture of *a* in Fig. 20.23 shows permafrost inside the surface. When the surface soil was removed, a frozen layer appeared (*whitish* central part of photograph)

Fig. 20.25 Bodont stream at 1 km downstream from the mining site of Figs. 20.23 and 20.24, Bodont, Tsenkher Sum, Arkhangai Aimag: there is no water in the stream

20.4 Lack of Legal Linkage Between the Natural Environment and Mining Land Use

20.4.1 Laxity of Mining Regulations

Mongolia is located on the Central Plateau of Asia, so the climate is notably dry and cold. Average precipitation is 241 mm per year (AQUASTAT of FAO), indicating a very low level of rainfall. In addition to permafrost, water in rivers and lakes has also played a significant role by providing drinking water and other daily necessities for nomadic people and their livestock, as well as maintaining the delicate ecological balance of the grasslands and forests. However, as we have already seen, Mongolia appears to have lax mining development regulations, despite the fragility of its natural environment.

With this concern in mind, Table 20.1 and Fig. 20.26 compare the regulated upper limit area for one exploration license and the annual average rainfall with those of foreign countries. Whenever a promising deposit of underground resources is found, the mining company has priority in obtaining the mining license. Given it is possible that an mining permit will be issued anywhere within the exploration license area, the scale of the regulated upper limit area for one exploration license can be considered an indicator of the laxity of mining regulations. Meanwhile, the annual average rainfall is considered an indicator of the ecosystem's recovery (it is assumed that lower levels of precipitation prolong the amount of time needed for the environment to recover).

The countries located at the lower right of the scatter plot in Fig. 20.26 are larger in the regulated upper limit area for one exploration license and are lower in annual rainfall. Saudi Arabia is placed at the far right in Fig. 20.26, showing the upper limit areas for one exploration license is the largest (at 1 million ha), and the precipitation is the lowest (at 59 mm/year; AQUASTAT of FAO).

Mongolia is located at the lower right, next to Saudi Arabia on the regulated upper limit area, on an area of 400,000 ha. Its annual precipitation is the fifth lowest at 241 mm. Saudi Arabia belongs exclusively to the desert climate zone, whereas Mongolia belongs to the desert, steppe, and cold wet climate zones. Grasslands account for 80% of the territory, and forests and permafrost layers are in the northern part of the country. Therefore, Mongolia's ecosystem is more delicate and fragile. Furthermore, taking into account underground resources, Saudi Arabia is focused on oil, which is not mined with open-pit mining. Conversely, Mongolia's major underground resources are copper, gold, coal, and zinc, which are generally extracted by open-pit mining, the excavation of grasslands, riverbeds, and forests.

Considering the aforementioned data and the ecological situation, it can be argued that Mongolia's current legal system is more lax than those of other countries.

Table 20.1 Comparison of the regulated upper limit area for one exploration license and annual average precipitation

Code	Country	Upper limit area for one exploration license hector	Annual precipitation (mm)
Sa	Saudi Arabia	1,000,000	59
Mo	Mongolia	400,000	241
Pn	Papua New Guinea	255,750	3,142
Mw	Malawi	250,000	1,181
Ni	Niger	200,000	151
Mt	Mauritania	150,000	92
Na	Namibia	100,000	285
Bo	Botswana	100,000	416
La	Laos	100,000	1,834
Id	Indonesia	100,000	2,702
Zi	Zimbabwe	65,000	3,107
Bv	Bolivia	62,500	1,146
So	Soloman	60,000	3,028
In	India	50,000	1,083
Co	Congo	40,000	1,646
Do	Dominican Republic	30,000	1,410
Bu	Burkina Faso	25,000	748
Pa	Panama	25,000	2,692
Ta	Tanzania	20,000	1,071
Vi	Vietnam	20,000	1,821
Ca	Cambodia	20,000	1,904
Ma	Mali	15,000	282
Gh	Ghana	15,000	1,187
Ar	Argentina	10,000	591
Ur	Uruguay	10,000	1,265
Gu	Guatemala	10,000	1,996
El	ElSalvador	5,000	1,724
Ec	Ecuador	5,000	2,087
Om	Oman	2,000	125
Cr	Costa Rica	2,000	2,926
Za	Zambia	1,600	1,020
Pe	Peru	1,000	1,738
No	Norway	1,000	1,414
Fi	Fiji	400	2,592
Mo	Morocco	400	346
Ja	Japan	350	1,668
Fn	Finland	100	536

Source: Upper limit area for one exploration license: Countries identified the data from "The environmental survey for resources development" (The Japan Oil, Gas and Metals National Corporation 2005). Annual precipitation: AQUASTAT of FAO

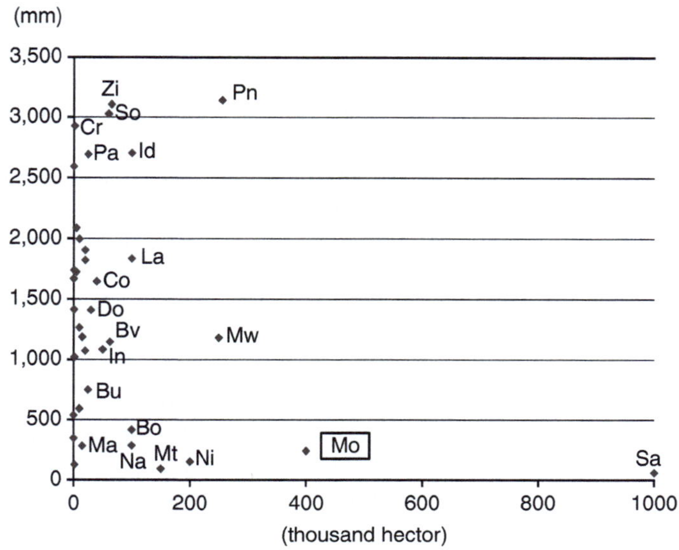

Fig. 20.26 Scatter plots of the regulated upper limit area for one exploration license and annual average precipitation of 37 countries. *Source*: Upper limit area for one exploration license: Counties identified the data from "The environmental survey for resources development" (The Japan Oil, Gas and Metals National Corporation 2005). Annual precipitation: AQUASTAT of FAO

20.4.2 Conflict Between Mining Rights and Pastureland Use

One more aspect of Mongolian mining regulations is worthy of comment. Nomadic livestock grazing has traditionally prevailed in Mongolia in such as way as to disperse grazing pressure over a wide area to utilize sustainably the very thin vegetation of the pasturelands. In addition, because of the harsh and unstable Mongolian climate, territory and rights over grassland use are not fixed, keeping pastureland use flexible. For this reason, there is no written legislation governing ownership or usage rights on pastureland.

On the other hand, mining rights are legislated clearly under the Minerals Law. This situation makes it difficult for nomadic herders to counter the mining rights. Furthermore, there is no provision for coordination between the surface ground use of nomadic herders and the mining rights of companies.

In the competition for access to pastureland between mining rights and grassland use by nomadic herders, the relationship is one sided, with all the power and strength on the side of the mining companies. To bring the two sides into balance, it is essential to add concrete provisions regulating mining rights to the Minerals Law or other laws to preserve pastureland, water resources, or forest for nomadic grazing and the environmental conservation of Mongolia.

At present there is only one way to protect good pasture areas, water reserves, or forests, and that is the provision of "land for special needs," as prescribed in Article 16 of the Land Law. If a Sum or Aimag assembly designates an area in their territory as "land for special needs," no mining authority can provide exploration licenses or mining licenses on that land, as prescribed in Articles 17 and 24 of the Minerals Law. Local assemblies can set aside "land for special needs" for several reasons, such as "reserve pastureland" (pastureland for "Otor" in Mongol) and "common hayfields for fodder reserves," as prescribed in Article 16 of the Land Law. For example, Zaamar Sum has designated the entire territory of the Sum other than that already licensed for exploration or mining, which is two-thirds (190,000 ha of a total of 280,000 ha) of Sum territory, preserve nomadic pastureland from mining development.

On the other hand, in Uyanga Sum in Ovorkhangai Aimag, the Sum assembly designated a valley area including the river basin of the Buuruljuut Stream as "land for special needs" in 2002. However the Aimag assembly, or prefectural parliament, canceled the Sum's designation in 2006. This case suggests that designating pasturelands according to the provisions of the Land Law may not be sufficient to ensure preservation.

20.4.3 The New Law to Regulate Mining

In 2009 a bill was passed to protect river basins and forests from mining development. Nongovernmental organizations (NGOs) consisting of local residents including herders affected by placer gold mining in the river basin petitioned parliament members to submit a bill. A bill prohibiting mining in river basins and forests was submitted to parliament in spring 2009 and was debated in the spring session of parliament. Following NGO demonstrations and hunger strikes in Ulaanbaatar in support of the bill, it was passed on 16 July 2009.

The outline of the law[5] is as follows:

- Exploration and mining in river basins and forests are prohibited.
- The government sets boundaries prohibiting exploration and mining.
- Licenses issued before the law will be invalidated, and the companies involved should restore the mined sites and be compensated.
- Restoration should be completed within 2 years, under the supervision of local administrations. Local administrations can delegate monitoring to environmental NGOs.
- The law does not apply to the Strategic Mineral Deposits (15 deposits; see Fig. 20.6).

[5] The text of the law: http://www.legalinfo.mn/insys/lawmain.php?vlawid=46655 (in Mongolian). http://rivermovements.org/pdf/JulyLaw.pdf (in English)

Under this law, new licenses will no longer be issued in prohibited areas, and licenses issued before the law was passed will be invalidated. Companies who have already carried out mining in the prohibited areas will be required to restore the mined sites. It is hoped that the law will be effective in conserving precious water resources and good pastureland in river basins and forests against the power of the mining companies. Several of the existing environmental conservation laws in Mongolia are rather ineffective, although they contain good provisions. It is important to ensure that this remarkable new law for Mongolian nomadism and natural environment is implemented effectively.

The government and related Sum administrations have worked on fixing the boundaries of prohibited areas. Meanwhile, the mining companies whose licenses will be invalidated are set to receive compensation from the government according to the law. There is some controversy involved in fixing the appropriate levels of compensation. To keep the law effective, it is important (1) to draw boundaries effectively, (2) to decide compensation according to the Mongolian legal system practically, and (3) to restore mined sites and surface vegetation appropriately.

The boundaries of the prohibited area initially be regulated according to Article 31.3 of Mongolia's Water Laws. This article states that "ordinary protection zones shall be set not less than 200 meters from the banks of a water reservoir area." Local governments set boundaries at least 200 m away from the banks of rivers or lakes.

Some people have pointed out that underground resources are national property according to the Mongolian Constitution, and that there might therefore be complications involving compensation for companies. Mining companies only have licenses to use national underground properties. There may well be some heated argument on this matter before the implementation of the law.

20.5 Conclusion

The mining industry in Mongolia will continue to develop at an accelerating pace. However, it is essential to coordinate this development with the natural environment, including the country's notably fragile water resources, as well as with the nomadic grazing that also depends on surface grounds, to keep Mongolia's two major industries sustainable.

The recent passing of the new law represents environmentally fragile Mongolia's first experience of dealing with the crucial conflicts between the two major sectors. Because this issue is specific to Mongolia, involving the natural environment and socioeconomic issues, it is difficult to apply foreign systems to Mongolia directly. It will be important to gather scientific data and ideas to find a way for the two major sectors to coexist.

In Mongolia, there is room for discussion and scope for systems or institutions that will coordinate these hardships. As a first step toward creating a sustainable and stable country, it is important to ensure that the new law regulating mining in environmentally fragile places is enforced and to coordinate issues between mining rights and nomadic herders' pastureland use.

Acknowledgments This research was supported by the Research Institute for Humanity and Nature (project number D-04) and the Mongol Law Study Project of the Nagoya University.

References

Korea Environment Institute and Mongolian Nature and Environment Consortium (2004, 2005) Joint Research between Korea and Mongolia on Water Quality and Contamination of Transboundary Watershed in Northern Mongolia. http://203.250.99.31/pub/docu/en/AN/BA/ANBA2006AAA/ANBA-2006-AAA.PDF. Accessed 6 March 2011

Mijiddorji R, Bayasgaralan S (2006) *Integrated Assessment on Drying Process in the Ongi River Basin*. Ulaanbaatar, pp 6–7 (in Mongol, with English abstract).

Mongolian Business Development Agency (2003) Ninja Gold Miners of Mongolia Assistance to Policy Formulation for the Informal Gold Mining Sub-sector in Mongolia. http://www.river-movements.org/pdf/Ninja_Report.pdf. Accessed 5 March 2011

The Japan Oil, Gas and Metals National Corporation (2005) The environmental survey for resources development. http://mric.jogmec.go.jp/mric_search/Search.do?akey=%E8%B3%87%E6%BA%90%E9%96%8B%E7%99%BA%E7%92%B0%E5%A2%83%E8%AA%BF%E6%9F%BB&dsel=98&fyear=&fmonth=&fdate=&tyear=&tmonth=&tdate=&syear=2005&smonth=10&sdate=7&check=7&okey=&pkey=&nkey=&fflg=false&psel=100&nowPage=1&recordSize=87&ssel=0&ccheck=7 + (in Japanese) Accessed 6 March 2011

Tumenbayar N (2002) Herders' Property Rights vs. Mining in Mongolia. http://www.uvm.edu/~shali/Mining%20Mongolia%20paper.pdf Accessed 6 March 2011

United Nations Human Rights Team Group (2006) Discussion Paper for the National Dialogue: Mining and Human Rights in Mongolia

U.S. Department of the Interior (2010) Mineral Commodity Summaries 2010. http://minerals.usgs.gov/minerals/pubs/mcs/2010/mcs2010.pdf Accessed 13 March 2011

World Bank (2006) Mongolia: A Review of Environmental and Social Impacts in the Mining Sector. http://siteresources.worldbank.org/INTMONGOLIA/Resources/Mongolia-Mining.pdf Accessed 6 March 2011

World Energy Council (2007) 2007 Survey of Energy Resources World Energy Council 2007 Uranium. http://www.worldenergy.org/documents/uranium_country_notes.pdf Accessed 5 March 2011

Chapter 21
Interference Impact of Global Warming and Globalization on the Society and Ecosystem in Mongolia

Zamba Batjargal and Batjargal Enkhjargal

Keywords Aerosol • Climate change • Data contamination • Development paradigm • Ecological footprint • Interference effect • Pastoralism • Pattern change • Tradition

21.1 Introduction

Climate change studies conducted in Mongolia so far have been mostly based on the outputs of global climate models (GCM), resulting in the very important message that the present global warming will in the long run lead to a shift of climate zones with dominance of arid and desert areas in this country (Dagvadorj et al. 1994; Batima and Dagvadorj 2000; Mijiddorj 2000; Batima et al. 2005; Gomboluudev 2006). The National Action Programme on Climate Change (NAPCC) of Mongolia (GoM 2000), approved by the Government of Mongolia in 2000, incorporated this message in the outline of possible actions as response measures to climate change. The Government of Mongolia has established an interagency and intersectoral National Climate Committee (NCC), led by the Minister in charge of the Environment, to coordinate and guide national activities and measures aimed to adapt to climate change and to mitigate Green House Gas (GHG) emissions. A series of comprehensive studies have been conducted in line with the NAPCC. Certain impact and risk analyses studies have been undertaken in relationship with different ecosystems and economy sectors, including agriculture with a strong focus on the livestock sector. However, leading experts on climate change in Mongolia recognized that the vulnerability and adaptation assessments of climate change for different natural zones and rangeland were defined so far by the judgments of experts rather than by quantitative estimation and

Z. Batjargal (✉)
World Meteorological Organization, UN, New York, NY, USA
e-mail: z_batjargal@yahoo.com

B. Enkhjargal
Mongolian Meteorological Society, Ulaanbaatar, Mongolia
e-mail: enkheb99@yahoo.com

N. Yamamura et al. (eds.), *The Mongolian Ecosystem Network: Environmental Issues Under Climate and Social Changes*, Ecological Research Monographs,
DOI 10.1007/978-4-431-54052-6_21, © Springer Japan 2013

integrated analyses. It is also true that most of the studies on climate change have covered only the economic and environmental aspects of the problem, avoiding the sensitive issues of food security and sustainable livelihood in Mongolia.

There is a need to further update the results of those studies by integrating new data and upgrading them in terms of spatial scale. In the downscaling exercise, there is a need to shift the focus from macro-scale frontal-type alteration to meso-scale pattern change and to evaluate the aftereffects of interaction of climate and human systems at a local level. It is important to conduct more detailed and localized analyses of current climate condition and projection of global climate change at the regional level for different ecozones at the territory of Mongolia, specifically, focusing on the current and expected responses of the local ecosystem, which is under pressure from the interference effect of the climate and human systems.

In respect to the methodological approach, there is a need to highlight the pattern behaviors of the climate system, the mosaic peculiarities of the land cover, and a mixed structure of the ecosystem with aggressive intervention of man-managed ecosystems where pastoralism plays a special role.

In practical terms, a high priority should be given to the near- and midterm incremental changes, adding certain long-term and low-probability but high-impact changes, such as the continued and increased aridity in some parts of Mongolia, rapid and deep alteration in moisture supply regime and related adaptation options, and possible livelihood alternatives for local communities.

21.2 Confidence in the Outcomes of the Climate Change Studies as a Basis for Policy Making

There still is a certain degree of uncertainty regarding confidence in the GCM-based projections to use them as a basis for nationwide policy-making exercises in categorical terms. Despite the fact that the emission scenarios used by the Intergovernmental Panel on Climate Change (IPCC) are sophisticated and relatively realistic (IPCC 2007), they missed some important factors such as the impact of climate on the carbon cycle, changes in ozone precursors, long-term changes in solar activity, and unpredictable but possible volcanic eruptions. The current GCMs did not capture all intrinsic components in the driving factors, such as the full range of aerosol impact and possible imperative nonlinear feedback effects related to land-cover change (resulting from extensive land use, deforestation, etc.), interaction with changing ocean parameters (sea ice coverage, ocean current alteration caused by water density variation), change in cloud patterns (caused by increased evapotranspiration and aerosol population as cloud condensation nuclei), and other associated factors.

It is also true that the interpretation of projections was mostly based on the "domino effect," resembling a linear approach rather than the nonlinear type transaction covering a full range of economic and environmental aspects of the problem with a clear target being the sustainable livelihood of people. It is essential that the study focuses on Mongolia, where social structure, lifestyle, and land use are unique and cannot fit the generalized picture common for most of the countries in Asia. The conclusions in this regard are that the outcomes of the current studies, particularly those related to future scenarios, can be considered in optional but not in categorical terms.

21.3 Scale and Level of Interaction Between Climate and Other Systems

Interaction of the climate system with other human and nonhuman systems is a very complex process, involving events of different scales in a wide range of timeframes. This process, in the case of Mongolia, has certain specifics because of the country's deep inland location with less direct influence from the world oceans and the fact that the livelihood approach of people there is to be adaptive to their surrounding living environment rather than attempting to conquer it. In Table 21.1 are highlighted some of those components of interaction that have had more pronounced impact on local ecosystem in Mongolia, if numerous transactions of cascade effects can be presented in a more simplified way as the "food chain."

Table 21.1 Interaction of the climate with social and ecological systems

Climate system	Social system	Ecological system
Global climate driving forces Orbital change, solar activity, volcanic eruption, GHG, aerosols.	*Global community* Globalization-induced interdependence, dilemma regarding production and consumption patterns, development polarization, homogenization of culture and lifestyle, burden, and risk sharing through trade, Foreign Direct Investment (FDI), Official Development Assistance (ODA).	*Planetary ecosystem* Cataclysm-like collapse of ecosystems resulting from exceeding a threshold of pressure caused by the interference effect of the natural system's evolutional and the human system's forced changes.
Regional climate factors Extra-tropical cyclone activity, alteration in strength of westerlies and East Asian monsoons, El Niño/ Southern Oscillation (ENSO).	*Regional community* Burden and risk sharing through regional integration, migration, vector-borne, and endemic disease-spreading risk.	*Regional ecosystem* Trans-boundary environmental pressure including desertification, air pollution with acid components, water sources depletion, forest and steppe fire risks, infestation of pests, insects, sand dune movement, wildlife migration.
Local climate-linked factors Geographic location, altitude, topography, albedo, soil properties.	*Local community* Accelerated urbanization, pastoralism under pressure, resource-based development policy, lifestyle and cultural value change.	*Local ecosystem* Increased shift from natural to man managed ecosystems, decreased ecosystem service, reduced biocapacity of adaptation to rapidly changing environment, increased cost for maintenance of ecosystems, and diversification of harmful effects on Nature as life-sustaining basis.

More detailed studies are needed at the level of localized patterns of key climate parameters, such as temperature and precipitation, to explore the specific advantages of global and regional models and interactive downscaling and upscaling exercises.

21.3.1 Localized Features of Temperature and Precipitation Trends

Within this study, certain analyses have been undertaken to demonstrate the local peculiarities of response to global climate change, revealed in the trends for temperature and precipitation. Standard observation records, which are included in the national meteorological database and kindly provided by the National Agency for Meteorology, Hydrology and Environmental Monitoring (NAMHEM), have been used. The selection criteria for the meteorological stations involved were based on geographic location, continued function, length of record, and presence of local factors that might contribute to the bias of representativeness or data "noise."

First of all, an analysis was completed to verify the differences in response to temperature trends in distinctive ecozones from a generalized picture for the entire territory. Four meteorological stations were chosen to represent (1) the Great Lakes area in the west—Ulaangom (lat. 29°48′, long. 92°05′; alt. 939 m); (2) the Khubsugul Mountain areas in the north—Murun (lat. 29°48′, long. 100°1′; alt. 1,285 m); (3) steppe areas in the east—Choibalsan (lat. 48°0′, long. 114°32′; alt. 747 m); and (4) the Gobi areas in the south—Dalanzadgad (lat. 43°35′, long. 104°25′; alt. 1,465 m).

The warming trend indicated in all ecological zones during the past four decades and the trends were more or less synchronous despite the great distance between the zones (Fig. 21.1).

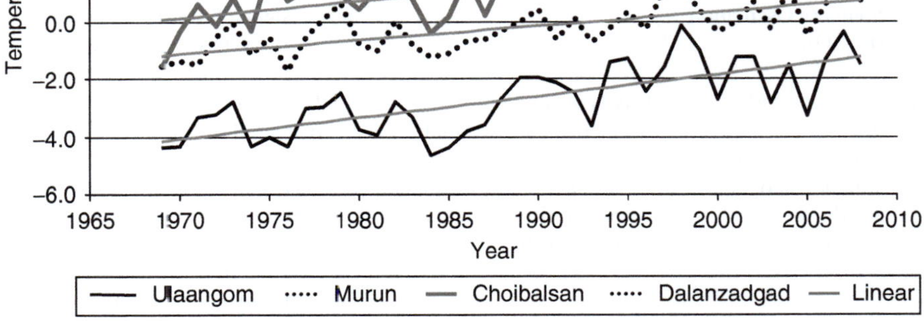

Fig. 21.1 Mean temperature trend in different distinct technological zones (West, North, East, and South) in Mongolia by year

With respect to precipitation, the trend was slightly different. Almost no change was indicated during this period except a slight decrease in the "open" areas such as the steppes in the east and the Gobi in the south (Fig. 21.2).

It was interesting to see differences in the precipitation trend in cold and warm periods of a year, especially in the areas of the country with most crop fields. Figure 21.3 illustrates the unfavorable conditions for cropping in this area (Station Baruunkharaa is located in the active crop cultivation area of Mongolia) resulting from a shift (although slight) of precipitation from the summer to winter season.

Moving south to the arid and semidesert areas, the difference in seasonal trend of precipitation is more apparent with an obvious misbalance, because a significant portion of increase in winter time cannot compensate for even a small portion of decrease in summer season, not mentioning its role in vegetation growth (Fig. 21.4).

It is well known that the SMH (Siberia-Mongolian High) plays a dominant role in the air temperature regime in winter seasons. Tosontsengel station is located at the edge of the Great Lakes Depression and therefore is more affected by the influence of the SMH compared to Galuut, located on the southwestern slope of Khangai Mountain. Figure 21.5 shows the variations of the annual mean temperature in both sites; they are identical in their phase, but the average temperature at the Tosontsengel is significantly lower than at the Galuut despite Tosontsengel's lower altitude above sea level (almost 400 m difference in their altitudes).

In the case of precipitation, the story is different from that of temperature. The average level of precipitation is roughly the same at both sites. Nonetheless, there is a slight asynchrony in phase, with a certain inconsistency in the magnitude of variations (Fig. 21.6), which means that the SMH, as a winter season circulation regime, has much less impact on the amount of precipitation. Instead, the dynamic and thermal effects of orography, obviously, have a greater contribution to precipitation scenarios in this area. Northern and southern slopes of mountains in Mongolia have more impacts but with opposite effects on the soil moisture regime through transpiration and convection mechanisms (Batjargal 2007b). The west and east sides of mountains play an important role depending on their orientation with respect to the dominant wind direction. However, precipitation, for instance, at the Baitag station, is much less (Fig. 21.7) despite its windward location compared to the Altai station, probably because the westerlies can bring much less air moisture to the Baitag site because of its location (Trans-Altai Gobi exposed to dry air mass from Central Asia). Increased precipitation at the Altai site can be partly explained by its high altitude (2,181 m above sea level, whereas the altitude of Baitag is 1,186 m). The average amount of precipitation gradually decreases moving away from the Altai site further to the east to the more open site of Mandalgobi. These outcomes are in good agreement with the generalized analysis for the entire territory (MNET 2009). Nevertheless, there is a need for additional arguments explaining the positive trend of precipitation to the west side of the Altai mountain range while there are sustained negative trends on the east side of the mountain range (Figs. 21.7 and 21.8).

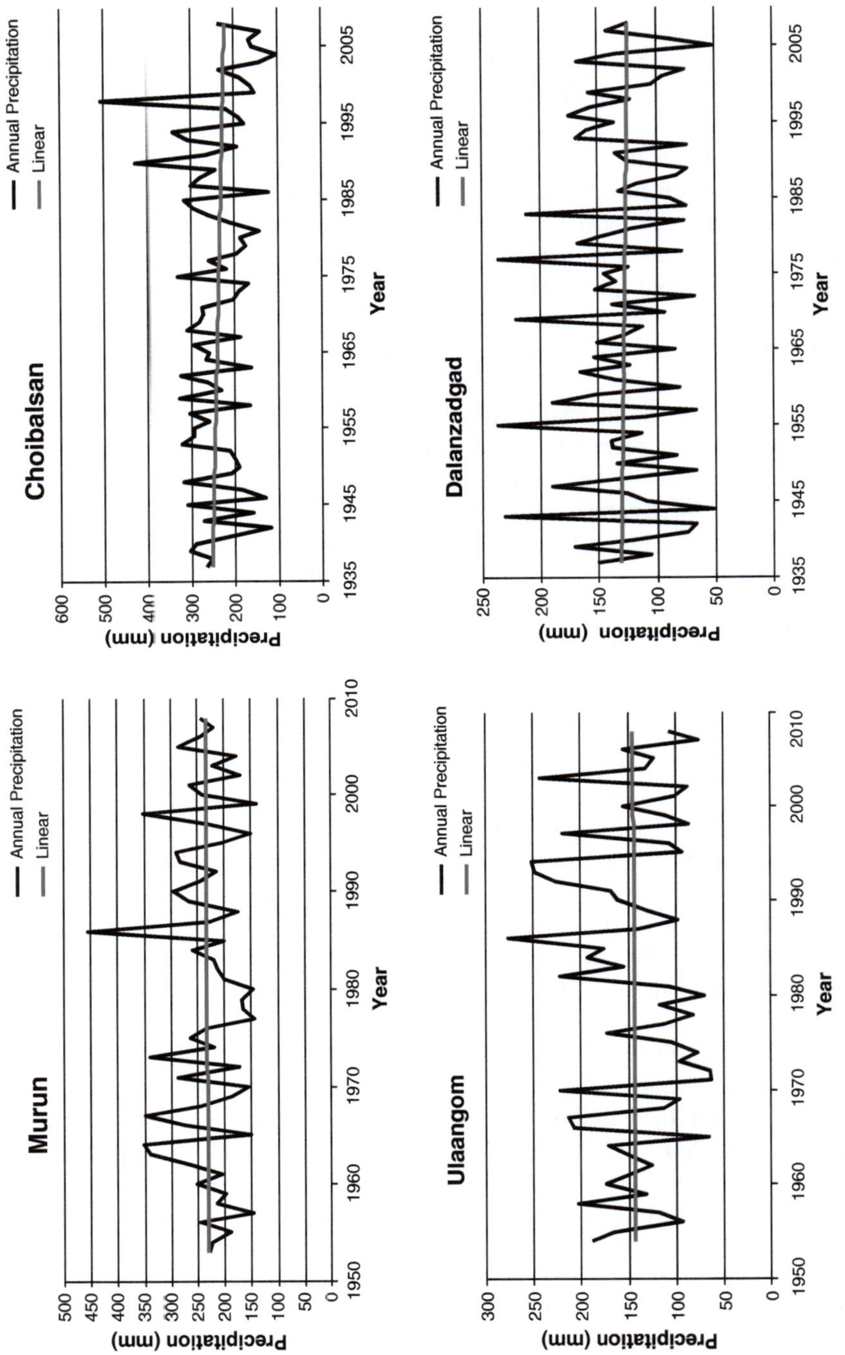

Fig. 21.2 Annual precipitation trend in different distinct ecological zones (west, north, east, and south) in Mongolia by year

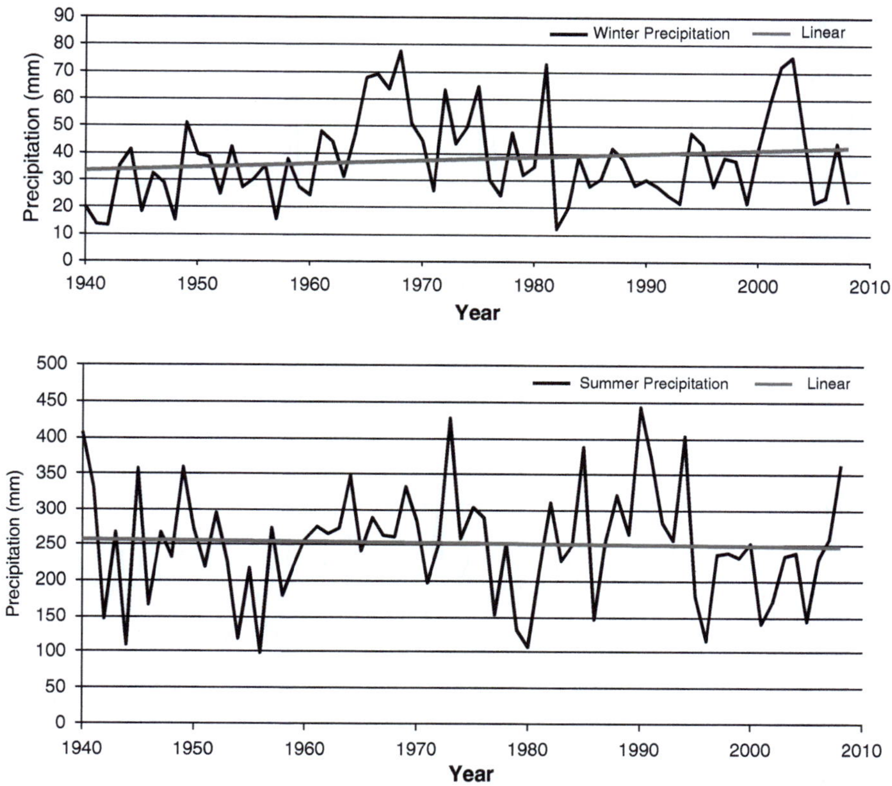

Fig. 21.3 Precipitation trend in cold (*top*) and warm (*bottom*) seasons (Baruunkharaa)

21.3.2 Urban Heat Island and Inversion Effects on Temperature Regime

Incomplete coverage of meteorological observation in Mongolia is an added obstacle to a more detailed study on local climate conditions that could match the complexity of topography and diversity of landscapes of the country. This lack contributes to some degree of uncertainty in the results of the study. Another source of errors can be associated with "urban warming" (caused by energy use and alterations of the natural environment), which has more systematic effects (Hansen et al. 2010).

Most meteorological stations in Mongolia can be considered as "free" from urban effects thanks to the absence of heavily populated mega-cities in this country. However, the long heating season and rapid urbanization in Mongolia are turning some key stations, with long periods of observation to category "not free" from urban influences. The urban effect in Mongolia reveals itself in a more complex way, in the form of mixed effects of heated air, "smog cap," and "moisture blanket."

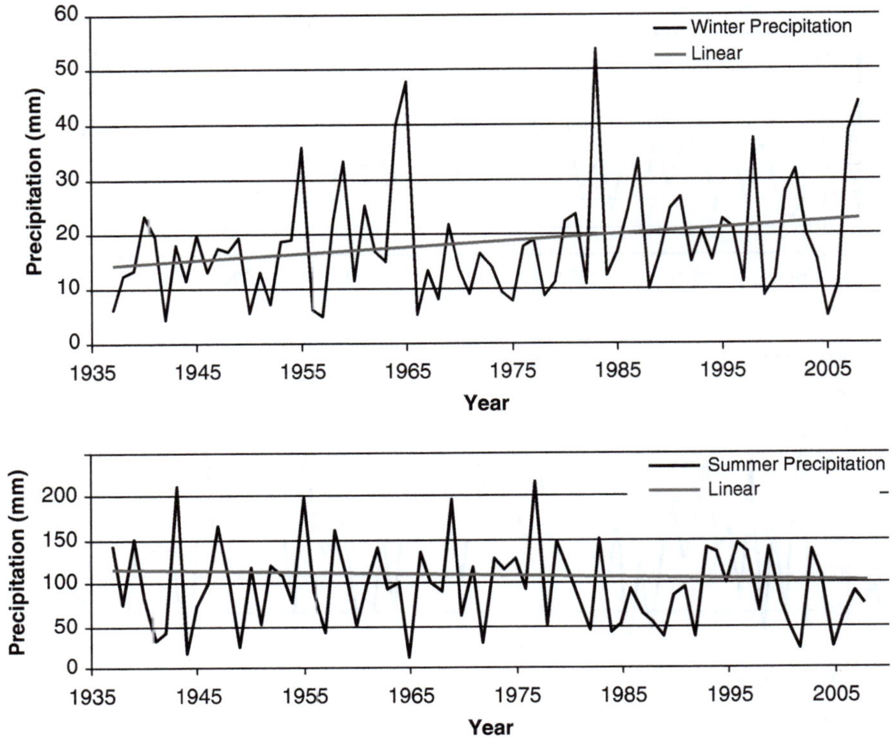

Fig. 21.4 Precipitation trend in cold (*top*) and warm (*bottom*) seasons (Dalanzadgad)

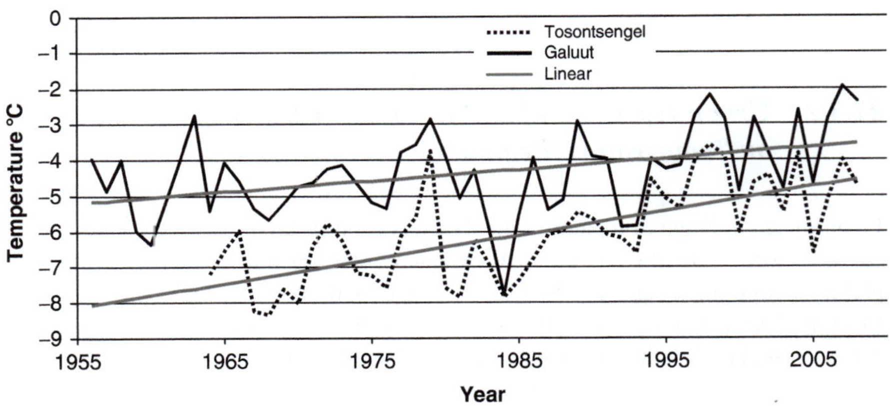

Fig. 21.5 Air temperature regime at the sites Tosontsengel and Galuut

Topography can add some "noise" in the data from certain micro-scale phenomena such as "shallow inversion," local "valley wind," etc., at the urban effect-free stations located in rural areas. For long-term trends, all the effects can be filtered using appropriate smoothing techniques.

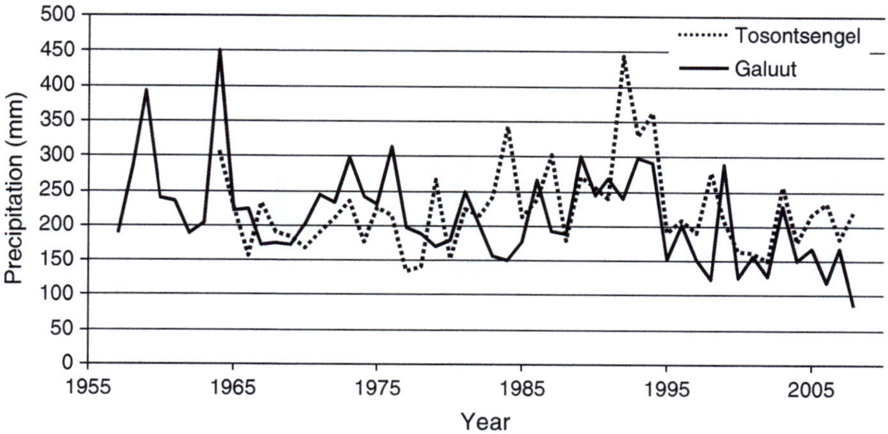

Fig. 21.6 Precipitation scenario at the Tosontsengel site (affected by SMH more strongly but located on the NW slopes of Hangai Mountain) and the Galuut site (located on the SW slopes of Khangai Mountain)

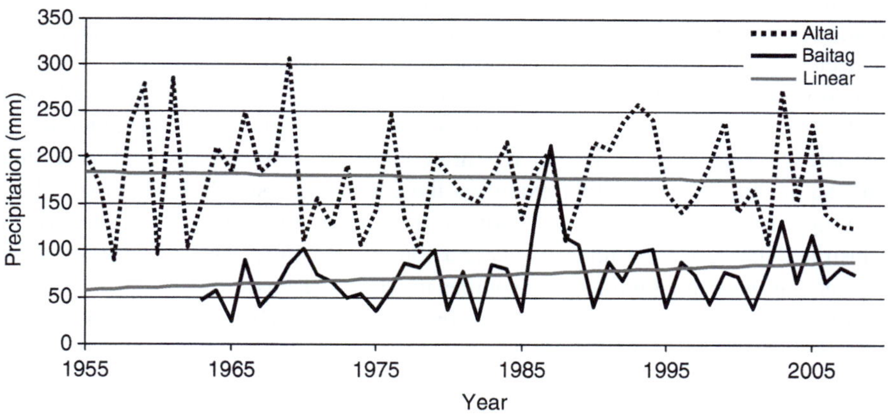

Fig. 21.7 Topography effect on precipitation on west (Baitag) and east (Altai) sides of Altai Mountain Range

Additional analyses have been undertaken to demonstrate the "reality" of the urban effect in general trends of climate parameters. In this exercise, four meteorological stations located at different parts of the Ulaanbaatar city were involved. First of all, trends in temperature have been recorded at the stations located within the capital city and beyond its parameters. Ulaanbaatar station has been gradually surrounded by residential dwellings since it was moved to a more open area in the 1970s. Ikhsurguuli station was established in the city center and thus was affected by ever-increasing city effects from the beginning of its function. In contrast to these two stations, the Buyantukhaa site has not experienced too much of the city effects thanks to its location by the national airport. It also has a relatively long-

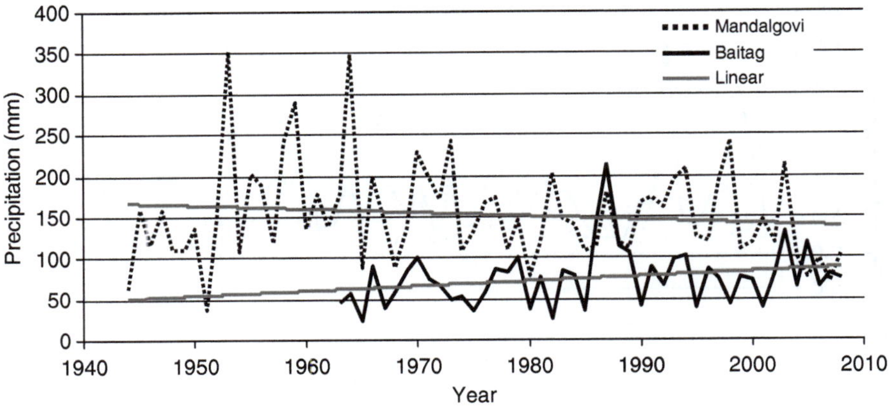

Fig. 21.8 Topography effect on precipitation on west or Trans-Altai Gobi (Baitag) versus east (Mandalgobi) close to the tip of the Altai Mountain range

period, uninterrupted record that can be used as a point of reference for this area. The difference in altitude between these three stations does not exceed 30 m, and distances between them are almost negligible for parameters such as temperature. The clear similarity in phase of temperature trend at these three stations proves this situation. Despite that, the average temperature at both city stations (Ulaanbaatar and Ikhsurguuli) is higher than at the Buyantukhaa site and has a notable tendency to increase faster (Fig. 21.9), which definitely leads to assumptions that this feature can be attributed to a certain extent to the urban heat island (UHI) effect. This assumption is supported by the results of the maximum temperature trend, which illustrated its great consistency at city and non-city stations, probably because of the absence of the nonnatural heating sources in warm seasons.

Intensive cooling of the land surface in winter seasons because of the high pressure dominance in Mongolia can be revealed differently in patterns of surface air temperature (SAT), depending on topography. The effect of inversion can be pronounced in the annual mean SAT. To illustrate this phenomenon, the SAT was recorded at the meteorological station located on top of Morin-uul hill (height about 240 m) near Buyantukhaa station. The direct projected distance between these two stations is no more than 5 km. It appears that the annual mean SAT at the top of Morin-uul hill is greater than at the level of Buyantukhaa station (Fig. 21.10), where temperature inversion occurs frequently in winter seasons.

The results of analysis of the local regime in temperature and precipitation trends can be summarized as follows:

- Temperature and precipitation at the local level are responding differently to global warming. Temperature increase is synchronized, with more pronounced changes in winter seasons and in the northern part of Mongolia. Precipitation responds differently in terms of both geography and season (with increases in winters).
- Topography plays a prevailing role in the precipitation pattern, which remains sensitive to external factors such as atmospheric circulation. A slight change in

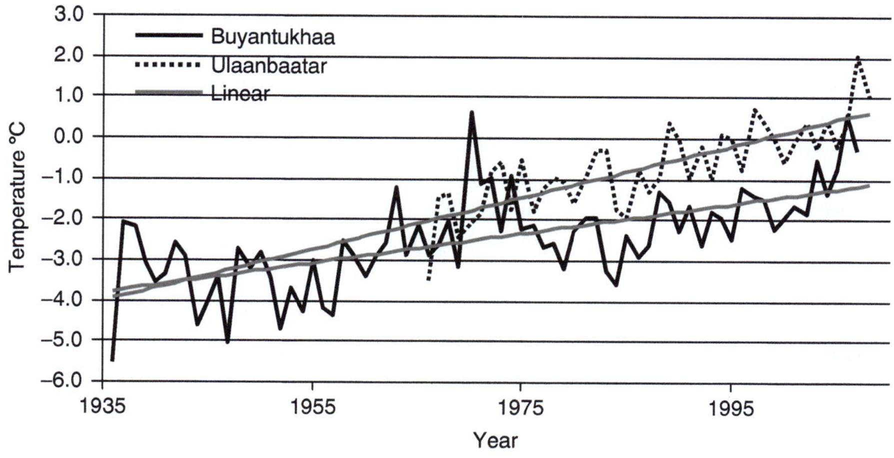

Fig. 21.9 Urban heat island (UHI) effect shown by average temperature at the city (Ulaanbaatar) and a non-city reference station (Buyantukhaa)

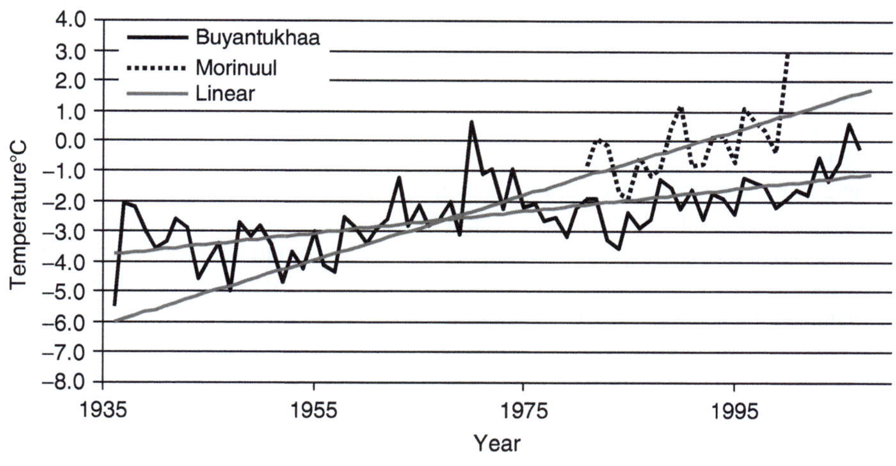

Fig. 21.10 Effect of inversion on annual mean temperature shown by average temperature at the Morinuul site (located on top of the hill; altitude 1,512.8 m) and at the Buyanukhaa site (altitude 1,272 m); distance between them is no more than 5 km

the dominant direction of the wind might cause dramatic changes in the local distribution of precipitation. Precipitation changes will have more impact on the local ecosystems and on the livelihood of the people.

- Maximum and minimum temperatures, particularly in winter seasons, might reflect the thermal condition of a tiny air mass by land surface rather than the entire area that the given station is supposed to represent, as a result of certain specifics of local landscapes, such as closed basins of lakes and mountain valleys where temperature inversion occurs frequently.

- The current trend of warming at the territory of Mongolia based on instrumental records needs to be evaluated taking into account location specifics of meteorological stations and the gradually increasing UHI effect in cold seasons in some key areas.
- Possible data "contamination" from incomplete coverage by meteorological standard observations, inadequate instrumental accuracy, and local effects, for instance, heat island and smog cap, should not be neglected in estimation of trends at the local level, in contrast to global trends in climate parameters.

21.4 Challenges for Mongolia to Cope with Global Warming and Globalization

21.4.1 Development Paradigm: Natural Resource or Human Knowledge

Mongolia has been struggling to find a way toward a more accelerated development as a nation in its path of history during the past several decades. It could not avoid failures experienced by many other, the so-called "resource-rich," developing countries, who tend to "enjoy" an easy way of "development" relying on their natural wealth rather than their human capacity. This tendency can be probed by referring to a recently developed "ecological footprint" concept (Ewing et al. 2009). Mongolia has greater per capita "biocapacity" compared to heavily populated countries in Asia, that is, China or India (Table 21.2). However, per person footprint by land-use type is exceeding the level of other Asian countries (Table 21.3). The total biocapacity is declining because of increase in population and individual consumption (Table 21.4). On the other hand, if we look at the change in biocapacity and progress in the human development index (HDI), it becomes clear that Mongolia is "burning" more natural resources for every component of HDI than any other country (Table 21.4).

Table 21.2 Per person biocapacity by land-use type in five countries

Country/ region	Total biocapacity (gha/person)	Cropland (gha/person)	Grazing land (gha/person)	Forest land (gha/person)	Fishing grounds (gha/person)	Built-up land (gha/person)
Mongolia	16.5	0.3	10.2	5.7	0.2	0.1
China	0.9	0.4	0.1	0.2	0.1	0.1
India	0.4	0.3	0.0	0.0	0.0	0.0
Japan	0.6	0.1	0.0	0.3	0.1	0.1
Korea	0.3	0.1	0.0	0.1	0.0	0.1

gha global hectares
Source: Ewing et al. (2009)

Table 21.3 Per person footprint by land-use type in five countries

Country/ region	Total ecological footprint (gha/person)	Cropland (gha/person)	Grazing land (gha/person)	Forest land (gha/person)	Fishing grounds (gha/person)	Carbon footprint (gha/person)	Built-up land (gha/person)
Mongolia	5.57	0.49	3.51	0.12	0.00	1.39	0.06
China	1.85	0.36	0.13	0.15	0.06	1.08	0.07
India	0.77	0.28	0.01	0.12	0.01	0.31	0.04
Japan	4.11	0.58	0.03	0.28	0.47	2.68	0.07
Korea	3.73	0.69	0.04	0.24	0.61	2.09	0.06

Source: Ewing et al. (2009)

Table 21.4 Change by country in footprint of consumption, biocapacity, and human development index (HDI)

Country/ region	Change (%) in population (millions)	Change (%) in per person ecological footprint of consumption	Change (%) in total ecological footprint of consumption	Change (%) in per person biocapacity	Change (%) in total biocapacity	HDI, 1980	HDI, 2007
Mongolia	165	−47	41	−69	−17	0.668 in 1985	0.727
China	97	1,659	423	−17	63	0.53	0.772
India	153	−12	122	−54	17	0.43	0.612
Japan	35	90	1,569	−41	−20	0.89	0.96
Korea	87	2,679	5,859	−49	−5	0.72	0.937

Source: Ewing et al. (2009)

Such kinds of development policy, as a rule, appeared to be unsustainable in terms of economic growth, partly because of price volatility in the international market. This policy also has side effects, for example, environmental damage. Moreover, there is a prerequisite for social tension attributed to the emerging disparity regarding access to available mineral resources and thus increased income polarization in society.

21.4.2 Dilemma: To Be or Not To Be—Pastoralism

At present, the adaptation deficit is a strong challenge in most countries of the world. However, Mongolia could be an exception, to some extent, as a nation that is still practicing pastoralism in its classical form, if human mobility is accepted as an essential option for survival as well as development (UNDP 2009). It is well known that the agriculture sector historically has shown an enormous capacity to adjust to social and environmental stimuli that are analogous to climate stimuli.

Historical success stories in coping with climate variability suggest that livestock producers are able to adjust to climate change with minimum damage to the sector's basic principles. To some extent, it is true that a pastoral subsistence lifestyle reduces the risks associated with irregular climate patterns (Batjargal 2000). In that sense, it can be assumed that Mongolia could propose more viable ways of adaptation to climate change, at least for itself. However, there is also a misperception (UNDP 2003) that, if pastoralism continues to exist in the near future, it will remain only in poor nations. Since the second half of the last century, governments in some countries have forced pastoralists to limit or stop migration in the name of prevention of overgrazing, not yet realizing fully that, in fact, the overgrazing problem emerged mostly because of the cash economy. Another strong negative attitude concerning the fate of the pasture-based livestock sector in Mongolia is associated with the occurrence of natural disasters with heavy loss of livestock (Batjargal 2001; NAMHEM 2010) and the ongoing and expected climate change (MNE 2006; START 2006).

These negative messages had some influence on public attitude, including some politicians who argue that pastoralism has no future in Mongolia and needs to transition to some form of settled farming. On the other hand, a counterargument can be put forward in favor of the traditional livestock system. First, a decision should be based on solid and numerous arguments. For instance, the possible reduction of animal weight linked only to expected increase of heat stress in accordance with predicted climate scenarios is an insufficient argument toward a conceptual decision.

In advocating for traditional livestock, other arguments can be put forward. (1) Pasture-based livestock is the only way to sustain the net level of production for Basic Human Needs (BHNs) in arid areas where vegetation cover is patchy and grass yield fluctuates in spatial and temporal scale, depending on various, mostly unpredictable, factors. (2) Pastoralism has served without failure for a long time as the principal component of the national economy. (3) It was the life-sustaining basis for a majority of, if not the entire, population of Mongolia for at least hundreds, if not thousands, of years. (4) Pastoralism has high-enough returns in terms of economic efficiency and in terms of ecosystem service as a man-managed but still seminatural ecosystem. (5) It will serve as a great contribution to the cultural heritage of mankind, being the last instance of classical nomadic civilization. (6) Pastoralism can make a priceless contribution to the natural heritage by preserving the gene pool of animals with high tolerance to harsh conditions and by preserving a portion of grasslands where the exercise of grazing is still controlled by the natural order.

21.4.3 Options for Survival of Mongols Under the Pressure of Such External Stresses as Global Warming and Globalization

The traditional lifestyle is not merely a subject of identity for Mongols or a subject for tourist attraction. It is a solid basis for the guaranteed survival of Mongols as a nation, especially because it is a principal economy sector not greatly affected directly by international market volatility. It is the most accessible option for BHN of Mongols, based on the guaranteed available resources, for instance, pastures and available capacity in the form of traditional knowledge tested and shaped throughout thousands of years of the nation's existence. The traditional lifestyle is unfailing as a life-sustaining option, withstanding many types of external stresses induced by climate variability, cultural invasion, political turbulence, and other pressing factors.

Speaking of tradition in a general term, we need to recognize that tradition is not static nor "frozen" and unyielding. It has its own dynamism. The best traditions can survive and continue to serve people when they are able to reconcile with evolution of the natural environment and with human history that is coping with the challenges of societal development. In this sense, tradition needs innovation, and the traditional lifestyle needs diversification.

21.4.3.1 Vulnerability of Tradition

The lessons from around the world show that the tradition might not "serve" to benefit the communities if left without necessary support at the society level. Mongols can compete with the world only by offering their own assets in maintenance of which they have inherent skills, and not by copying others with whom they need to compete.

Complete elimination of the traditional lifestyle, particularly as linked to pastoralism, might lead to a loss of guaranteed income source and, more importantly, to cutting off the roots of the cultural background. Traditionally, life in Mongolia in rural areas was sustained on the tireless effort and contribution of every member of family, both junior and senior. People in Mongolia said "Every new family member will add own portion to family foodstuffs" and welcomed every child, in contrast to some cultures where newcomers were considered an additional burden for a family's limited living expenses.

There is a concern with regard to some emerging issues such as cultivation of attitudes to live off unearned income, for example, "Grant of Motherland" or " Share of Treasure" (referring to election promises made by politicians regarding the possible future dividends from major mining explorations) and to rely on good fortune, on the power of government, on generosity of donors, etc.

21.4.3.2 Globalization-Associated Shock

Conventional wisdom implies that globalization-related stresses might have a more disturbing impact on the Mongolian society, particularly on the traditional livelihood of the people, compared to the effect from global warming, as a consequence of the nation's long-lasting isolation in the past from the broad spectrum of cultural domains in the world. There is no question that the adaptive capability of people practicing nomadic or pastoral lifestyles to new circumstances, with their tolerance to new cultural elements, can play a positive role with respect of adaptation to variable climate conditions. However, these adaptive behaviors and high tolerance might not bring a positive outcome on every occasion, that is, in terms of preservation of their unique indigenous traditional culture.

The rate of tolerance and acceptance can be defined mostly by the aggressiveness of invading nonnative cultures rather than by its benefits for local communities. Alcoholism, drug addiction, gambling, highly commercialized "dirty" entertainment, cross-border economic crimes, etc., are only a few examples of the side effects of globalization that could be more disturbing to society than infrequent extreme weather events.

Cultural shocks might have longer-lasting effects on society than natural disasters like dzud and, therefore, might require more effort, funds, and time for recovery. Such shocks could jeopardize the sustainable livelihood of people for long periods of time with irreversible consequences.

Therefore, the globalization-induced changes in society need to be harmonized with natural capital and traditional cultural values to avoid possible "shocks" that might trigger negative "domino effects," shaking the sustainability of the entire society. Moreover, it can bring more stress on local ecosystems, which had not yet experienced such society-generated pressures since the human colonization of these areas.

Society and ecosystems in Mongolia can be affected by interference effects not only generated within its territory but also those originating at great distances. Climate change-induced shocks, for instance, having occurred in some parts of Europe can be transferred to Mongolia with amplified magnitude through banking system networks, trade, FDI, ODA, and so on. Some contagious diseases originating in distant areas can easily reach Mongolia where people are not immune to disease endemic to other areas, for instance, tropical diseases. On the other hand, this negative factor can also be positive because broad and intensive contact with such diseases can result in an improved immune response system, which is not the case for people who live in isolation with limited or no outside contact and who, when challenged by such unfamiliar diseases, will be affected more severely.

21.4.3.3 Tradition and Innovation

Tradition is not static, as was emphasized before. The best tradition is that of human behavior that is streamlined with the modern tendency of evolution of human civilization. Tradition has its own dynamism based on the continued process of lessons learned and best practices. The traditional lifestyle in Mongolia can be enriched by adapting a broader spectrum of innovations and technologies. A list of such should include the following: modern communication technologies with wireless and mobile connections; renewable energy facilities with flexible energy supply opportunities; water purification equipment; Internet access with remote education and healthcare services; modern compact machinery to process livestock products for family-based and small-scale businesses; and a modern commercial network that promotes access to markets in a city without the people having to migrate into urbanized areas.

Crop cultivation, based on modern science and agricultural technology, would be essential for diversification of income sources for local communities. Crop cultivation can be developed as a complementary business activity, circumventing the conflicts that existed in the past between livestock and the state's large crop production entities during the socialist modernization period. In some parts of Mongolia, rich experience has been gained on how to combine livestock and crop cultivation with maximum efficiency within traditional way of life (Batjargal 2007a). Such kinds of progressive practices can be included in the list of innovative approaches and management practices.

These and other innovative approaches, put into practice with certain cautions, should be an essential part of the adaptation strategies for Mongolia to cope with challenges emerging from global warming and globalization.

21.5 Concluding Remarks

- At present Mongolia is facing multiple challenges in relationship to the current and expected climate conditions. Human activities are making an increasing contribution to the destabilization of ecosystems in Mongolia.
- In the long history of human society, the ecosystem in the territory of Mongolia has been controlled by direct impacts of the global climate system through its regional patterns. Since the second half of the twentieth century, human activities in Mongolia have increasingly intervened in the natural order and harmony, with a notable change in the ecosystem structure.
- Human intervention in the future might aggravate the effects of forcing generated by the global climate system in a different direction, resulting in negative and positive outcomes, in terms of benefits for the ecosystem functions and for society.
- A science-based policy is needed in respect to a development paradigm of the country to prevent any collapse-level degradation in ecosystems caused by the interference effects between human and natural systems.
- Mongolia has certain advantages because it can benefit from its ecosystem services, thanks to the dominance of natural and seminatural ecosystems in the country and its people's unique lifestyle based on human mobility in the form of classical pastoralism.
- Traditional pastoralism, enriched by the advances of modern technology and knowledge and diversification, involving cropping, ecological tourism, and other income source-related activities in a complementary and environmentally sound manner, can be the best possible option in the adaptation to climate change strategy to minimize stress on society caused by global warming.
- Globalization-related stresses might have a more disturbing impact on the Mongolian society in comparison with the effects of global warming as a consequence of Mongolia's long-lasting isolation in the past from the wide spectrum of cultural domains in the world. Therefore, globalization-induced changes in society need to be harmonized with natural capital and traditional cultural values to help avoid possible "shocks" that might trigger a negative "domino effect" such as consequences for local ecosystems.
- In terms of adaptation strategy timeframe, a high priority should be given to the near- and midterm incremental changes with special focus on high-impact events,

such as change in precipitation patterns with the increase of its winter proportion and lesser-frequency but higher-intensity rains in summer.

- In the midterm plan, focus is necessary on possible long-lasting aridity in some parts of the territory of Mongolia, with increased lack of water for livelihood and production activities.
- In the long-term plan, some uncertain or low-probability but high-impact events should be kept in the field of vision, that is, a complete shift of climate zones with intensified rate of desertification, which results in dramatic social consequences such as mass displacement of people within and beyond their national border, including flows of people in both direction as ecological refugees.
- International cooperation is needed to identify the "tipping points" in relationship to the aerosol impact on regional climate systems, which in some areas could reach a level that may counteract with the warming effect of the GHG, as a result of intensified land-use change. In some parts of Mongolia, focus of the study should be made on the role of the so-called black carbon or other albedo effective aerosols in acceleration of glacier melting and its matching rate with possible above-"normal" accumulation of snow in winter seasons.
- The traditional way of livestock husbandry can be threatened by a multitude of factors such as global warming-caused cold waves with heavy snowstorms, and early establishment of long-lasting snow coverage of the territory in winter. These factors in combination with increased soil moisture deficit, caused by a possible decrease in summer rainfall and intensified evopotranspiration, can prevail as pressures on livestock, with heat stress no longer being the single stress factor for animals.
- Increased incidents of "white" and "iron" dzud situations in winter seasons will have serious negative impacts on livestock as the economy sector and on the livelihood of the local people with limited options for income sources for life. It might require a change of infrastructure development concept, which was based, so far, on the past climate background and norms. Improved concepts should cover the full range of transport, telecommunication, and energy supply networks, including road construction standards, energy gridding with autonomous backup based on local and renewable energy sources, wireless and satellite communications, rehabilitation of air services with access to remote areas, etc. Adequate government policy would be needed to bring medical and educational systems to rural areas using the advancements of modern Information and Communications Technology (ICT) to provide equal access for everyone to a high standard of service.

References

Batima P, Dagvadorj D (2000) Climate change impacts in Mongolia. JEMR Publishing, Ulaanbaatar

Batima P, Natsagdorj L, Gombluudev P, Erdenetsetseg B(2005) Observed climate change in Mongolia. AIACC Working Paper No. 12

Batjargal Z (2000) Climate and Pastoralism. Presentation at the international symposium on "Nomads and the use of pastures today," December 1999. Nomadic IISNC Newsletter Nos. 4 and 5. Ulaanbaatar, Mongolia

Batjargal Z(2001). Lessons learnt from consecutive dzud disaster of 1999–2000 in Mongolia. Presentation at the symposium "Change and sustainability of pastoral land use systems in temperate and central Asia." IISNC, Ulaanbaatar, Mongolia

Batjargal Z (2007a) Climate condition and human activities are the principal factors for fragility of ecosystem and desertification risk in Mongolia. In: Fragile environment, vulnerable people and sensitive society. Kaihatu-sha, Tokyo, Japan, pp 137–153

Batjargal Z (2007b) Eco-politics and nature conservation in Mongolia. Presentation at the Conference "Asian green belt: its past, present and the future," 30–31 October. RIHN, Kyoto, Japan

Dagvadorj D, Mijiddorj R, Natsagdorj L (1994) Climate change in Mongolia. Papers on Meteorology, No. 17. Ulaanbaatar, pp 3–10

Ewing B, Goldfinger S, Oursler A, Reed A, Moore D, Wackernagel M (2009). The ecological footprint atlas 2009. Global Footprint Network, Oakland

GoM (2000) The National Action Programme on Climate Change (NAPCC) of Mongolia. Ulaanbaatar

Gomboluudev P (2006) Future climate change of Mongolia under Special Report Emission Scenarios (SRES). Proceedings of Fifth Mongolia-Korea Joint Seminars on Environmental Changes of Northeast Asia, October 10–14. Ulaanbaatar, Mongolia

Hansen J, Ruedy R, Sato M, Lo K (2010) Current GISS Global Surface Temperature Analysis. NASA Goddard Institute for Space Studies, New York

IPCC (2007) Climate Change 2007: Synthesis Report. Contribution of Working Groups. IPCC, Geneva, Switzerland

Mijiddorj R(2000) Climate change and sustainable development. Ulaanbaatar

MNE (2006) Climate change and sustainable livelihood of rural people in Mongolia. Ministry of Nature and the Environment of Mongolia (MNE), Netherlands climate change studies assistance programme. Project Report, Ulaanbaatar

MNET (2009) Mongolia Assessment Report on Climate Change 2009. Ministry of Environment, Nature and Tourism of Mongolia (MNET). Ulaanbaatar

NAMHEM (2010) Information note on weather condition of winter (2009–2010). Ulaanbaatar (in manuscript)

START (2006) Climate change vulnerability and adaptation in the livestock sector of Mongolia. A final report submitted to Assessments of Impacts and Adaptations to Climate Change (AIACC), Project No. AS 06. International START Secretariat, Washington, DC

UNDP (2003) Pastoralism and mobility in the drylands. The Global Drylands Imperative. United Nations Development Programme (UNDP), www.undp.org/drylands/docs/

UNDP (2009) Human Development Report 2009. Overcoming barriers: human mobility and development. United Nations Development Programme (UNDP), New York

Index

N. Yamamura et al. (eds.), *The Mongolian Ecosystem Network: Environmental Issues Under Climate and Social Changes*, Ecological Research Monographs,
DOI 10.1007/978-4-431-54052-6, © Springer Japan 2013

K

L

M

N

O

P

MIX
Papier aus verantwortungsvollen Quellen
Paper from responsible sources
FSC® C105338

If you have any concerns about our products,
you can contact us on
ProductSafety@springernature.com

In case Publisher is established outside the EU,
the EU authorized representative is:
Springer Nature Customer Service Center GmbH
Europaplatz 3, 69115 Heidelberg, Germany

Printed by Libri Plureos GmbH
in Hamburg, Germany